ADVANCES IN
QUANTUM CHEMISTRY

VOLUME 35

ADVANCES IN
QUANTUM CHEMISTRY

PROPAGATING INSIGHT: A TRIBUTE TO YNGVE ÖHRN

EDITOR-IN-CHIEF

PER-OLOV LÖWDIN

PROFESSOR EMERITUS

DEPARTMENT OF QUANTUM CHEMISTRY
UPPSALA UNIVERSITY
UPPSALA, SWEDEN

AND

QUANTUM THEORY PROJECT
UNIVERSITY OF FLORIDA
GAINESVILLE, FLORIDA

EDITORS

JOHN R. SABIN
MICHAEL C. ZERNER

QUANTUM THEORY PROJECT
UNIVERSITY OF FLORIDA
GAINESVILLE, FLORIDA

ERKKI BRÄNDAS

DEPARTMENT OF QUANTUM CHEMISTRY
UPPSALA UNIVERSITY
UPPSALA, SWEDEN

GUEST EDITORS

J. V. ORTIZ

DEPARTMENT OF CHEMISTRY
KANSAS STATE UNIVERSITY
MANHATTAN, KANSAS

HENRY KURTZ

DEPARTMENT OF CHEMISTRY
UNIVERSITY OF MEMPHIS
MEMPHIS, TENNESEE

VOLUME 35

ACADEMIC PRESS

San Diego London Boston New York Sydney Tokyo Toronto

Academic Press Rapid Manuscript Reproduction

This book is printed on acid-free paper. ∞

Academic Press
a division of Harcourt Brace & Company
525 B Street, Suite 1900, San Diego, California 92101-4495, USA
http://www.apnet.com

Academic Press
24-28 Oval Road, London NW1 7DX, UK
http://www.hbuk.co.uk/ap/

Library of Congress Catalog Card Number:

International Standard Book Number: 0-12-034835-7

PRINTED IN THE UNITED STATES OF AMERICA
99 00 01 02 03 04 EB 9 8 7 6 5 4 3 2 1

Contents

Contributors

Numbers in parentheses indicate the pages on which the authors' contributions begin.

Jean-Marie André (95), Laboratoire de Chimie Théorique Appliquée, Facultés Universitaires Notre-Dame de la Paix, B-5000 Namur, Belgium

Rodney J. Bartlett (149), Quantum Theory Project, University of Florida, Gainesville, Florida 32611

R. Cabrera-Trujillo (175), Chemistry Department, Odense University, 5230 Odense M, Denmark; and Department of Physics, Universidad Autonoma Metropolitana–Iztapalapa, 09340 Mexico D.F., Mexico

L. S. Cederbaum (77), Lehrstuhl für Theoretische Chemie, Institut für Physikalische Chemie, Universität Heidelberg, D-69120 Heidelberg, Germany

Benoît Champagne (95), Laboratoire de Chimie Théorique Appliquée, Facultés Universitaires Notre-Dame de la Paix, B-5000 Namur, Belgium

Marshall G. Cory (357), Quantum Theory Project, University of Florida, Gainesville, Florida 32611

M. S. Deleuze (77), Theoretische Scheikunde, Departement SBG, Limburgs Universitair Centrum, Universitaire Campus, B-3590 Diepenbeek, Belgium

Erik Deumens (21), Quantum Theory Project, University of Florida, Gainesville, Florida 32611

Joseph G. Fripiat (95), Laboratoire de Chimie Théorique Appliquée, Facultés Universitaires Notre-Dame de la Paix, B-5000 Namur, Belgium

J. W. Gauld (339), Department of Quantum Chemistry, Uppsala University, S-75120 Uppsala, Sweden

Aage E. Hansen (193), Chemical Laboratory IV, Ørsted Institute, University of Copenhagen, DK-2100 Copenhagen Ø, Denmark

Christof Hättig (111), Department of Chemistry, University of Aarhus, DK-8000 Aarhus C, Denmark

Y. Itagaki (339), Department of Physics and Measurement Technology, Linköping University, S-58183 Linköping, Sweden

Morten Østergaard Jensen (193), Chemical Laboratory IV, Ørsted Institute, University of Copenhagen, DK-2100 Copenhagen Ø, Denmark

Poul Jørgensen (111), Department of Chemistry, University of Aarhus, DK-8000 Aarhus C, Denmark

Vandana K. (261), Department of Chemistry, Indian Institute of Technology Bombay, Powai, Mumbai 400 076, India

R. M. Kadam (339), Department of Physics and Measurement Technology, Linköping University, S-58183 Linköping, Sweden

Jeffrey L. Krause (249), Quantum Theory Project, University of Florida, Gainesville, Florida 32611

Jan Linderberg (1), Department of Chemistry, University of Aarhus, DK-8000 Aarhus C, Denmark

Ricardo L. Longo (53), Departamento de Química Fundamental, Universidade Federal de Pernambuco, 50740-540 Recife, PE, Brazil

A. Lund (339), Department of Physics and Measurement Technology, Linköping University, S-58183 Linköping, Sweden

S. Lunell (339), Department of Quantum Chemistry, Uppsala University, S-75120 Uppsala, Sweden

David A. Micha (317), Quantum Theory Project, University of Florida, Gainesville, Florida 32611

Manoj K. Mishra (261), Department of Chemistry, Indian Institute of Technology Bombay, Powai, Mumbai 400 076, India

David H. Mosley (95), Laboratoire de Chimie Théorique Appliquée, Facultés Universitaires Notre-Dame de la Paix, B-5000 Namur, Belgium

J. Oddershede (175), Chemistry Department, Odense University, 5230 Odense M, Denmark; and Quantum Theory Project, University of Florida, Gainesville, Florida 32611

J. V. Ortiz (33), Department of Chemistry, Kansas State University, Manhattan, Kansas 66506

Hideo Sekino (149), Quantum Theory Project, University of Florida, Gainesville, Florida 32611

John R. Sabin (175), Chemistry Department, Odense University, 5230 Odense M, Denmark; and Quantum Theory Project, University of Florida, Gainesville, Florida 32611

Stephan P. A. Sauer (175), Chemistry Laboratory IV, Ørsted Institute, University of Copenhagen, DK-2100 Copenhagen Ø, Denmark

Jack Simons (283), Henry Eyring Center for Theoretical Chemistry, Chemistry Department, University of Utah, Salt Lake City, Utah 84112

Krassimir K. Stavrev (357), Quantum Theory Project, University of Florida, Gainesville, Florida 32611

S. B. Trickey (217), Quantum Theory Project, Department of Physics and Chemistry, University of Florida, Gainesville, Florida 32611

B. Weiner (217), Department of Physics, Pennsylvania State University, DuBois, Pennsylvania 15801

Michael C. Zerner (357), Quantum Theory Project, University of Florida, Gainesville, Florida 32611

Preface

This volume of *Advances in Quantum Chemistry* honors the many accomplishments of Dr. N. Yngve Öhrn on the occasion of his 65th birthday. The authors invited to contribute to this volume include past and present students, postdoctoral associates, and collaborators of Dr. Öhrn.

Yngve's research has concentrated on the development of theoretical techniques and computational applications of these innovations to experimental problems. Examples of techniques most closely associated with him are the electron and polarization propagator methods (starting in the 1960s), the antisymmetrized geminal power (AGP) wavefunction (1970s and 1980s), and recent work on electron nuclear dynamics (1980s and 1990s). Yngve has over 130 publications on these and many other topics. His versatility is exemplified by two consecutive publications from his vita—"Spin Projection of Ferminon Coherent States" in the *Journal of Mathematical Physics* and "Semiempirical MO-Study of Charge Transfer in (4-Aminophenyl)borane and Related Compounds—Molecular Paddle-Wheels" in the *Journal of Molecular Structure.*

One of the most important contributions Yngve Öhrn has made is embodied in the title for this volume, "Propagating Insight." This teaching role includes many years of courses and lectures at the University of Florida, where he has been a faculty member since 1966 and has taught students in the Department of Chemistry and the Quantum Theory Project. He has recently been recognized for excellence in undergraduate teaching. His lectures have also been steadily in demand beyond the confines of his university. Yngve has introduced many scientists to group theory and propagator methods at the International Summer Institutes in Norway and Sweden and the Winter Institutes in Florida. He has lectured extensively in Sweden, Denmark, and Belgium. He held the Chaire Francqui Interuniversitaires d'Etranger at the Universities of Namur, Antwerp, and Liege in 1995.

Yngve's role in service to his profession is also exceptional. He not only has played the important roles in research and teaching outlined above, but also has devoted much time to academic administration. He was chairman of the Department of Chemistry at the University of Florida from 1977 to 1983. This was a period of active growth for the department and a time of its emergence as one of the strongest departments in the United States. Yngve has also played a central role in the development of the Quantum Theory Project. He started as one of the initial contingent of Swedes from 1961 to 1963 and served as Associate Director from 1967 to 1977 and as Director from 1983 to the present.

We are honored to assemble this volume, as Yngve Öhrn's influence and example have informed, inspired, and guided each of us. His integrity, dedication, and insight have benefited a broad community of scientists. We look forward to many more years of association with Yngve and to the fruits of his work.

H. A. KURTZ
J. V. ORTIZ
J. R. SABIN

YNGVE ÖHRN, SCIENTIST, LEADER, AND FRIEND

Jan Linderberg
Department of Chemistry
Aarhus University
DK-8000 Aarhus C, Denmark

1. PREAMBLE.

Bergslagen, the Swedish mining district, covers part of Värmland, Dalarna, Västmanland, and Uppland. Copper was mined at Falun since the thirteenth century, silver at Sala, and iron was plentiful and of good metallurgical quality at numerous places. It was the backbone for an economy that allowed Swedish kings to implement an expansive international policy in the seventeenth century and for a technological and scientific development with prominent results in

ADVANCES IN QUANTUM CHEMISTRY, VOLUME 35

0065-3276/99 $30.00

engineering and natural sciences with people such as Swedenborg, Linné, Polhem, Scheele, and Bergman making pioneering discoveries. Jöns Jacob Berzelius and his students discovered and characterized a number of elements in the minerals from this region. Berzelius initiated the classification, in the spirit of Linné's sexual system for the flora, of chemical elements that led to Mendelejev's periodic system and he rose to become a most influential chemist in the first half of the nineteenth century.

Mining and iron extraction through blast-furnaces based on charcoal from the extensive forests became a major occupation in Bergslagen in the seventeenth century when most farmers had a share of a mine and a furnace. Development of the technical use of the iron was accelerated by foreign blacksmiths, principally from the Walloon part in present day Belgium, and there grew a population of skilled and respected members of the many villages and towns of the region. Knowledge of the processes involved in steel manufacturing developed mainly as an intellectual capital in the people occupied in the production, there was little instrumentation to tell about composition and temperature during the operation.

Traditions and societal structures from the emerging Swedish steel industry were still very much present in the first half of this century. Yngve Öhrn was born in 1934 as the son of Nils and Elsa Öhrn at Avesta in the heart of Bergslagen. Nils had his occupation in the steel rolling mill in a town dominated by the Avesta Järnverk, one of several exponents of the Swedish "bruks"-tradition (1) that had grown out of the original farmer shared operations. Respect for knowledge and skills was still manifest and schooling was encouraged at all levels. The low birth rate in Sweden in the 1930's (2) gave incentive to government initiatives in the realm of education as well as in other fields. A concept of an "intellectual reserve" was established to promote further education throughout the Swedish population.

Sports activities was a significant part of the social life in the growing industrial towns, track and field, football (soccer), cross country skiing and bandy, a team sport on ice with a small light ball and hooked sticks, were in the focus of attention and rivalry between towns. Yngve Öhrn showed his prowess in school and in track and field, he earned a baccalaureate from Avesta Gymnasium with high grades and a relay championship in 4x100m at the national school track meet.

Military service was mandatory for able as well as some not so able male Swedes and Yngve Öhrn subscribed to reserve officer's training in the signal corps. He served a first installment of fifteen months, then started studies in mathematics at Uppsala University. This was followed by physics and theoretical physics while he took summers off to complete the military training as an ensign. He married Ann in August 1957 and completed the undergraduate studies with a master's degree, well within the standard allotted time and by 1958 he was looking into the options for graduate studies.

Per-Olov Löwdin established the Uppsala Quantum Chemistry Group in 1956 and acquired the first of a series of contracts with the Aerospace Research Laboratories, OAR, through the European Office of Aerospace Research, United States Air Force. This funding extended substantially the base provided by the King Gustaf VI Adolf's 70-Years fund for Swedish Culture and the Knut and Alice Wallenberg's Foundation and Löwdin could offer opportunities for several students to pursue research and studies towards graduate degrees. Yngve Öhrn learned about this possibility through his friend Jan Nordling, who had joined the effort in connection with the acquisition by the Group of the first electronic computer at Uppsala University in late 1957. He approached Löwdin and was quickly introduced to the rather special and quite international ambiance of "research in Atomic, Molecular and Solid State Theory". Thus I, who had joined Löwdin in 1957, got to know, respect, and earn the friendship of Yngve Öhrn.

Graduate studies in quantum chemistry at Uppsala in the late fifties were not formally recognized by the authorities of the University. The Quantum Chemistry Group was associated with the University but was to a large degree autonomous, having for instance its own fiscal arrangements with granting agencies. Löwdin's position as an employee of the research council gave no rights of examination or degree granting privileges. It was indicated that a solution to this calamity might be forthcoming, possibly by an American option. Löwdin was negotiating with institutions "over there" about a cooperative project. This came into being in 1960 when the University of Florida invited Löwdin to establish the Quantum Theory Project as a joint effort between the departments of chemistry and physics. The same year he ascended to the new chair of quantum chemistry at Uppsala whereby the field of study was recognized and the privileges of curriculum definition and examination were established for the new Department of Quantum Chemistry. We could then strive for the degree of *filosofie licentiat* in quantum chemistry. A doctorate required that the candidate had the *fil. lic.*, a published thesis, which could be either a summary of published papers or a monograph and which was submitted to the university library in some 300 copies, and that some further formalities by the college were observed. The successful defense of the thesis in a public ceremony against an opponent appointed by the college was also required. The right to teach, *jus docendi,* at the University was bestowed on the candidate who passed with honors. No remuneration followed the title of *docent* .

Florida appeared to fledgling Swedish scientists to be an exciting place and quite a few of us were given the opportunity for an extended stay in Gainesville with the Quantum Theory Project (3). My first sojourn there was from May 1960 to May 1961. Yngve Öhrn and his family arrived in March 1961. Yngve had then gotten well into the project of electronic correlation in small atoms and while we had struck up a friendship from his first day with the Uppsala group we had not been engaged in the same project. Both of us felt that we could benefit from a closer collaboration. Our outlooks on the field of quantum chemistry, on society, sports, and family were kindred and it was anticipated that we should get on to a common project eventually.

Formal structures for the acquisition of academic degrees in quantum chemistry were established in the year following Löwdin's appointment to the chair in May 1960. Anders Fröman became the first to earn the *fil. lic.* degree in December 1961 and thus started the "institutionalization" of the field. It had until then a flavor of novelty and exclusivity which was embraced by some and repelled others. The new status changed the prerequisites for graduate studies as well, undergraduate chemistry as well as experimental physics should now be a part of the baggage of new students.

Yngve Öhrn returned to Uppsala in May 1963 after nearly two years at Gainesville and a month with Hirschfelder's group at Wisconsin. His work with Jan Nordling on the awkward integrals for three electron wave functions with interparticle coordinates was published (4) and was the basis for his *fil. lic.* degree, which is translated as equivalent to the *Ph. D.* from a U. S. academic institution. Yngve and I initiated our long and close collaboration during the academic year 1963-64. The last paper for my doctorate was submitted in late February 1964 and then we had started a project with Anders Fröman on "Penetration Effects in the ^{2}F-Series of Cs I" (5) which gave a reasonable description of the quantum defects determined by Kjell Bockasten.

2. PROPAGATORS.

Yngve Öhrn's research profile added an essential element in early 1964. A paper of lasting influence in theoretical chemistry and physics was published by John Hubbard in the November 26, 1963 issue of the Proceedings of the Royal

Society (6). The model carrying the author's name was defined and a particular approximation was introduced in the hierarchy of equations for the one-particle Green function. Obvious formal similarities relate Hubbard's model hamiltonian with the prominently featured Pariser-Parr-Pople formalism (7) as extensions of the tight-binding schemes by Hückel (8) for conjugated hydrocarbons and Bloch (9) for crystalline systems. The Green function usage tied in with the elegant work on the Hückel model by Coulson (10) and it became imperative to try and apply this technique to incorporate correlation effects in the molecular situation. Ball and McLachlan (11) published their application of the Random Phase Approximation, with exchange, also termed the linearized time-dependent Hartree-Fock method, to the Pariser-Parr-Pople hamiltonian in 1964. Yngve and I felt that an improvement of the one-particle propagator was essential and used the Hubbard form as an element in the construction of a polarization propagator equation. The effective interaction was chosen as a single center parameter. Results were quite satisfactory and Charles Coulson was sympathetic towards our attempt and communicated the first paper (12).

Delight filled Yngve's and my senses and we pursued the initial ideas. A follow-up paper (13) extended the calculations from the ethylene and benzene cases to larger alternant hydrocarbons. The writing was done at Umeå University in Northern Sweden where I spent a year. This was when I realized that Yngve was my contra-author rather than co-author. The way that we worked out the sentences in a constant interchange of formulations set the pattern for our future cooperation. I still find this paper to be the most readable and convincing piece with my name on it. It was later reprinted in Japan (14). We prepared to present our results at the Molecular Spectroscopy Conference in Copenhagen in August 1965 and considered some heterosystems, among them aniline. Transition energies were excellent but here we began to realize that there were problems with the oscillator strengths and the symmetry classification. Nevertheless we got a most generous "pat-on-the-back" from Coulson who chaired our session.

Yngve completed his work towards the doctorate during some of the coldest months ever recorded at Uppsala in early 1966. A major part was the work on atoms with wave functions in terms of interparticle coordinates (15) but the two papers on propagators (12, 13) were included as well (16). The Swedish procedure at the public thesis defense required that the "respondent" first gave the floor to the official opponent, who had been assigned by the Faculty of Natural Sciences, then there should be an opportunity for opposition "ex auditorio" before the two opponents chosen by the respondent were given the word. Thus, when the faculty opponent Jean-Louis Calais had concluded his perusal of Yngve's thesis, Per-Olov Löwdin asked about the *N-representability* (17) of the Green function methods. Yngve had to admit, of course, that no proof was available on whether the approximations in the chain of equations of motion preserved this feature and that no ground state vector could be constructed. The problem stayed with us and it was a decade later that we could address the problem and obtain a partial solution (18).

Florida and Denmark offered greener pastures to Yngve and me respectively and we left Uppsala in the summer of 1966 to pursue academic careers on different shores of the Atlantic Ocean. Cooperation did not cease, frequent mutual visits and longer stays made it possible to actively build on the propagator formulation. The symmetry and sum rule problems that were apparent in the original Green function decoupling needed attention as well as resolution and we concluded that the original Hubbard truncation of the chain of equations could be generalized through an operator algebra (19) with many implications (20).

Some notes (21) that I prepared for a summer school in Debrecen and Balatonszeplak in 1967 were seeds for our intense efforts to provide a textbook (22) where Green functions should be introduced as the preferred vehicle for the interpretation, calculation, and development of molecular theory. Most of the writing took place during the academic year 1970-71 when Yngve was a visiting professor at Aarhus University, whereas some material had been developed during my stay at Gainesville from April through June 1968 and at the summer school on many-body methods at Batelle, Seattle arranged by Bill Kern in 1969.

Yngve inspired several of his students to explore the electron propagator. George Purvis (23), Henry Kurtz (24), Greg Born (25), and Lynn Tyner (26) furthered the computational as well as the formal development. Manoj Mishra (27) examined the dilated hamiltonian and the related propagator for scattering problems. Vince Ortiz (28) has carried the one-electron Green function studies to an efficient and accurate means for the examination of molecular and atomic anions and so called Rydberg states.

Polarization propagators became a specialty for students at Aarhus (29, 30) and the quest for *N-representability* gave the first result for the Random Phase Approximation where excitations are described by particle-hole operators in a Hartree-Fock reference. Such operators are related to the generators of the unitary group on the one-electron basis (31). Their usefulness in many-electron theory were exploited well by Matsen (32) and Paldus (33). A generator coordinate representation (34) that had been used with advantage in nuclear theoretical physics was brought closer to Yngve through his contacts with the Belgian group around Piet Van Leuven (35). It proved to be applicable in establishing the nature of a ground state that was consistent with the annihilation features of the deexcitation operators (18). Detailed examination of this ground state demonstrated that it was the Antisymmetrized Geminal Power state (36) so prominently featured in the theory of superconductivity (37) and density matrix analysis (38). The state was further analyzed (39) and several calculations and formal improvements came about from work in Yngve's group (40).

3. POTENTIAL CURVES.

Significant contributions to electronic structure theory were initiated by Yngve Öhrn upon arrival in Florida in 1966. He assembled a system of programs, using the Harris-Michels integral code (41) for a Slater type orbital basis in a diatomic frame, and explored a number of systems with his associates. Yngve exhibits, through this effort, his military training. The goals were set out (42):

A computational method which is suitable for studies of this nature should fulfill certain basic requirements: (a) it should be sufficiently economical to allow computation of full potential-energy curves for comparatively large number of states, (b) the calculated potential curves for bound states should give rise to vibrational and rotational constants which are in reasonable agreement with experiment when a comparison is possible, (c) the calculated total energies of all the states should be of comparable accuracy, and (d) the ordering of the states should be correct.

and he accomplished this program through a natural orbital valence-configuration-interaction approach which is a *consolidation of two current techniques* as it is expressed in the first of a series of papers.

There followed, after the initial study of imidogen (42), papers on boron carbide (43), nitric oxide and its ions (44), as well as fluoronitride and its cation (45). Results from these calculations are still useful since they offer mutually consistent representations of potential curves for a considerable number of states. This successful campaign was furthermore beneficial to the development at Aarhus since several Danes were given the opportunity to participate in the work and to learn to appreciate Yngve's way of scientific quest.

Pursuit of scientific accomplishment seems to have many similarities to artistic activities, first comes the desire to find out, then follows the hard work to establish the necessary skills. The latter cannot be secured without careful planning and by persistent following a plan. Öhrn demonstrated the ability to create and keep in motion a demanding research effort in the field of diatomic molecule problems while actively contributing to several other projects.

4. LEADERSHIP.

Per-Olov Löwdin established both the Quantum Chemistry Group at Uppsala and the Quantum Theory Project at Gainesville through his personal and scientific stature and he was unquestionably the Director of both operations. Several prominent scientists were invited during the first years to be Associate Directors and to maintain the spirit when "the Force" was absent. Yngve Öhrn was invited to join the faculty at Florida as Associate Professor in 1966 and he chose to become a member of the Quantum Theory Project. He assumed the Associate Directorship of the Quantum Theory Project in 1967 and proved quickly to have the qualities that allowed him to deal effectively with the university administration at all levels as well as with issues within the rather diverse and extensive body of scientists and students of the Project. Fairness, compassion, and persistence are words that come to my mind when I characterize Yngve's leadership.

Administrative competence and scientific merit made Yngve the choice for chairman of the Department of Chemistry at Florida in 1977. He served two terms and stepped down in 1983. The six years that Öhrn was busy in the chair of the department resulted in substantial changes. An endowment was secured to bring Alan Katritsky to Gainesville and two senior theoreticians, Rod Bartlett and Mike Zerner, were convinced to join the faculty. According to the surveys by C&E News there was an increase in research and development in chemistry at the University of Florida from $769,000 in 1977 to $2,302,000 in 1981 (46). Federally financed spending was nearly constant during the period and the increase was engineered by a department with an aggressively active chairman. Yngve initiated a new resource for the department by inviting persons with a Ph. D. degree in chemistry from Florida to form an advisory panel. These people provide contacts with the market for chemistry graduates and they can, from prominent positions in industry and business, indicate trends and developments as seen from outside the ivory tower of the academic world.

The successful chairmanship and the apparent qualities of personnel management landed several offers of deanships in Yngve's lap. He decided, to my delight, to remain a dedicated scientist. Administrative obligations remained, he assumed the role of Director of the Quantum Theory Project in 1983 and led this rather informal association of faculty members from Chemistry and Physics to be established, in 1987, as the "Institute for Theory and Computation in Molecular and Materials Sciences" with a well defined position in the University system. This was another demonstration of the ability to create coherence and commitment amongst a body of strong minded people with diverse interests. It is a consequence that Florida remains a world renowned hub of quantum chemistry activities.

5. END.

Theoretical challenges were changing dramatically during the seventies with the advent of ever more powerful computational devices. Accurate calculations became feasible for reasonably large systems and the opportunity arose to utilize quality potential energy surfaces for the study of reactive scattering in its quantum mechanical form. The papers by Schatz and Kuppermann (47) herald the beginning of two decades of intense development and progress. Electronic structure methods have, more or less, converged on the use of particular approaches: atomic orbital basis sets in terms of primitive Gaussians, superposition of configurations, possibly with elements of orbital optimization, and sometimes propagators and/or coupled cluster techniques. No consensus is yet on hand in dynamics studies. The search continues and Yngve Öhrn decided that a directly time-dependent scheme has advantages that can be economically explored.

Electron Nuclear Dynamics (48) departs from a variational form where the state vector is both explicitly and implicitly time-dependent. A coherent state formulation for electron and nuclear motion is given and the relevant parameters are determined as functions of time from the Euler equations that define the stationary point of the functional. Yngve and his group have currently implemented the method for a determinantal electronic wave function and products of wave packets for the nuclei in the limit of zero width, a "classical" limit. Results are coming forth: protons on methane (49), diatoms in laser fields (50), protons on water (51), and charge transfer (52) between oxygen and protons.

Dedication to science and commitment to excellence in his work is manifest in Yngve's Electron Nuclear Dynamics effort. Novel use has been made of available techniques in order to establish an effective framework for the exploration of the molecular quantum theoretical thicket by means of inspired choices away from the well trodden paths.

6. A FRIEND.

Compatibility is necessary for a fruitful collaboration in science, as well as in other enterprises. Friendship is not a requirement, thus it is the more rewarding when present. There is a similarity between Yngve's and my background, both of us are an only child, our mothers stayed at home, our fathers were proud providers and skilled workmen, the households encouraged reading, sports, and traditional values of decency. We were trained in a school system where scholastic prowess was rewarded. This system was to be considered elitist and "unfair" to the youth from the working class, such as we were, and became the target of an endless series of misconstrued reforms by nominally socialist governments.

Integrity emanates from Yngve and to win his friendship has been a source of great satisfaction for me. The development of our relation has been eased by the affinity between our wives and children which has allowed us to spend extended periods in each others homes without strain.

Companionship is brought upon you when you travel together and Yngve and I have a number of fine memories to share from joint excursions. We were young "greenhorns" when Löwdin took us around some downtown bars in Chicago during a conference at Northwestern in August 1961, but we escaped unscathed.

The two of us set out for the UK in October 1963 to visit Belfast, Oxford, Keele, and to take part in a meeting "On the teaching of quantum chemistry" which was arranged at Barton Manor northwest of Chester by J. M. Wilson of Loughborough. SAS botched up our flight to London and we were late for our dinner appointment with the Cohens at Belfast but we both had opportunity to present a seminar to David Bates and his group, where Alec Dalgarno was a member. Ronald McCarroll introduced us to the Guinness at a pub and we got a fine start on our week. Mathematical training served us well at the inn at Oxford where the gas heater should be fed coins. We observed the motion of the meter for ten minutes and extrapolated to seven hours with a perfect result. Charles Coulson took us to lunch at Wadham College and we felt quite privileged. The following night we were exposed to sounds beyond description from a plumbing system in a small hotel in Birmingham while Roy McWeeny provided a comfortable expanse in the mansion at the University College, Keele.

Carved wooden bears in all shapes and sizes overwhelmed Yngve and me at our visit to Noboribetsu in 1976. There was an afternoon to spare before the opening of the "Oji International Seminar on Theories and Ab Initio Computations of Molecular Electronic Structure" at Tomakomai, Hokkaido in the fall of 1976 so we wished to experience the hot springs. The train left us with a choice of buses, the destinations of which were clearly indicated in Japanese writing. We found the right one and came to a city in a canyon where the sulfur fumes and hot water let themselves out. The kind reception by Kimio Ohno and Fukashi Sasaki at Hokkaido University remains a vivid memory.

Numerous trips have brought us together at more or less remote places, with or without our families, and I expect that the future will bring many close encounters as well. Denmark houses many friends and admirers of Yngve's and he spends extended periods of time here every year. He is probably the only person that has been invited to share Carl Ballhausen's office at the H. C. Ørsted Institute, an indication of a special recognition by a discerning scientist. Yngve has also a special relation to Belgium where he is welcome and has ongoing projects with both Flemish and Walloon universities. We have both developed long lasting relationships with Utah through Frank Harris and Jack Simons. Ann and Yngve's annual ski vacation at Snowbird is combined with scientific exploits on the campus.

7. CONCLUSION.

The editors asked me to write a personal appreciation of Yngve Öhrn and I am grateful to Henry Kurtz, Vince Ortiz, and Jack Sabin for offering the opportunity. Scientific details of his works can be gathered from the bibliography and very likely from other contributions to this volume. Awards and honors have been left out by me here. A teacher's greatest satisfaction is the continuing respect from his former students and his peers, just as a parent derives his joy from his children's affection.

REFERENCES

(1) Huntford, R. "The new totalitarians"; Allen Lane The Penguin Press: London, 1971; p42ff.

(2) Myrdal , A. ; Myrdal, G. "Kris i befolkningsfrågan"; Stockholm, 1934.

(3) Löwdin, P.-O. In "Partners in Progress", Kastrup, A.; Olsson, N. W.,Eds. Swedish Council of America, 1977; p. 255

(4) Öhrn, Y.; Nordling, J. *J. Chem. Phys.* **1963**, *39* ,1864.

(5) Fröman, A.; Linderberg, J.; Öhrn, Y. *J. Opt. Soc. Am.* **1964**, *54,* 1064.

(6) Hubbard, J. *Proc. Roy. Soc.* (London) **1963**, *A276* , 238.

(7) Pariser, R.; Parr, R. G. *J. Chem. Phys.* **1953,** *21,* 446, 767, Pople, J. A. *Trans. Faraday Soc.* **1953**, *42*, 1375.

(8) Hückel, E. *Z. Physik* **1931**, *70,* 204; **1931**, *72,* 310; **1932**, *76,* 628.

(9) Bloch, F. *Z. Physik* **1929**, *52,* 555; **1929**, *57,* 545.

(10) Coulson, C. A. *Proc. Camb. Phil. Soc.* **1940**, *40,* 201; Coulson, C. A.; Longuet-Higgins, H. C. *Proc. Roy. Soc.* (London) **1947**, *A191*, 39; **1948**, *A192,* 16.

(11) Ball, M. A.; McLachlan, A. D. *Rev. Mod. Phys.* **1964**, *36,* 944; *Mol. Phys.*.**1964**, *7,* 501.

(12) Linderberg, J.; Öhrn, Y. *Proc. Roy. Soc.* (London) **1965**, *A285*, 445.

(13) Öhrn, Y.; Linderberg, J. *Phys. Rev.* **1965**, *A139,* 1063.

(14) "Series of Selected Papers in Physics" by the Physical Society of Japan, *Theory of Molecular Structure* **1966**, *II,* 131.

(15) Öhrn, Y.; Nordling, J. *Arkiv Fysik* **1966**, *31*, 471, Öhrn, Y.; McWeeny, R. *Arkiv Fysik* **1966**, *31,* 561.

(16) Öhrn, Y., "Quantum Mechanical Studies of Eletronic Spectra of Atomic and π-Electron Systems" *Acta Universitatis Upsaliensis* **1966**, *68.*

(17) For a review see Davidson, E. R. "Reduced Density Matrices in Quantum Chemistry"; Academic Press: New York, 1966.

(18) Linderberg, J.; Öhrn, Y., *Int. J. Quant. Chem.* **1977**, *12*, 161.

(19) Linderberg, J.; Öhrn, Y., *Chem. Phys. Lett.* **1967**, *1,* 295.

(20) Goscinski, O. *Int. J. Quant. Chem.* **1968**, *2,* 761; Roth, L. M. *Phys. Rev. Lett.* **1968**, *20,* 431; Rowe, D. J. *Rev. Mod. Phys.* **1968**, *40,* 153.

(21) Linderberg, J. *Magyar Fizikai Folyóirat* **1968**, *16,* 1.

(22) Linderberg, J.; Öhrn, Y. "Propagators in Quantum Chemistry"; Academic Press: London, 1973.

(23) Purvis, G. D.; Öhrn, Y. *J. Chem. Phys.* **1976**, *65***,** 917.

(24) Kurtz, H.; Öhrn, Y. *J. Chem. Phys.* **1978**, *69,* 1162.

(25) Öhrn, Y. ;. Born, G. *Adv. Quant. Chem.* **1981**, *13,* 1.

(26) Tyner Redmon, L.; Purvis, G. D.; Öhrn, Y. *J. Chem. Phys.* **1975**, *63,* 5011.

(27) Mishra, M.; Goscinski, O.; and Öhrn, Y. *J. Chem. Phys.* **1983**, *79,* 5494, 5505. See also Mishra, M. K.; Medikeri, M. N. *Adv. Quant. Chem.*. **1996**, *27,* 223.

(28) Ortiz, J. V.; Öhrn, Y. *Chem. Phys. Lett.* **1981**, *77,* 548. See also Ortiz, J. V. *J. Chem. Phys.* **1998**,*108,* 1008.

(29) Jørgensen, P. *Ann. Rev. Phys. Chem.* **1975**, *26,* 359.

(30) Oddershede, J. *Adv. Quant. Chem.* **1978**, *11,* 275.

(31) A fine introduction is offered by Louck, J. D. *Am. J. Phys.* **1970**, *38,* 3.

(32) Matsen, F. A. *Int. J. Quant. Chem.* **1976**, *10,* 525.

(33) Paldus, J. In "Theoretical Chemistry, Advances and Perspectives", Eyring, H.; Henderson, D., Eds.; Academic Press: New York, 1976; p.131.

(34) Hill, D. L.; Wheeler, J. A. *Phys. Rev.* **1539**, *89,* 1102.

(35) Deumens, E.; Öhrn, Y.; Lathouwers, L.; Van Leuven, P. *J. Chem. Phys.*. **1986**, *84,* 3944, Erratum: *ibid.* **1986**, *85,* 3138.

(36) Coleman, A. J. *Int. J. Quant. Chem.* **1978**, *13,* 67.

(37) Cooper, L. N. *Phys. Rev.* **1956**, *104,* 1189.

(38) Coleman, A. J. *Rev. Mod. Phys.* **1963,** *35,* 668, Sasaki, F. *Phys. Rev.* **1965**,*138,* B1338.

(39) Öhrn, Y.; Linderberg, J., *Int. J. Quant. Chem.* **1979**, *15,* 343.

(40) Ortiz, J. V.; Weiner, B.; Öhrn, Y. *Int. J. Quant. Chem.* **1981**, *S15,* 113; Lozes, R. L.; Weiner, B.; Öhrn, Y., *ibid.* **1981**, *S15,* 129; Jensen, H. J. Aa.; Weiner, B.; Ortiz, J. V.; Öhrn, Y. *ibid.* **1982**, *S16,* 615; Jensen, H. J. Aa.; Weiner, B.; Öhrn, Y. *ibid.* **1982**, *23,* 66.

(41) Harris, F. E.; *J. Chem. Phys.* **1960**, *32,* 3; Harris, F. E.; Michels, H. H. *Adv. Chem. Phys.* **1967**, *13,* 205.

(42) Kouba, J.; Öhrn, Y. *J. Chem. Phys.* **1970**, *52,* 5387.

(43) Kouba, J.; Öhrn, Y. *J. Chem. Phys.* **1970**, *53,* 3923.

(44) Kouba, J.; Öhrn, Y. *Int. J. Quant. Chem.* **1971**, *5,* 539; Thulstrup, E. W.; Öhrn, Y. *J. Chem. Phys.* **1972**, *57,* 3716; Thulstrup, P. W.; Thulstrup, E. W.; Andersen, A.; Öhrn, Y. *J. Chem. Phys.* **1974**, *60,* 3975.

(45) Andersen, A.; Öhrn, Y. *J. Mol. Spectroscopy* **1973**, *45,* 358.

(46) *Chem. Eng. News,* **1983**, *XX,* 63.

(47) Schatz, G. C.; Kuppermann, A. *J. Chem. Phys.* **1976**, *65,* 4642, 4668.

(48) Deumens, E.; Öhrn, Y. *J. Phys. Chem.* **1988**, *92,* 3181; Deumens, E.; Diz, A.; Longo, R.; Öhrn, Y. *Rev. Mod. Phys.* **1994**, *66,* 917.

(49) Jacquemin, D.; Morales, J. A.; Deumens, E.; Öhrn, Y. *J. Chem. Phys.* **1997**, *107,* 6146.

(50) Broeckhove, J.; Coutinho-Neto, M. D.; Deumens, E.; Öhrn, Y. *Phys. Rev.* **1997**, *A56*, 4996.

(51) Hedström, M.; Morales, J. A.; Deumens, E.; Öhrn, Y. *Chem. Phys. Lett.* **1997**, *279,* 241.

(52) Hedström, M.; Deumens, E., Öhrn, Y., *Phys. Rev.* **1998**, *A57,* 2625.

Yngve Öhrn

Teaching Quantum Mechanics

Erik Deumens
Quantum Theory Project
University of Florida
Gainesville, Florida 32611–8435

ABSTRACT

The axiomatic way of teaching quantum mechanics (QM) is analyzed in the light of its effectiveness in making students ready to understand and use QM. A more intuitive method of teaching QM is proposed. An outline of how a course implementing that method could be structured is presented.

Contents

1. Introduction

It is a pleasure to contribute to this volume in honor of Prof. Yngve Öhrn. For those familiar with Yngve's work, this chapter may appear light: it has almost no equations in it. However, the subject tackled lies at the heart of why Yngve likes to use many equations. Yngve likes simplicity and clarity, but not if accuracy of presentation or precision of concepts will be sacrificed. Over the years Yngve and I have often discussed teaching quantum mechanics (QM), in particular, that most courses seem to fail to establish a solid foundation of understanding and skill in the students. This seems to be true for the brightest students as well as for average students. Nevertheless, there are many excellent scientists around who clearly have a full grasp

0065-3276/99 $30.00

of QM. The question then arises: Why does it take a long period of study, including graduate school and significant work beyond, to master QM? Other fundamental theories of Physics, like Newtonian mechanics (NM), Maxwell theory (MT), and the special and the general theories of relativity seem more readily mastered.

I do not propose any new theory, nor do I introduce a new interpretation of QM. The purpose of this chapter is to analyze of the process of teaching QM. I propose a reordering of the stages students are taken through while learning QM with the goal of arriving at the "confident and skilled" level more quickly. No course yet exists following this new sequence, but we present an outline for such a course.

The argument can be sketched as follows: Everyday life forms the intuitive basis for NM. Teaching NM proceeds axiomatically to organize and formalize this knowledge. The teaching process employs only a few "token" experiments. Teaching QM traditionally follows the same path, but the students have no intuitive knowledge whatsoever about the new realm: condensed matter physics, atomic, molecular, and optical physics. QM itself is well understood by those who master it, and they are confident that it is correct. The conventional process of teaching from a small selection of experiments fails for two reasons: First, the students have no prior knowledge and therefore the experiments provide the only experience the students have in the new realm. Second, the commonly used experiments involve macroscopic measuring devices, the area where QM is least understood, most hotly debated and for which no rigorous mathematical solutions are available.

The proposal is to first make students familiar with the new realm of QM by showing them how atoms behave, molecules react, light interacts with atoms and molecules, etc. This should be done without introducing measurements, but rather by using modern computer simulation and visualization technology. Then wave functions and the Schrödinger equation governing them can be introduced at the appropriate mathematical level, depending on whether the course has a requirement of calculus or not. After that, the "principles of quantum mechanics" can be presented as the rules that allow interpretation of experiments with macroscopic measuring devices. Finally. real experiments can be studied in detail to confirm and solidify the new knowledge.

2. Analysis

Before turning to the teaching of QM, let us look at some aspects of teaching classical theories throughout history.

2.1. Historical prelude: Kepler's laws

The axiomatic teaching of Newtonian mechanics (NM)(1, 2), with or without calculus, is very effective. The abstract concepts are quickly connected by students to the practical experience and intuitive insight they have developed throughout their lives. This access to their own direct experience allows each student to make the necessary adjustments and corrections to the abstract concepts if the way they are presented by one particular teacher following one particular textbook does not optimally match the particular student's style and background. As a result, the majority of students of NM gain an effective level of understanding and skill within the time of the course.

In any course of NM, one of the first applications is the derivation of Kepler's laws of planetary motion. Historically this is one of the great triumphs of NM. Kepler's laws state that the orbits of the planets around the sun are ellipses with the sun in one of the focal points and that the speed of the planets is such that equal areas inside of the ellipse are swept in equal times.

Now consider the hypothetical problem of trying to teach the physics of space flight during the period in time between the formulation of Kepler's laws and the publication of Newton's laws. Such a course would introduce Kepler's laws to explain why all spacecraft proceed on elliptical orbits around a nearby heavenly body with the center of mass of that heavenly body in one of the focal points. It would further introduce a second principle to describe course corrections, and define the "orbital jump" to go from one ellipse to another. It would present a table for each type of known spacecraft with the burn time for its rockets to go from one tabulated course to another reachable tabulated course. Students completing this course could run mission control, but they would be confused about what is going on during the "orbital jump" and how it follows from Kepler's laws.

Real understanding of "orbital jumps" comes with NM and Newton's laws, in particular

$$\vec{F} = m\vec{a} \tag{1}$$

the force equals mass times acceleration. The knowledge that Kepler's laws follow from this, that the expression for the force of gravitational attraction between massive bodies is

$$\vec{F} = -\frac{Gm_1m_2}{|\vec{r}_2 - \vec{r}_1|^3}(\vec{r}_2 - \vec{r}_1) \tag{2}$$

and that firing the rockets produces just another force, unifies the mechanics of orbital motion and "orbital jumps". This makes the relevant processes in the realm of the physics of space flight understandable.

The principles of Kepler's laws and "orbital jumps" in isolation would leave students confused. Alternatively suppose students were taught that only free-falling space flight can be understood from Kepler's laws, and that the tables for course corrections had been constructed from careful experimentations and observations. In this case, students would not be confused either. The confusion comes from stating that everything will be explained theoretically and then only explaining half.

2.2. Historical prelude: Maxwell Theory

Similarly, the axiomatic teaching of Maxwell's theory (MT) of electromagnetism (EM) is very effective. Students again can draw from their practical experience and intuitive insight, most of which has now been formalized and made rigorous by NM. This time they focus on a special class of phenomena related to electrical and magnetic forces and on waves, all of which they have experienced before the course starts. Elastic waves in water and sound waves in air, water and solids are familiar and have been formally described as applications of NM. The abstract concepts of the electromagnetic fields can therefore be grasped and mastered effectively.

In a modern course, no one tries to explain the electromagnetic waves as elastic waves of some special medium in an attempt to reduce the physics of EM to an application of NM. On the contrary, the EM field is introduced as an irreducible entity defined and described fully and exhaustively by Maxwell's equations. The electromagnetic field is one of the four fundamental forces of Nature and is a cornerstone of modern.Physics.

However, in Maxwell's days everyone assumed that there had to be a mechanical underpinning for the theory of EM. Many researchers worked on very detailed "hidden variable" theories for the EM field, in an attempt to prove that the laws of EM were in fact a theorem in NM, just like Kepler's laws are a theorem in NM. No one noticed that it was impossible to do this, since Maxwell's equations are not Galilei invariant and Newton's laws are. That includes Lorentz who discovered around 1900 that the Maxwell equations are invariant under another transformation that now bears his name.

Michelson and Morley tried to establish the absolute motion of the Earth through space by measuring the difference in the velocity of light, a form of EM waves, parallel and perpendicular to the Earth's path. It is well-known that they failed to observe any difference. Since we know that the Earth is moving, this experiment implied a fundamental flaw in their reasoning. They conceived their experiment from the intuition of EM fields as elastic waves in some mechanical medium called "Ether", not from a thorough mathematical solution of Maxwell's equations. Such an analysis, impossible in practice, would have lead them to the correct conclusion. One of the consequences of Michelson and Morley failing to observe the motion of the Earth, was that the quest for a mechanical foundation for the EM field, a "hidden variable" theory for MT, was abandoned. The fact that Michelson and Morley posed a valid question is evident from the fact that the Cosmic Background Explorer satellite, launched by NASA in 1989, provided accurate data to yield a Solar System velocity of 370.6 $\pm$ 0.5 km/s in the rest frame of the Cosmic Microwave Background.(3)

The Michelson and Morley experiment shows the critical importance of intuitive concepts and understanding in the progress of Physics. It is essential that the abstract concepts and intuitive notions of a theory are accurate, precise and correct. A necessary condition is that they correspond exactly to the mathematical formulation of these concepts. In the case of MT during the last century and the beginning of this century, that correspondence was flawed. This example demonstrates the importance of teaching students both the concepts and the mathematics and to make sure that the relationship between the two is fully understood.

2.3. Axiomatic teaching of Quantum Mechanics

The special theory of relativity (STR), describing the correction to NM for motions with high velocity comparable to the speed of light, and the general theory of relativity (GTR), describing the extension and correction to Newton's theory of the gravitational force for time varying and strong fields, can both be taught very effectively using the axiomatic approach. The reason is, as with MT, that students have a large set of experiences and intuitions to which a precise modification must be made. Within the framework of each new theory, the concepts and axioms of NM can be recovered by a well defined and clear process. That process can be applied during any analysis or derivation in the new theory, so that the deviations from the familiar and intuitive world can be clearly exhibited. Thus, the students can methodically and effectively extend their intuition and experiences to include phenomena described by STR and GTR. This typically happens quickly and with limited confusion, even though the students do not have any personal experience with relativistic phenomena that are in contradiction with NM.

In view of the success of the axiomatic method for teaching NM, MT, STR and GTR, it should be no surprise that QM is also taught this way. The works of Dirac(4), Von Neumann(5) and Pauli(6) all were written by masters in the field who were completely familiar with all the phenomena relating to QM. To them and their colleagues these works are a clean axiomatic summary of what they knew.

Most, if not all, textbooks and courses of QM proceed in a similar axiomatic way. At the end of the introduction to QM, most, if not all, students are confused. It takes a long time, several advanced courses, graduate school and often more to master QM. It is ironic that current masters of QM also internalize the paradigm of QM instruction that initially confused and frustrated them, and thus instruct new students in the same way. As a result, those students who take QM as part of a general education and not as the first step of a career in Physics, are left confused forever. I believe this constitutes a failure of the Physics community to communicate to humanity its most precious, successful and beautiful accomplishment.

What goes wrong in the process of conveying the knowledge about QM from the master to new students has three aspects:

1. There is a total lack of reference points for the students regarding the subject.
2. The wording of the physical principles of QM, which guides thinking about QM problems, is imprecise.
3. The domain of applicability of some of the most crucial postulates is not clearly established.

We examine these aspects in detail.

2.4. Problem: lack of reference points

The realm of phenomena into which the axioms of QM are bringing structure and order is completely unknown to the students. Only after many years of intense work does one accumulate a meaningful collection of experiences, similar to the experience with the operation of NM gathered by every child during the first years of life. The collection of experiments used to show the peculiar nature of QM is very small and it is mainly chosen to surprise rather than to teach. Worst of all, these experiments do not deal with QM systems but with the interaction between QM systems and classical measuring devices. This subject of measurement theory is the most controversial part of QM and it is usually dealt with either in a superficial manner or by a long-winded philosophical exposé, but rarely in a scientific manner. In the last fifteen years a more detailed approach to measurement centered around the concept of "decoherence" has emerged(7, 8), but it is not yet clear whether it will become a rigorous theory of measurement.

2.5. Problem: imprecise boundaries

The second problem with the axiomatic teaching of QM is analogous to teaching the physics of space flight with only Kepler's laws. Most QM courses start out with one or more mystifying experiments to surprise and intrigue the student, like the beginning of a good detective story. Most detective stories, however, do explain the mystery before the novel ends, whereas students of QM find the explanation of the experiments lacking. Then the axioms of QM are exhibited in an attempt to explain the experiment. The initial experiment is like the full trajectory of a spacecraft that is launched from Earth and sent to land on the Moon. This trajectory cannot be explained from Kepler's laws alone, since it does not explain the course corrections. It would be clearer to talk about the free-falling parts of the trajectory and the course corrections separately. Similarly, in teaching QM, the dynamics of QM should be introduced separately from the process of experimental verification to eliminate confusion about what explains what.

2.6. Problem: inaccurate formulation

The third problem is like the confusion caused in MT by maintaining the concept of the Ether. Most practitioners of QM think about microscopic systems in terms of the principles of QM: probability distributions, superposition principle, uncertainty relations, complementarity principle, correspondence principle, wave function collapse. These principles are an approximate summary of what QM really is, and following them without checking whether the Schrödinger equation actually confirms them does lead to error.

The new delightful book by Greenstein and Zajonc(9) contains several examples where the outcome of experiments was not what physicists expected. Careful analysis of the Schrödinger equation revealed what the intuitive argument had overlooked and showed that QM is correct. In Chapter 2, "Photons", they tell the story that Einstein got the Nobel Prize in 1922 for the explaining the photoelectric effect with the concept of particle-like photons. In 1969 Crisp and Jaynes(10) and Lamb and Scully(11) showed that the quantum nature of the photoelectric effect can be explained with a classical radiation field and a quantum description for the atom. Photons do exist, but they only show up when the EM field is in a state that is an eigenstate of the number operator, and they do not reveal themselves in the photoelectric effect.

Another example of slight conceptual inaccuracy is given by the Wigner function(12) and Feynman path integral(13). Both are useful ways to look at the wave function. However, because of the prominence of classical particles in these concepts, they suggest the view that QM is a variant of statistical mechanics and that it is a theory built on top of NM. This is unfortunate, since one wants to convey the notion that NM can be recovered as an integral part of QM pertaining to for macroscopic systems.

A final example is the concept of QM state. It is often stated that the wave function must be square integrable because the modulus square of the wave function is a probability distribution. States in QM are rays in Hilbert space, which are equivalence classes of wave functions. The equivalence relation between two wave functions is that one wave function is equal to the other multiplied by a complex number. The space of QM states is then a projective space, which by an infinite stereographic projection is isomorphic to a sphere in Hilbert space with any radius, conventionally chosen as one. Hence states can be identified with normalized wave functions as representatives from each equivalence class. This fact is important for the probability interpretation, but it is not a consequence of the probability interpretation.

The inaccuracies in the formulation have a more severe impact in teaching QM than in any other axiomatic instruction, because students have no direct knowledge of their own to properly integrate and assimilate the principles and postulates of QM.

2.7. Solution: reference points from a journey

Traditionally, because of the history of the discovery of QM, experience has been equated to experiments. Experiments with macroscopic devices involve measurement theory, which most physicist agree is the most difficult and least understood part of QM. These are therefore often unclear, confusing and contradictory. They are the wrong place to provide experience. They are absolutely essential to verify the correctness of QM, but as the most difficult part of QM they should be treated at the end of the course, not at the beginning. Also it is not sufficient to provide a few experiences, but a thorough immersion is required. Hence my choice of the word "journey"; a short "visit" will not do.

With time-dependent computer simulation and visualization we can give the novices to QM a direct "mind's eye" view of many elementary processes. The simulations can include interactive modes where the students can apply forces and radiation to control and manipulate atoms and molecules. They can be posed challenges like trapping atoms in laser beams. These simulations are the "inside story" of real experiments that have been done, but without the complexity of macroscopic devices. The simulations should preferably be based on rigorous solutions of the time dependent Schrödinger equation, but they could also use proven approximate methods to broaden the range of phenomena to be made accessible to the students. Stationary states and the dynamical transitions between them can be presented as special cases of the full dynamics. All these experiences will create a sense of familiarity with the QM realm. The experiences will nurture accurate intuition that can then be made systematic by the formal axioms and concepts of QM.

2.8. Solution: precise boundaries

Because of our inability to analyze the interaction of microscopic QM systems and macroscopic measuring devices to a sufficient degree, we make use of a set of empirical rules that are known as measurement theory. Some day, measurement theory will become a proven set of theorems in QM, , as the proponents of the decoherence theory, among others, claim. Until such time, it is beneficial to introduce the measurement process, and the principles associated with it, separately from the dynamics described by the Schrödinger equation.

This separation will allow the students to properly assess the measurement process, which plays a special and complex role in QM that is different from its role in any classical theory. Just as Kepler's laws only cover the free-falling part of the trajectories and the course corrections, essential as they may be, require tabulated data, so too in QM, it should be made clear that the Schrödinger equation governs the dynamics of QM systems only and measurements, for now, must be treated by separate rules. Thus the problem of inaccurate boundaries of applicability can be addressed by clearly separating the two incompatible principles governing the change of the wave function: the Schrödinger equation for smooth evolution as one, and the measurement process with the collapse of the wave function as the other.

2.9. Solution: accurate formulation

Even though historically the formulation of all QM phenomena in terms of macroscopic phenomena was very significant and useful, it seems counter productive to insist on making every student go through the same laborious process.

The Ether is not useful to teach MT. The EM field is most effectively viewed as an irreducible entity completely defined by Maxwell's equations. (If one wants to make the interaction with point charges in NM or QM explicit, one can add the Lorentz force or the minimal coupling.) All physical properties of the EM field and its interaction with matter follow from Maxwell's equations and the matter equations.

Similarly, after 70 years of successful application of QM, it is clear that the wave function does not describe the properties of something else. There is no Ether, there are no hidden variables. The effect on the wave function of every interaction between QM systems is described by the Schrödinger equation. The wave function does not describe something unknown and eluding(14, 15), that is sometimes a wave and sometime a particle.

The wave function is an irreducible entity completely defined by the Schrödinger equation and this should be the core of the message conveyed to students. It is not useful to introduce any hidden variables, not even Feynman paths. The wave function is an element of a well defined state space, which is neither a classical particle, nor a classical field. Its nature is fully and accurately defined by studying how it evolves and interacts and this is the only way that it can be completely and correctly understood. The evolution and interaction is accurately described by the Schrödinger equation or the Heisenberg equation or the Feynman propagator or any other representation of the dynamical equation.

2.10. Intuitive teaching of Quantum Mechanics

It is beyond the scope of this chapter to describe in detail the contents of an introductory QM course based on the rearranged sequence of material proposed here. However, based on the analysis we propose an approach to teaching QM that consists of three sections:

I. **Introduction to quantum phenomena.** This section takes the students into the realm of atoms and molecules and uses mathematical modeling and computer simulation, animations and visualization to give them experience in the phenomena that must be described by QM and cannot be described by NM. The simulations could cover the following processes among others:

   a. Photoelectric effect: Observe electronic wave function of a surface or cluster as light shines on the surface, the light can be treated classically (10).
   b. Spectroscopy: Light radiating atoms and molecules causes excitation to excited rotational, vibrational and electronic states, which then decay emitting radiation. The intensity of the emitted radiation shows peaks at the excitation energies corresponding to transitions between stationary states.
   c. Black body radiation and Planck's Law.
   d. Chemical reactions: Atoms and molecule interact when they get sufficiently close and form new molecules. Electronic charge distributions change during the reaction.
   e. Metastable states, ionization, tunneling: Atoms have long lived states that decay by emission of radiation or by ionization. And nuclei have metastable states that can lives thousands of years before radioactive decay breaks them apart or turns them into the nucleus of a different chemical element.

f. Crystal structure: Matter does not shrink to ever smaller size, but the lowest-energy state has a finite size which accounts for the stability of matter.
g. Superconductivity and Josephson junction.
h. Trapping of single atoms with light pressure.

In these simulations the focus should be on the wave function by displaying modulus squared of the wave function as a three dimensional field changing in time, without mentioning the probability interpretation. Concepts like position and energy and momentum and angular momentum should be used, without going into formal definitions, as pertaining globally to wave functions. Thus students can acquire reference points as to how wave functions evolve and interact.

II. **Principles of Quantum Mechanics.** This section defines the state of a system, the wave function, the Schrödinger equation, the superposition principle and the different representations. It can be given with or without calculus and with or without functional analysis, depending on the mathematical preparation of the students. Additional topics include:

a. Exactly solvable systems: harmonic oscillator, rigid rotor, hydrogen atom, Morse oscillator.
b. Classical dynamics is studied as a special case by analyzing the Ehrenfest theorem, coherent states (16) and systems with quasi classical dynamics like the rigid rotor for molecules (17) and the oscillator (18) for various particle systems and for EM field in a laser.
c. Analysis of dynamics in terms of eigenstates, both for discrete and continuous spectra.
d. Time dependent perturbation theory.
e. Infinite and extended condensed matter systems.
f. The incomplete and active subject of describing macroscopic systems.
g. Quantum statistical mechanics with the concepts of mixed states, density operators and the Liouville equation.

III. **Experimental observation of Quantum Mechanics.** Only this final section should address the rules that govern interpretations of experiments measuring properties of QM systems with macroscopic devices. This includes probability interpretation, uncertainty relations, complementarity and correspondence. Then experiments can be discussed to show how the wave functions manipulated in section I can be used to predict the probabilistic outcome of experiments.

3. Conclusion

In this chapter, we analyzed the process of teaching QM from a set of axioms in an attempt to identify why so many students are confused and dissatisfied at the end of the typical introductory course. Although students majoring in Physics or Chemistry eventually do master the subject and become effective practitioners of QM, those students who have an introductory course as their only contact with QM are left with an unsatisfactory state of understanding. We identify three problems:

1. Lack of reference points. The postulates of QM are given to the students without giving them any experience with the new realm that is supposedly governed by these principles. The only experience that is provided is in the form of experiments for which the interpretation is not straightforward.
2. Imprecise boundaries. The basic concept of the state of a system is governed by two mutually incompatible laws, namely the Schrödinger equation for normal dynamics and the measurement process for interactions with macroscopic devices. It is not made clear where the applicability of one ends and the other begins.
3. Inaccurate formulation. The principles of QM are, because of their complex nature, not manifestly equivalent to the mathematical formulation. This leads to incorrect expectations for experiments when following the principles without doing the mathematical computation.

To avoid the confusion caused by the above problems, a three-step process is proposed:

1. Familiarize the student extensively with time dependent dynamics of QM by using computer simulation, animation and visualization.
2. Introduce the wave function and the Schrödinger equations to formalize and organize this experience.
3. Make the connection with experiments to allow verification of the theory of QM.

References

(1) Goldstein, H. *Classical Mechanics*, 2nd ed.; Addison-Wesley: Reading, 1980.

(2) Abraham, R.; Marsden, J. E. *Foundations of mechanics*, 2nd ed.; Benjamin Cummings: Reading, 1978.

(3) Bennet, C. L.; Turner, M. S.; White, M. *Physics Today* **Nov 1997**, *50*, 32–38.

(4) Dirac, P. A. M. *The principles of Quantum Mechanics*, 4 ed.; Clarendon: Oxford, 1974.

(5) von Neumann, J. *Methemahiesche Grundlagen der Quantenmechanik*; Springer: Berlin, 1968.

(6) Pauli, W. *General Principles of Quantum Mechanics*; Springer: New York, 1980.

(7) Omnès, R. *Rev. Mod. Phys.* **1992**, *64*, 339–382.

(8) Zurek, W. H. In *The wave particle dualism*; Dimer, S., Ed.; Reidel: Dordrecht, 1984, p 515.

(9) Greenstein, G.; Zajonc, A. G. *The Quantum Challenge: Modern Research on the Foundations of Quantum Mechanics*; Jone: Sudbury, MA, 1997; Vol. 1 of *The Challenge Series*.

(10)Crisp, M. D.; Jaynes, E. T. *Phys. Rev.* **1969**, *179*, 1253–1261.

(11)Lamb, Jr., W. E.; Scully, M. O. In *Polarisation, Matière et Rayonnemt*; Presses University France, 1969.

(12)Wigner, E. P. *Phys. Rev.* **1932**, *40*, 749.

(13)Feynman, R. P.; Hibbs, A. R. *Quantum Mechanics and Path Integrals*; McGraw-Hill: New York, 1965.

(14)Aharanov, Y.; Anandan, J. *Phys. Rev. A* **1993**, *47*, 4616.

(15)Aharanov, Y.; Vaidman, L. *Phys. Lett. A* **1993**, *178*.

(16)Klauder, J. R.; Skagerstam, B.-S. *Coherent States, Applications in Physics and Mathematical Physics*; World Scientific: Singapore, 1985.

(17)Janssen, D. *Soviet Journal of Nuclear Physics* **April 1977**, *25*, 479–484.

(18)Cohen-Tannoudji, C.; Diu, B.; Laloe, F. *Quantum Mechanics*, second ed.; Jon Wiley: New York, 1977.

Toward an Exact One–Electron Picture of Chemical Bonding

J. V. Ortiz
Department of Chemistry
Kansas State University
Manhattan, Kansas 66506–3701
U. S. A.

This article is dedicated to the author's graduate advisor, Professor Yngve Öhrn.

Contents

ADVANCES IN QUANTUM CHEMISTRY, VOLUME 35

0065-3276/99 $30.00

1 Conceptual Goals for One–Electron Theories

One–electron pictures of molecular electronic structure continue to inform interpretations of structure and spectra. These models are the successors of qualitative valence theories that attempt to impose patterns on chemical data and to stimulate experimental tests of predictions. Therefore, in formulating a one–electron theory of chemical bonding, it is desirable to retain the following conceptual advantages.

One–electron Energies Orbital energies should be equal to ionization energies and electron affinities (electron binding energies). They should be the eigenvalues of a one–electron operator with an effective, many–electron potential. All ionization energies should be predicted, without artificial distinctions between core versus valence or principal versus shakeup descriptions of final states. Bound and unbound electron–attached states should correspond to negative and positive orbital energies, respectively.

One–electron States For each electron binding energy, there should be a corresponding orbital that is an eigenfunction of the same effective, one–electron operator. Orbitals should display phase relationships that permit interpretations in terms of kinetic energy and interference effects. Transition probabilities corresponding to the electron binding energies should be related to the same orbitals. Effects of electron detachment or attachment on nuclear positions should depend on forces associated with the orbitals.

Density Matrices One–electron density matrices of initial and final states should be related to the orbitals used to interpret electron binding energies. Their eigenvalues should lie between zero and unity and their traces should equal the number of electrons in each state. One–electron properties should be size–extensive.

Total Energies Total energies should be related to orbitals and their energies. They should be size–extensive as well.

Numerical Representation The theory should be systematically improvable with respect to basis sets or integration schemes.

Independence There should be no parameters which have been adjusted to reproduce certain data from experiment or from unrelated sources.

Potential Energy Surfaces All potential energy surfaces generated with this theory should be smooth. No discontinuities due to symmetry reduction should occur.

Electron Correlation The theory should have a limiting case of exact total energies, electron binding energies and corresponding transition probabilities.

Hartree–Fock (HF), molecular orbital theory satisfies most of the criteria, but qualitative failures and quantitative discrepancies with experiment often render it useless. Methods that systematically account for electron correlation, employed in pursuit of more accurate predictions, often lack a consistent, interpretive apparatus. Among these methods, electron propagator theory [1] is distinguished by its retention of many conceptual advantages that facilitate interpretation of molecular structure and spectra [2, 3, 4, 5, 6, 7, 8, 9].

2 Electron Propagator Poles and Residues

The physical meaning of the electron propagator rests chiefly in its poles (energies where singularities lie) and residues (coefficients of the terms responsible for the singularities) [1]. In its spectral form, the r, s element of the electron propagator matrix is

$$G_{rs}(E) = \lim_{\eta \to 0} \{ \sum_n \frac{\langle N | a_r^\dagger | N-1, n \rangle \langle N-1, n | a_s | N \rangle}{E + E_n(N-1) - E_0(N) - i\eta} + \sum_m \frac{\langle N | a_s | N+1, m \rangle \langle N+1, m | a_r^\dagger | N \rangle}{E - E_m(N+1) + E_0(N) + i\eta} \}. \quad (1)$$

The propagator matrix is energy-dependent; poles occur when E equals an ionization energy, $E_0(N) - E_n(N-1)$, or an electron affinity, $E_0(N) - E_m(N+1)$. Dyson orbitals (DOs) for ionization energies are defined by

$$\phi_n^{Dyson,IE}(x_1) = \int \Psi_N(x_1, x_2, x_3, \ldots, x_N) \Psi_{N-1,n}^*(x_2, x_3, x_4, \ldots, x_N) \, dx_2 dx_3 dx_4 \cdots dx_N, \quad (2)$$

and are related to propagator residues (also known as Feynman–Dyson amplitudes) by

$$\phi_n^{Dyson,IE}(x) = \sum_r \phi_r(x) \langle N-1, n | a_r | N \rangle. \quad (3)$$

For electron affinities, the corresponding relationships are

$$\phi_n^{Dyson,EA}(x_1) = \int \Psi_N^*(x_2, x_3, x_4, \ldots, x_{N+1}) \Psi_{N+1,n}(x_1, x_2, x_3, \ldots, x_{N+1}) \, dx_2 dx_3 dx_4 \cdots dx_{N+1} \quad (4)$$

and

$$\phi_n^{Dyson,EA}(x) = \sum_r \phi_r(x) \langle N+1, n | a_r^\dagger | N \rangle. \quad (5)$$

Many kinds of transition probabilities depend on DOs. Photoionization cross sections, σ^{PI}, are proportional to the absolute squares of matrix elements between DOs and continuum orbitals, or

$$\sigma^{PI} = \kappa^{PI} |\langle \phi^{Dyson} | \hat{T} \phi^{Continuum} \rangle|^2, \tag{6}$$

where κ^{PI} is a constant and $\hat{T}$ is a transition operator describing the interaction between electrons and the radiation field [10]. DOs also are useful in computing cross sections for various electron scattering processes [11, 12].

For the final state, n, the pole strength, P_n, is defined by

$$P_n = \int |\phi_n^{Dyson}(x)|^2 dx. \tag{7}$$

Normalized DOs therefore read

$$\psi_n^{Dyson}(x) = P_n^{-\frac{1}{2}} \phi_n^{Dyson}(x). \tag{8}$$

In the uncorrelated limit, where the many–electron Fock operator replaces the full electronic Hamiltonian, familiar objects of HF theory are recovered as special cases. $|N\rangle$ becomes a HF, determinantal wavefunction for N electrons and $|N \pm 1\rangle$ states become the frozen–orbital wavefunctions that are invoked in Koopmans's theorem. Poles equal canonical orbital energies and DOs are identical to canonical orbitals.

It is possible to use full or limited configuration interaction wavefunctions to construct poles and residues of the electron propagator. However, in practical propagator calculations, generation of this intermediate information is avoided in favor of direct evaluation of electron binding energies and DOs.

3 Operator Space Diagonalization

Through introduction of superoperators and a corresponding metric [13], the propagator may be represented more compactly [2, 6]. Superoperators act on field operator products, X, where the number of annihilators exceeds the number of creators by one. The identity superoperator, $\hat{I}$, and the Hamiltonian superoperator, $\hat{H}$, are defined by

$$\hat{I}X = X \tag{9}$$

and

$$\hat{H}X = [X, H]_-, \tag{10}$$

respectively. The superoperator metric, defined by

$$(\mu|\nu) = \langle N|[\mu^\dagger, \nu]_+|N\rangle, \tag{11}$$

depends on the choice of the N–electron reference state, $|N\rangle$. Consideration of ionization energy and electron affinity poles in a single propagator leads to the anticommutator, $\mu^\dagger\nu + \nu\mu^\dagger$, contained in the metric definition. (Had this discussion considered either the first or second summations in equation 1, the anticommutator would have been abandoned in favor of $\mu^\dagger\nu$ or $\nu\mu^\dagger$, respectively.) With this notation, one may write

$$G_{rs}(E) = (a_r|(E\hat{I} - \hat{H})^{-1}a_s). \tag{12}$$

Thus the matrix elements of the electron propagator are related to field operator products arising from the superoperator resolvent, $(E\hat{I} - \hat{H})^{-1}$, that are evaluated with respect to $|N\rangle$. In this sense, electron binding energies and DOs are properties of the reference state.

In matrix notation, equation 12 is rewritten as

$$\mathbf{G}(E) = (\mathbf{a}|(E\hat{I} - \hat{H})^{-1}\mathbf{a}), \tag{13}$$

where the rank of the propagator matrix equals the number of spin–orbitals. After inner projection,

$$\mathbf{G}(E) = (\mathbf{a}|\mathbf{u})(\mathbf{u}|(E\hat{I} - \hat{H})\mathbf{u})^{-1}(\mathbf{u}|\mathbf{a}), \tag{14}$$

where $\mathbf{u}$ is the vector of all X field operator products. An inverse matrix instead of an inverse superoperator is considered henceforth.

If $\mathbf{u}$ is partitioned into the primary space, $\mathbf{a}$, and an orthogonal space of product operators, $\mathbf{f}$, the partitioned form of the propagator matrix reduces to

$$\mathbf{G}(E) = \begin{bmatrix} \mathbf{1} & \mathbf{0} \end{bmatrix} \begin{bmatrix} E\mathbf{1} - (\mathbf{a}|\hat{H}\mathbf{a}) & -(\mathbf{a}|\hat{H}\mathbf{f}) \\ -(\mathbf{f}|\hat{H}\mathbf{a}) & E\mathbf{1} - (\mathbf{f}|\hat{H}\mathbf{f}) \end{bmatrix}^{-1} \begin{bmatrix} \mathbf{1} \\ \mathbf{0} \end{bmatrix}. \tag{15}$$

Poles of the propagator therefore occur at values of E that are equal to eigenvalues, ω, of the superoperator Hamiltonian matrix:

$$\omega_n \begin{bmatrix} \mathbf{U_{a,n}} \\ \mathbf{U_{f,n}} \end{bmatrix} = \begin{bmatrix} (\mathbf{a}|\hat{H}\mathbf{a}) & (\mathbf{a}|\hat{H}\mathbf{f}) \\ (\mathbf{f}|\hat{H}\mathbf{a}) & (\mathbf{f}|\hat{H}\mathbf{f}) \end{bmatrix} \begin{bmatrix} \mathbf{U_{a,n}} \\ \mathbf{U_{f,n}} \end{bmatrix} \tag{16}$$

or

$$\mathbf{U}\omega = \hat{\mathbf{H}}\mathbf{U}. \tag{17}$$

In the new basis of operators,

$$\mathbf{G}(E) = \begin{bmatrix} \mathbf{1} & \mathbf{0} \end{bmatrix} [\mathbf{U}(E\mathbf{1} - \omega)^{-1}\mathbf{U}^\dagger] \begin{bmatrix} \mathbf{1} \\ \mathbf{0} \end{bmatrix}. \tag{18}$$

The DO corresponding to the pole, ω_n, is

$$\phi_n^{Dyson} = \sum_r \phi_r U_{r,n}^* \tag{19}$$

and the pole strength reads

$$P_n = \sum_r |U_{r,n}|^2. \tag{20}$$

Note that contributions from the secondary sector of the eigenvectors, $\mathbf{U_f}$, do not appear in the residues, for the summation index, r, pertains to spin–orbitals only.

Because this route to poles and residues requires only solutions of equation 17, the usual matrix diagonalization techniques characteristic of CI calculations may be applied [14, 15]. The chief conceptual difference between diagonalization of $\hat{\mathbf{H}}$ and diagonalization of the Hamiltonian matrix in Hilbert space is that operators, not many–electron configurations, form the basis. In addition, solutions correspond not to state energies, but to electron binding energies. For each of these energy differences, there corresponds an operator expressed as a linear combination of X components.

Simultaneous treatment of the ionization energy and electron affinity components of the electron propagator allows operator mixings that are not present in Hilbert space methods for energy differences. For ionization energies, operators corresponding to virtual (particle or p) orbitals, shakeon (two particle, one hole or 2p–h) and other (3p–2h, 4p–3h, 5p–4h, et cetera) processes may contribute to the eigenvector, $\mathbf{U}$, in addition to the usual CI–like operators for occupied (hole or h) orbitals, shakeup (two hole, one particle or 2h–p) and other (3h–2p, 4h–3p, 5h–4p, et cetera) processes that generate (N–1)–electron states in Hilbert space when operating on a reference configuration. Electron affinity operators also have h, p, 2h–p, 2p–h and higher–operator–product constituents.

4 Solving One–Electron Equations

Partitioning the operator manifold can lead to efficient strategies for finding poles and residues that are based on solutions of one–electron equations with energy–dependent effective operators [16]. In equation 15, only the upper left block of the inverse matrix is relevant. After a few elementary matrix manipulations, a convenient form of the inverse–propagator matrix emerges, where

$$\mathbf{G}^{-1}(E) = E\mathbf{1} - (\mathbf{a}|\hat{H}\mathbf{a}) - (\mathbf{a}|\hat{H}\mathbf{f})\left[E\mathbf{1} - (\mathbf{f}|\hat{H}\mathbf{f})\right]^{-1}(\mathbf{f}|\hat{H}\mathbf{a}). \tag{21}$$

Because

$$(a_r|\hat{H}a_s) = h_{rs} + \sum_{tu}(rs||tu)\rho_{tu}, \tag{22}$$

where ρ is the one–electron density matrix, the primary operator space block of $\hat{\mathbf{H}}$ may be considered to be a generalized Fock matrix, $\mathbf{F}$. It is possible to separate the correlated and uncorrelated contributions to the $(\mathbf{a}|\hat{H}\mathbf{a})$ block. In the canonical MO basis,

$$(a_r|\hat{H}a_s) = \epsilon_r\delta_{rs} + \sum_{tu}(rs||tu)\rho^c_{tu} = F_{rs}, \tag{23}$$

where the correlation contribution to the one–electron density matrix is ρ^{c} and

$$\rho = \rho^{HF} + \rho^c. \tag{24}$$

Elements of the zeroth–order, inverse–propagator matrix are

$$\mathbf{G}_0^{-1}(E)_{rs} = (E - \epsilon_r)\delta_{rs}. \tag{25}$$

(The poles correspond to Koopmans's theorem.) The inverse–propagator matrix and its zeroth–order counterpart therefore are related through

$$\mathbf{G}^{-1}(E) = \mathbf{G}_0^{-1}(E) - \mathbf{\Sigma}(\infty) - \mathbf{\Sigma}'(E) \tag{26}$$

where

$$\mathbf{\Sigma}(\infty)_{rs} = (a_r|\hat{H}a_s)_{correlation} = \sum_{tu}(rs||tu)\rho^c_{tu} \tag{27}$$

and

$$\mathbf{\Sigma}'(E) = (\mathbf{a}|\hat{H}\mathbf{f})\left[E\mathbf{1} - (\mathbf{f}|\hat{H}\mathbf{f})\right]^{-1}(\mathbf{f}|\hat{H}\mathbf{a}). \tag{28}$$

Corrections to the zeroth–order, inverse propagator in equation 26 are gathered together in a term known as the self–energy matrix, $\mathbf{\Sigma}(E)$. The Dyson equation may be written as

$$\mathbf{G}^{-1}(E) = \mathbf{G}_0^{-1}(E) - \mathbf{\Sigma}(E). \tag{29}$$

In the self–energy matrix, there are energy–independent terms and energy–dependent terms:

$$\mathbf{\Sigma}(E) = \mathbf{\Sigma}(\infty) + \mathbf{\Sigma}'(E). \tag{30}$$

In the limit of $|E| \to \infty$, $\mathbf{\Sigma}(E)$ approaches its energy–independent component, $\mathbf{\Sigma}(\infty)$.

When $|det\mathbf{G}(E)| \to \infty$, E is a pole. It is equivalent to require $\mathbf{G}^{-1}(E)$ to have a vanishing eigenvalue at the pole energy, where

$$\mathbf{G}^{-1}(E)\mathbf{C}(E) = 0\mathbf{C}(E). \tag{31}$$

This condition implies that

$$[\epsilon + \mathbf{\Sigma}(E)]\,\mathbf{C}(E) = E\mathbf{C}(E). \tag{32}$$

The latter expression may be rewritten as

$$[\mathbf{F} + \mathbf{\Sigma}'(E)]\,\mathbf{C}(E) = E\mathbf{C}(E). \tag{33}$$

Here, an effective one–electron operator matrix has Fock and energy–dependent, self–energy terms. From this matrix expression, one may abstract one–electron equations in terms of the generalized Fock and energy–dependent, self–energy operators:

$$[F + \Sigma'(E)]\,\phi^{Dyson} \equiv \Gamma(E)\phi^{Dyson} = E\phi^{Dyson}. \quad (34)$$

When E is an eigenvalue of $\mathbf{\Gamma}(E)$, E is a pole. The corresponding operator, $\Gamma(E)$, is nonlocal and energy-dependent. In its exact limit, it incorporates all relaxation and differential correlation corrections to canonical orbital energies.

A normalized DO is determined by an eigenvector of $\Gamma(E_{pole})$ according to

$$\psi^{Dyson}(x) = \sum_r \phi_r(x) C_r(E_{pole}), \quad (35)$$

and satisfies

$$\langle \psi^{Dyson} | \psi^{Dyson} \rangle = 1 \quad (36)$$

provided $\mathbf{C}^\dagger \mathbf{C} = 1$. The normalization factor, $\sqrt{P}$, occurring in

$$\phi^{Dyson}(x) = \sqrt{P}\psi^{Dyson} \quad (37)$$

is related to the pole strength, P, such that

$$P = \left[1 - \mathbf{C}^\dagger(E_{pole}) \frac{d\Sigma(E)}{dE} \,|_{E=E_{pole}}\, \mathbf{C}(E_{pole}) \right]^{-1}. \quad (38)$$

When $\mathbf{\Sigma}(E)$ is neglected, P equals unity for each Koopmans final state.

Results on valence ionization energies of closed–shell molecules generally indicate that off–diagonal elements of the self–energy matrix in the canonical basis are small and have a negligible effect on poles and DOs. Diagonal self–energy approximations explicitly neglect these matrix elements and, as a consequence of equation 35, constrain the DOs to be equal to canonical orbitals. The associated pole search becomes especially easy, for the zeros of the diagonal elements of the Dyson equation can be found by solving

$$E = \epsilon_p + \Sigma_{pp}(E). \quad (39)$$

The usual initial guess, $\epsilon_p + \Sigma_{pp}(\epsilon_p)$, usually leads to convergence in three iterations. Relationships between diagonal self–energy approximations, the transition operator method, the ΔSCF approximation and perturbative treatments of electron binding energies have been analyzed in detail [17, 18].

5 Approximations

5.1 Perturbative $\Sigma(E)$ for Large Molecules

The usual choice of superoperator metric starts from a HF wavefunction plus perturbative corrections [4, 5]:

$$(Y|Z) = \langle HF|(1 + T^\dagger)[Y^\dagger, Z]_+(1 + T)|HF\rangle \quad (40)$$

where

$$T = T_2^{(1)} + T_1^{(2)} + T_2^{(2)} + T_3^{(2)} + T_4^{(2)} + \cdots. \tag{41}$$

The level of excitation in $T_e^{(f)}$ is indicated by the subscript, e, and the order is defined by the superscript, f. For example, second-order, triple excitations are represented by $T_3^{(2)}$. Coupled-cluster parametrizations of this metric [19] suggest an alternative form:

$$(Y|Z) = \langle HF|e^{-T}[Y^\dagger, Z]_+ e^T|HF\rangle. \tag{42}$$

This choice produces asymmetric superoperator matrices. A simplified final form for the self-energy matrix that does not require optimization of cluster amplitudes is sought for large molecules; the approximation

$$e^T \approx 1 + T_2^{(1)} \tag{43}$$

therefore is made.

With this choice, several third-order terms that appeared with the usual metric are eliminated. The new self-energy matrix in third order is asymmetric and is expressed by

$$\begin{aligned}\boldsymbol{\Sigma}(E) = (\mathbf{a}|\hat{H}\mathbf{f}_3)^{(1)}\{E\mathbf{1} - (\mathbf{f}_3|\hat{H}\mathbf{f}_3)^{(0)}\}^{-1}(\mathbf{f}_3|\hat{H}\mathbf{a})^{(1)} \\ +(\mathbf{a}|\hat{H}\mathbf{f}_3)^{(1)}\{E\mathbf{1} - (\mathbf{f}_3|\hat{H}\mathbf{f}_3)^{(0)}\}^{-1}(\mathbf{f}_3|\hat{H}\mathbf{a})^{(2)} \\ +(\mathbf{a}|\hat{H}\mathbf{f}_3)^{(1)}\{E\mathbf{1} - (\mathbf{f}_3|\hat{H}\mathbf{f}_3)^{(0)}\}^{-1}(\mathbf{f}_3|\hat{V}\mathbf{f}_3)^{(1)}\{E\mathbf{1} - (\mathbf{f}_3|\hat{H}\mathbf{f}_3)^{(0)}\}^{-1}(\mathbf{f}_3|\hat{H}\mathbf{a})^{(1)},\end{aligned} \tag{44}$$

where $\mathbf{f}_3$ is a vector whose elements are 2h–p or 2p–h operators. Note that energy-independent terms in the third-order, self-energy matrix are not retained.

Two observations suggest additional economies. First, numerical results for ionization energies show that third-order, 2p–h terms in equation 44 are small relative to their 2h–p counterparts. Terms arising from these operators are important in second order, however. Second, evaluation of the third-order, 2p–h terms requires integrals with four virtual indices. Because of the large number of these integrals that typically is generated, their storage is often avoided through semidirect algorithms [20]. Contractions involving integrals with four virtual indices remain the bottleneck in third-order calculations.

Neglect of third-order, 2p–h terms produces this self-energy matrix:

$$\begin{aligned}\Sigma(E)_{pq} = \frac{1}{2}\sum_{iab}\frac{\langle pi||ab\rangle\langle ab||qi\rangle}{E+\epsilon_i-\epsilon_a-\epsilon_b} \\ +\frac{1}{2}\sum_{aij}\frac{\langle pa||ij\rangle W_{qaij}}{E+\epsilon_a-\epsilon_i-\epsilon_j} + \frac{1}{2}\sum_{aij}\frac{U_{paij}(E)\langle ij||qa\rangle}{E+\epsilon_a-\epsilon_i-\epsilon_j}\end{aligned} \tag{45}$$

where i, j, k are occupied indices, a, b, c are virtual indices, p, q are general indices,

$$W_{qaij} = \langle qa||ij\rangle + \frac{1}{2}\sum_{bc}\frac{\langle qa||bc\rangle\langle bc||ij\rangle}{\epsilon_i + \epsilon_j - \epsilon_b - \epsilon_c} + (1 - P_{ij})\sum_{bk}\frac{\langle qk||bi\rangle\langle ba||jk\rangle}{\epsilon_j + \epsilon_k - \epsilon_a - \epsilon_b} \quad (46)$$

and

$$U_{paij}(E) = -\frac{1}{2}\sum_{kl}\frac{\langle pa||kl\rangle\langle kl||ij\rangle}{E + \epsilon_a - \epsilon_k - \epsilon_l} - (1 - P_{ij})\sum_{bk}\frac{\langle pb||jk\rangle\langle ak||bi\rangle}{E + \epsilon_b - \epsilon_j - \epsilon_k}. \quad (47)$$

This partial third-order expression has been designated by the abbreviation P3 [21].

Comparison of the self-energy matrix elements of equation 45 with older, related methods [7, 15] reveals the advantages of the P3 approximation. Among the intermediates required in third order is

$$V_{piab}(E) = \frac{1}{2}\sum_{cd}\frac{\langle pi||cd\rangle\langle cd||ab\rangle}{E + \epsilon_i - \epsilon_c - \epsilon_d} + (1 - P_{ab})\sum_{jc}\frac{\langle pj||bc\rangle\langle ic||ja\rangle}{E + \epsilon_j - \epsilon_b - \epsilon_c}. \quad (48)$$

The first summation requires electron repulsion integrals with four virtual indices. Efficient algorithms that avoid the storage of these integrals have been discussed in detail [20]. For every orbital index, p, this OV^4 contraction must be repeated for each energy considered in the pole search; it is usually the computational bottleneck.

In the diagonal, P3 approximation, nondiagonal elements of the self-energy matrix are neglected. The first contraction in equation 46 is the most demanding, for it has an arithmetic scaling factor of O^2V^3. This step also requires electron repulsion integrals with one occupied and three virtual indices. The W intermediate is energy-independent and must be evaluated once only for each ionization energy of interest.

The diagonal, P3 self-energy was compared with older propagator methods for 19 ionization energies of six closed-shell molecules with the correlation-consistent, triple ζ basis [21]. The average absolute errors in eV were: 1.34 for Koopmans's theorem, 0.25 for 3+ (a method similar to ADC(3) [7]), 0.25 for OVGF-B (a method based on a scaled, third-order self-energy [7]) and 0.19 for P3. The P3 procedure exhibits accuracy at least as good as that of other methods, superior arithmetic scaling and no need for electron repulsion integrals with four virtual indices.

5.2 Renormalized Methods

According to equation 15, eigenvalues of the superoperator Hamiltonian matrix, $\hat{\mathbf{H}}$, are poles (electron binding energies) of the electron propagator. Several renormalized methods can be defined in terms of approximate $\hat{\mathbf{H}}$ matrices. The latter are defined by the operator manifold and the reference state employed in the superoperator metric. Renormalized spin–orbitals may be generated according to the so–called Brueckner doubles (BD) recipe [22]. Here, a coupled–cluster singles and doubles wavefunction is reduced to the form

$$|BD\rangle = e^{T_2}|Brueckner\rangle \tag{49}$$

by rotation of the orbitals in the reference determinant, $|Brueckner\rangle$. A convenient reference metric is given by

$$(Y|Z) = \langle Brueckner|\left[Y^\dagger, Z\right]_+ e^{T_2}|Brueckner\rangle. \tag{50}$$

To take advantage of procedures used for configuration interaction calculations, eigenvalues of the symmetrized matrices, $\frac{1}{2}\{\hat{\mathbf{H}} + \hat{\mathbf{H}}^\dagger\}$, are computed.

This choice requires only one major modification in programs written for canonical, HF orbitals. In the Brueckner orbital basis, elements of the $\hat{\mathbf{H}}_{h,p}$ and $\hat{\mathbf{H}}_{p,h}$ blocks of the superoperator Hamiltonian matrix no longer vanish. Double replacement amplitudes from T_2 replace their first–order counterparts that are used in the P3 method. Orbitals that diagonalize the $\hat{\mathbf{H}}_{h,h}$ and $\hat{\mathbf{H}}_{p,p}$ blocks may be chosen without altering the BD ansatz. Electron repulsion integrals and diagonal elements of the generalized Fock matrix in the approximate Brueckner orbital basis therefore replace integrals and orbital energies in the canonical, HF basis. The $\hat{\mathbf{H}}$ matrix employed in calculations described below has the form

$$\hat{\mathbf{H}} = \begin{bmatrix} \hat{\mathbf{H}}_{h,h} & \hat{\mathbf{H}}_{h,p} & \hat{\mathbf{H}}_{h,2hp} & \hat{\mathbf{H}}_{h,2ph} \\ \hat{\mathbf{H}}_{p,h} & \hat{\mathbf{H}}_{p,p} & \hat{\mathbf{H}}_{p,2hp} & \hat{\mathbf{H}}_{p,2ph} \\ \hat{\mathbf{H}}_{2hp,h} & \hat{\mathbf{H}}_{2hp,p} & \hat{\mathbf{H}}_{2hp,2hp} & \hat{\mathbf{H}}_{2hp,2ph} \\ \hat{\mathbf{H}}_{2ph,h} & \hat{\mathbf{H}}_{2ph,p} & \hat{\mathbf{H}}_{2ph,2hp} & \hat{\mathbf{H}}_{2ph,2ph} \end{bmatrix}. \tag{51}$$

An additional approximation is introduced here: elements of the $\hat{\mathbf{H}}_{2hp,2ph}$ block are neglected. Since this block vanishes identically when HF reference states are used, the present approximation may be regarded as an improvement to the so–called 2p–h TDA [7, 23, 24] method with orbital and reference–state renormalizations [25, 26, 27].

6 Applications

6.1 Benzopyrene

Benzo[a]pyrene, a molecule with five, fused, hexagonal rings, is among the most carcinogenic of the polycyclic aromatic hydrocarbons (PAHs). Such biological activity may be related to the electronic structure of benzo[a]pyrene and its metabolites. Ionization energies of these molecules therefore have been investigated with photoelectron spectroscopy [28].

The diagonal, P3 self–energy is ideal for interpreting these experiments [29]. All occupied MOs except for 1s–like core orbitals were included in electron propagator calculations performed with the 6-311G(d,p) basis [30]. Pole strengths (P), listed immediately beneath P3 ionization energies in Table I, indicate that the Koopmans description of each final state is qualitatively valid, for shakeup character is minor. A symmetry–adapted, semidirect algorithm was employed [20]. Programs are incorporated in a modified version of GAUSSIAN-94 [31]. Because of limitations on memory and disk storage, virtual orbitals were dropped in the P3 calculations. A total of 310 occupied and virtual valence orbitals was retained; only virtual orbitals with energies above 2.71 atomic units were dropped.

Table I: Benzo[a]pyrene Ionization Energies (eV)

Orbital	KT	P3 P	Expt. [32]	Orbital Type
10a″	6.90	7.02 0.87	7.12	π
9a″	7.94	7.93 0.86	8.05	π
8a″	9.04	8.67 0.85	8.79	π
7a″	9.43	8.95 0.86	8.97	π
6a″	10.18	9.49 0.84	9.51	π
5a″	10.77	9.89 0.83	9.95	π
56a′	12.51	10.84 0.88		σ

In the photoelectron spectrum [32, 33], the sharp peak at 7.12 eV is followed by four regularly spaced peaks of lesser intensity up to 7.7 eV. This vibrational structure is due to the delocalized nature of the corresponding DO. There are no especially large amplitudes in the so–called bay region. A similar pattern of results obtains for the second cationic state. In the spectrum, a sharp peak at 8.00 eV exhibits a subsidiary peak at 8.2 eV that is succeeded by less distinct shoulders up to 8.4 eV. These features also are due to vibrational excitation in the second electronic state. Larger discrepancies between uncorrelated and correlated results occur for the third final state. P3 calculations with the full complement of virtual orbitals will approach 8.7 eV, in excellent agreement with experiment. Correlation effects are quantitatively important for the fourth and fifth final states as well. In the experimental report, the highest energy feature that was assigned occurred at 9.95 eV. No attempt was made therein to assign higher energy features to specific electronic states. Around 10.5 eV, the lowest discernible features appear. A clearly separable local maximum occurs around 10.9 eV. This feature is part of a broad system with many maxima up to 12.0 eV. This system, in turn, is followed by several, broad, jagged features up to 16 eV. According to the P3 results, the first final state with a σ hole occurs between 10.9–11.0 eV. (This estimate is based on the observation that inclusion of more virtual orbitals increases the predicted ionization energies.) The extensive nuclear rearrangements induced by removal of a σ electron may account for the structure seen on the low energy side of the 10.9–11.0 eV estimate.

6.2 1,10 Phenanthroline

Diaza derivatives of PAHs possess unique chelating properties, display extensive biological activity and are widely used in analytical chemistry, pharmacology and molecular biology [34]. These molecules have been thoroughly scrutinized for antitumor activity, carcinogenicity and mutagenicity. The ability of some phenanthrolines to inhibit electron transfer in biological systems has attracted attention. Derivatives of phenanthrolines have been studied for their potential anti–viral (including anti–HIV) properties.

Ionization energies from molecules with adjacent nitrogen lone pairs usually are assigned with the aid of a simple, two–level splitting model. Here, two lone–pair hybrids produce two combinations, the out–of–phase σN_- and the in–phase σN_+. Relative positions are determined by "through–space" or "through–bond" interactions. Photoelectron spectra of molecules with two aza centers usually are assumed to have two bands corresponding to ionizations from nitrogen lone–pair orbitals. Unfortunately, this concept may lead to erroneous assignments, especially when bands overlap. In azabenzenes, numerous final state misorderings from Koopmans's theorem have been revealed [35].

Because the two nitrogen atoms in 1,10 phenanthroline are separated by less than 3 Å, electron correlation has important consequences for the order of final states. The first ionization energy relates to a π MO with a distribution pattern similar to that of the parent hydrocarbon, phenanthrene. The P3 result shown in Table II is very close to the experimental value [34]. The second ionization band with the experimental maximum at ~8.8 eV represents the overlapping of two cationic states, π_2 $^2A_2$ and the out–of–phase, nitrogen lone–pair hybrid combination, $^2B_2$. A strong resemblance obtains between the former's MO and its π_2 counterpart in phenanthrene. The MO $19b_2$, corresponding to the third ionization, while having some C–C bonding character, consists chiefly of non–bonding, lone–pair hybrids. In this case, the MO with an antibonding combination of lone–pair hybrids lies above its bonding counterpart. P3 ionization energies obtained for these states are 8.67 and 8.87 eV, respectively. In the spectrum, there is a large peak at 8.8 eV with a shoulder at 8.6–8.7 eV. Both features are explained by these calculations. The fourth ionization occurs from an MO dominated by an in–phase combination of lone–pair hybrids,

$21a_1$. P3 results are in excellent agreement with experiment. An assignment made on the basis of INDO calculations is incorrect [36]. The peak at ~8.4 eV was assigned to an out-of-phase combination of nitrogen lone-pair hybrids; the in-phase combination was assigned to a feature at 11.76 eV. Our calculations predict the energy splitting of the two lone-pair levels to be only 0.4–0.5 eV and place these levels much closer to the first two π levels. Correlation corrections to Koopmans's theorem must be included in calculations that are performed to interpret this spectrum. The next three ionizations pertain to π cationic states and the calculated energies are in good agreement with the experimental values.

Table II: 1,10–Phenanthroline Ionization Energies (eV)

State	KT	P3	Expt. [36]	Orbital
$^2B_1$	8.29	8.39	8.35	π_1
$^2A_2$	8.62	8.67	8.82	π_2
$^2B_2$	10.82	8.87		σN_-
$^2A_1$	11.28	9.33	9.39	σN_+
$^2A_2$	10.62	10.17	10.11	π_3
$^2B_1$	10.86	10.37	10.47	π_4
$^2B_1$	12.49	11.47	11.16	π_5

6.3 Anions

Anion photoelectron spectroscopy [37, 38] and photodetachment techniques [39] provide accurate information on electron detachment energies of negative ions. Ten closed-shell anions considered here exhibit sharp peaks, indicative of minor or vanishing final-state nuclear rearrangements, in their photoelectron spectra. Comparisons between theory and experiment are straightforward, for differences between vertical and adiabatic electron detachment energies (VEDEs and AEDEs, respectively) are small.

These data are ideal tests for renormalized *ab initio* methods. Perturbative propagator methods have yielded poor agreement with experiment for F^- and OH^- [40]. For example, OVGF predictions for F^- and OH^- with a polarized, triple ζ basis augmented with diffuse functions are 5.00 and 2.86 eV, respectively.

Table III displays VEDEs obtained with the Brueckner-reference methods discussed in Section 5.2 and augmented, correlation-consistent, triple-ζ basis sets [41]. AEDEs include zero-point energy differences and relaxation energies pertaining to geometrical relaxation on the neutral's potential energy surface. The average absolute error with respect to experiment is 0.05 eV [26].

Table III: Electron Detachment Energies (eV)

Anion	Final State	VEDE	AEDE	Expt. [26]
F^-	^2P	3.54		3.40
OH^-	$^2\Pi$	1.85	1.85	1.83
NH_2^-	$^2B_1$	0.68	0.70	0.77 ± 0.005
Cl^-	^2P	3.61		3.61
SH^-	$^2\Pi$	2.29	2.30	2.32 ± 0.01
PH_2^-	$^2B_1$	1.20	1.22	1.27 ± 0.01
BO^-	$^2\Sigma^+$	2.57	2.54	2.51 ± 0.01
CN^-	$^2\Sigma^+$	3.83	3.83	3.86
AlO^-	$^2\Sigma^+$	2.72	2.69	2.60 ± 0.03
AlS^-	$^2\Sigma^+$	2.74	2.68	2.60 ± 0.03

For F^-, still the most difficult case, the pole strength is 0.90. The 2p orbital in the reference determinant dominates the normalized DO with a coefficient of 0.9997. In the **U** vector of equation 17, the $a^\dagger_{3p\beta}a_{2p\alpha}a_{2p\beta}$ contribution ≈ 0.1.

For OH^-, the pole strength is 0.89 and the 1π coefficient in the normalized DO is 0.9994. A 2h–p operator, $a^\dagger_{2\pi\beta}a_{1\pi\alpha}a_{1\pi\beta}$, also has a **U** element that is approximately 0.1.

Similar results follow for the remaining anions with eight valence electrons. Pole strengths are between 0.88 and 0.90 for NH_2^-, SH^- and PH_2^-. A somewhat larger value, 0.95, obtains for Cl^-. For NH_2^- and PH_2^-, there are 2h–p **U** elements for both final states with absolute values between 0.1 and 0.2. The operators in question have the form $a^\dagger_{tb_1\beta}a_{nb_1\alpha}a_{nb_1\beta}$ or $a^\dagger_{tb_1\beta}a_{nb_1\alpha}a_{ma_1\beta}$, where n and m are labels for occupied orbitals and t is an unoccupied orbital label.

For diatomics with ten valence electrons, pole strengths lie between 0.86 and 0.89. DOs are dominated by a single occupied orbital in all cases. In the normalized DO for the $^2\Sigma^+$ state of AlO, there are other contributions with coefficients near 0.02. For the $^2\Sigma^+$ states of BO and AlO, certain $a^\dagger_{t\sigma\beta}a_{n\sigma\alpha}a_{n\sigma\beta}$ operators have **U** elements that are approximately 0.1. Recent experimental work has produced a revised figure, 2.508 ± 0.008 eV, for the electron affinity of BO [42] and the entry in Table III is in excellent agreement. Similar agreement occurs for the electron affinities of CN, AlO and AlS.

6.4 Ozone

Applications of electron propagator methods with a single–determinant reference state seldom have been attempted for biradicals such as ozone, for operator space partitionings and perturbative corrections therein assume the dominance of a lone configuration in the reference state. Assignments of the three lowest cationic states were inferred from asymmetry parameters measured with Ne I, He I and He II radiation sources [43].

Correlated variational calculations with small basis sets provided qualitative descriptions of the cationic states in terms of symmetry–adapted or localized orbitals. Hay, Dunning and Goddard's GVB–CI description of ground state O_3 consists of a $6a_1^2 4b_2^2 1b_1^2 1a_2^2$ reference and a $1a_2^2 \rightarrow 2b_1^2$ double excitation [44]. Here, a singlet–coupled, biradical pair of electrons is placed on two p_π orbitals centered on terminal oxygens. Application of $4b_2$, $6a_1$ or $1a_2$ annihilation operators to this state produces a valence–bond description of the cations where holes occur in terminal–atom π or lone–pair σ orbitals. Kosugi, Kuroda and Iwata emphasized the importance of 2h–p configurations in describing the $^2A_1$ and $^2B_2$ states [45]. Malmquist, Ågren and Roos observed the importance of single and double replacements from $1b_1$ to $2b_1$ for the $^2A_2$ state [46].

OVGF calculations reversed the order of the $^2A_2$ and $^2B_2$ states [47]. Fock–space, multi–reference, coupled–cluster calculations with a 5s4p2d basis obtained excellent agreement with experiment [48]. Recent multiconfigurational, spin–tensor, electron propagator calculations were based on an active orbital space defined by the irreducible representations of the three highest occupied and three lowest unoccupied HF orbitals [49]. Six electrons are assigned to an active space consisting of $6a_1$, $4b_2$, $1a_2$, $2b_1$, $7a_1$ and $5b_2$ orbitals. (Symmetry adaptation of 2p atomic orbitals produces a space that also contains $5a_1$, $1b_1$ and $3b_2$ constituents.) The results are relatively insensitive to basis improvements and are in excellent agreement with experiment.

The Brueckner–reference method discussed in Section 5.2 and the cc–pvqz basis set without g functions were applied to the vertical ionization energies of ozone [27]. Errors in the results of Table IV lie between 0.07 and 0.17 eV; pole strengths (P) displayed beside the ionization energies are approximately equal to 0.9. Examination of cluster amplitudes and elements of **U** vectors for each ionization energy reveals the reasons for the success of the present calculations. The cluster operator amplitude for the double excitation to $2b_1^2$ from $1a_2^2$ is approximately 0.19. For each final state, the most important operator pertains to an occupied spin–orbital in the reference determinant, but there are significant coefficients for 2h–p operators. For the $^2A_2$ case, a balanced description of ground state correlation requires inclusion of a 2p–h operator as well. The $2b_1$ orbital's creation or annihilation operator is present in each of the 2h–p and 2p–h operators listed in Table IV. Pole strengths are approximately equal to the square of the principal h operator coefficient and contributions by other h operators are relatively small.

Table IV: Ozone Ionization Energies (eV) and Operators

n	ν	ν Type	$\lvert U_{\nu,n}\rvert$	P	Pole	Expt. [43]
$^2A_1$	$(6a_1\alpha)$	h	0.93	0.88	12.66	12.73
	$(2b_1\beta)^\dagger(4b_2\alpha)(1a_2\beta)$	2h–p	0.14			
$^2B_2$	$(4b_2\alpha)$	h	0.93	0.88	12.83	13.00
	$(2b_1\beta)^\dagger(6a_1\alpha)(1a_2\beta)$	2h–p	0.14			
$^2A_2$	$(1a_2\alpha)$	h	0.93	0.87	13.65	13.54
	$(2b_1\beta)^\dagger(1b_1\alpha)(1a_2\beta)$	2h–p	0.14			
	$(1a_2\beta)^\dagger(2b_1\alpha)(2b_1\beta)$	2p–h	0.18			

7 Conclusions

Electron propagator theory generates a one–electron picture of electronic structure that includes electron correlation. One–electron energies may be obtained reliably for closed–shell molecules with the P3 method and more complex correlation effects can be treated with renormalized reference states and orbitals. To each electron binding energy, there corresponds a Dyson orbital that is a correlated generalization of a canonical molecular orbital. Electron propagator theory enables interpretation of precise *ab initio* calculations in terms of one–electron concepts.

8 Acknowledgments

For many years, the lectures of Yngve Öhrn on the theory of chemical bonding have been models of clarity and incisiveness to graduate students at the University of Florida and at various topical schools. Their success in introducing the assumptions and conclusions of molecular orbital theory, group theory, electron correlation methods and related subjects has engendered a critical, but liberal attitude toward competing doctrines.

Students who join the Öhrn research group soon discover that graduate school, in Yngve's view, is a license to learn. The scientific individuality of each of Yngve's present and former students has been built on this foundation of freedom and on the patient friendship of an exemplary advisor.

This work was supported by the National Science Foundation under grants CHE–9873897 and CHE–9796095 and by Gaussian, Incorporated.

References

[1] Linderberg, J.; Öhrn, Y. "Propagators in Quantum Chemistry"; Academic Press: New York, 1973.

[2] Pickup, B.; Goscinski, O. *Mol. Phys.* **1973**, *26*, 1013.

[3] Cederbaum, L. S.; Domcke, W. In "Advances in Chemical Physics", Prigogine, I.; Rice, S. A., Eds.; Wiley: New York, 1977; Vol. *36*, 205–344.

[4] Simons, J. In "Theoretical Chemistry: Advances and Perspective", Eyring, H.; Henderson, D., Eds.; Academic: New York, 1978; Vol. *3*, 1–13.

[5] Herman, M. F.; Freed, K. F.; Yeager, D. L. In "Advances in Chemical Physics", Prigogine, I.; Rice, S.A., Eds.; Wiley: New York, 1981; Vol. *48*, 1–69.

[6] Öhrn, Y.; Born, G. In "Advances in Quantum Chemistry", Löwdin, P.O., Ed.; Academic: New York, 1981; Vol. *13*, 1-88.

[7] von Niessen, W.; Schirmer, J.; Cederbaum, L. S. *Comput. Phys. Rep.* **1984**, *1*, 57–125.

[8] Ortiz, J. V. In "Computational Chemistry: Reviews of Current Trends", Leszczynski, J., Ed.; World Scientific: Singapore, 1997; Vol. *2*, 1–61.

[9] Ortiz, J.V.; Zakrzewski, V.G.; Dolgounitcheva, O. In "Conceptual Trends in Quantum Chemistry", Kryachko, E.S., Ed.; Kluwer: Dordrecht, 1997; Vol. *3*, 465–517.

[10] Bethe, H.A.; Salpeter, E.E. "Quantum Mechanics of One and Two Electron Atoms"; Academic Press: New York, 1957.

[11] Csanak, G.; Taylor, H.S.; Yaris, R. In "Advances in Atomic and Molecular Physics" Bates, D.R.; Esterman, I. Eds.; Academic Press: New York, 1971; Vol. *7*, 288–361.

[12] Coplan, M.A.; Moore, J.H.; Doering, J.P. *Rev. Mod. Phys.* **1994**, *66*, 985.

[13] Goscinski, O.; Lukman, B. *Chem. Phys. Lett.* **1970**, *7*, 573.

[14] Baker, J.; Pickup, B.T. *Chem. Phys. Lett.* **1980**, *76*, 537.

[15] Ortiz, J.V. *J. Chem. Phys.* **1993**, *99*, 6716.

[16] Purvis, G.D.; Öhrn, Y. *J. Chem. Phys.* **1974**, *60*, 4063.

[17] Born, G.; Kurtz, H. A.; Öhrn, Y. *J. Chem. Phys.* **1978**, *68*, 74.

[18] Kurtz, H. A.; Öhrn, Y. *J. Chem. Phys.* **1978**, *69*, 1162.

[19] Bartlett, R.J. *J. Phys. Chem.* **1989**, *93*, 1697.

[20] Zakrzewski, V.G.; Ortiz, J.V. *Int. J. Quant. Chem., Quant. Chem. Symp.* **1994**, *28*, 23; Zakrzewski, V.G.; Ortiz, J.V. *Int. J. Quant. Chem.* **1995**, *53*, 583.

[21] Ortiz, J.V. *J. Chem. Phys.* **1996**, *104*, 7599.

[22] Cizek, J.; Paldus, J. *Physica Scripta* **1980**, *21*, 251; Bartlett, R.J.; Purvis, G.D. *Physica Scripta* **1980**, *21*, 255; Chiles, R.A.; Dykstra, C.E. *J. Chem. Phys.* **1981**, *74*, 4544.

[23] Born, G.; Öhrn, Y. *Chem. Phys. Lett.* **1979**, *61*, 307.

[24] Mishra, M.; Öhrn, Y. *Chem. Phys. Lett.* **1980**, *71*, 549.

[25] Ortiz, J.V. *J. Chem. Phys.* **1998**, *109*, 5741.

[26] Ortiz, J.V. *Chem. Phys. Lett.* **1998**, *296*, 494 and references therein.

[27] Ortiz, J.V. *Chem. Phys. Lett.* **1998**, *297*, 193.

[28] Fetzer, S.M.; Huang, C.; Harvey, R.G.; LeBreton, P.R. *J. Phys. Chem.* **1993** *97*, 2385.

[29] Zakrzewski, V.G.; Dolgounitcheva, O.; Ortiz, J.V. *J. Chem. Phys.* **1997**, *107*, 7906.

[30] Krishnan, R.; Binkley, J.S.; Seeger, R.; Pople, J. A. *J. Chem. Phys.* **1980**, *72*, 650.

[31] Frisch, M. J.; Trucks, G. W.; Schlegel, H. B.; Gill, P. M. W.; Johnson, B. G.; Robb, M. A,; Cheeseman, J. R.; Keith, T. A.; Petersson, G. A.; Montgomery, J. A.; Raghavachari, K.; Al-Laham, M. A.; Zakrzewski, V. G.; Ortiz, J. V.; Foresman, J. B.; Cioslowski, J.; Stefanov, B. B.; Nanayakkara, A.; Challacombe, M.; Peng, C. Y.; Ayala, P. Y.; Chen, W.; Wong, M. W.; Andres, J. L.; Replogle, E. S.; Gomperts, R.; Martin, R. L.; Fox, D. J.; Binkley, J. C.; Defrees, D. J.; Baker, J.; Stewart, J. J. P.; Head-Gordon, M.; Gonzalez, C.; Pople, J. A. *Gaussian 94, Revision B.3*; Gaussian, Inc.; Pittsburgh PA, 1995.

[32] Akiyama, I.; Li, K.C.; LeBreton, P.R.; Fu, P.P; Harvey, R.G. *J. Phys. Chem.* **1979**, *83*, 2997.

[33] Boschi, R.; Murrell, J.N.; Schmidt, W. *Discuss. Faraday Soc.* **1972**, *54*, 116.

[34] Dolgounitcheva, O.; Zakrzewski, V.G.; Ortiz, J.V. *J. Phys. Chem. A* **1997**, *101*, 8554 and references therein.

[35] Ortiz, J. V.; Zakrzewski, V. G. *J. Chem. Phys.* **1996**, *105*, 2762.

[36] Hush, N. S.; Cheung, A. S.; Hilton, P. R. *J. Electron Spectrosc. Relat. Phenom.* **1975**, *7*, 385.

[37] Ervin, K.M.; Lineberger, W.C. In "Advances in Gas Phase Ion Chemistry", Adams, M.G.; Babcock, L.M., Eds.; JAI: Greenwich, 1992; Vol. *1*, 121–66.

[38] Arnold, S.T.; Eaton, J.G.; Patel-Misra, D.; Sarkas, H.W.; Bowen, K.H. In "Ion and Cluster Ion Spectroscopy and Structure", Maier, J.P., Ed.; Elsevier: Amsterdam, 1988; 147–69.

[39] Drzaic, P.S.; Marks, J.; Brauman, J.I. In "Gas Phase Ion Chemistry", Bowers, M.T., Ed.; Academic Press: New York, 1984; Vol. *3*, 167–211.

[40] Dolgounitcheva, O.; Zakrzewski, V.G.; Ortiz, J.V. *Int. J. Quant. Chem.* **1997**, *65* 463.

[41] Dunning, T.H. *J. Chem. Phys.* **1989**, *90*, 1007; Kendall, R.A.; Dunning, T.H.; Harrison, R.J. *J. Chem. Phys.* **1992**, *96*, 6796.

[42] Wenthold, P.G.; Kim, J.B.; Jonas, K.-L.; Lineberger, W.C. *J. Phys. Chem.* **1997**, *101*, 4472.

[43] Katsumata, S.; Shiromaru, H.; Kimura, T. *Bull. Chem. Soc. Jpn.* **1984**, *57*, 1784.

[44] Hay, P.J.; Dunning, T.H.; Goddard, W.A. *J. Chem. Phys.* **1975**, *62*, 3912.

[45] Kosugi, N.; Kuroda, H.; Iwata, S. *Chem. Phys.* **1981**, *58*, 267.

[46] Malmquist, P.A.; Ågren, H.; Roos, B. *Chem. Phys. Lett.* **1983**, *98*, 444.

[47] Cederbaum, L.S.; Domcke, W.; von Niessen, W.; Kraemer, W.P. *Mol. Phys.* **1977**, *34*, 381.

[48] Barysz, M.; Rittby, M.; Bartlett, R.J. *Chem. Phys. Lett.* **1992**, *193*, 373.

[49] McKellar, A.J.; Heryadi, D.; Yeager, D.L.; Nichols, J.A. *Chem. Phys.*, **1998**, *238*, 1.

Spin Density Properties from the Electron Propagator: Hyperfine and Nuclear Spin-Spin Couplings

Ricardo L. Longo

Departamento de Química Fundamental
Universidade Federal de Pernambuco
50740-540 Recife, PE - Brazil

Contents

ADVANCES IN QUANTUM CHEMISTRY, VOLUME 35

0065-3276/99 $30.00

1. Introduction

Electron-spin resonance (ESR) techniques have been widely used in chemistry and biochemistry. For instance, the molecular structure of organic and organometallic reaction intermediates as well as the dynamics of ion pairs in solution have been inferred from ESR spectra (1-3). ESR spectroscopy has also been employed to track down electron transfer in biological and model systems (4). The development of low-temperature matrix isolation techniques have enable the studies of unusual open-shell species by probing their spin properties (5-7). These spin properties are basically the isotropic (Fermi contact term) and anisotropic (dipolar term) hyperfine coupling constants (5-7). Their origin is related to the interaction between the unpaired electron(s) and the nuclear spin ($I > 0$), thus yielding information about the electronic spin density. In addition to ESR, high resolution microwave (MW) spectroscopic techniques have also been applied to polyatomic transient species in gas phase and yielded hyperfine coupling constants (8). Models based on frontier orbitals, such as the McConnell relation (9) for hydrogen and Karplus-Fraenkel relation (10) for ^{13}C in pi-radicals, have helped the qualitative interpretation of ESR spectra (1-3,5-7). However, in order to obtain the most information from ESR or high resolution MW spectra a quantitative theory for calculating spin properties is necessary.

There are several theoretical approaches designed to calculate the spin density and its properties. These approaches range from *ab initio* to semi-empirical and from Hartree-Fock (HF), e.g., ROHF (Restricted Open-Shell HF), UHF (Unrestricted HF) and PUHF (Projected UHF) (11), to highly correlated methods, e.g., MBPT (Many-Body Perturbation Theory) - CC (Coupled-Cluster) (12,13), MR-CI (Multi-reference - Configuration Interaction) (14,15), SAC-CI (Symmetry Adapted Cluster - CI) (16,17), including also density functional theory (DFT) based approaches (18,19). Even with a large number of configurations included in the MR-SDCI (MR-Singles and Doubles CI) with very large uncontracted basis sets, the isotropic hyperfine coupling constants for ^{11}B, ^{17}O, and ^{19}F are still being calculated with an error larger than 10% when compared to experimental data (14,15). The use of DFT with large basis sets has not improved the results, in fact, the calculated values show a very large dependence with the exchange and correlation functionals, as well as, with the corrections to the local spin density approximation (18,19). The treatment of molecules is much more demanding, so that, smaller atomic basis sets and lower levels of correlation effects have to be employed. As a result, the search for more accurate and less demanding theoretical approaches is very important. The approaches mentioned above seem to present two important problems, namely, *i*) the spin density is a small quantity and is obtained from the difference between two large quantities (spin density of the spin-α and spin-β electrons); and *ii*) the reference state is an open shell which has been known to be very difficult to handle in a balance way at either the HF or correlated levels. These problems are similar to those encountered in the calculation of the electron binding energies, which are obtained as the energy difference between the ionic and the neutral states, and at least one of them is open shell. The electron propagator or one particle Green's function seems to be the appropriate tool to solve the electron binding problem. Thus, it is reasonable to assume that such an approach would also be appropriate to apply to the hyperfine coupling constant problem. It will be shown that main advantage of electron propagator approach is that it enables the calculation of the hyperfine coupling constant directly, that is, without the need to perform any difference between large quantities. In addition, the reference state can be chosen to be a closed shell from which the open shell state is obtained by adding or detaching an electron.

The development and application of propagators in quantum chemistry has been pioneered by Linderberg and Öhrn (20). In fact, the use of electron propagator to calculate the hyperfine interaction in states with one unpaired electron has already been proposed in the final pages of their seminal book (21,22). The second order expansion of the electron propagator was first used by Reinhardt and Doll (23,24) to calculate electron binding energies of atoms and by Cederbaum et al. (25,26) for molecules. Higher order expansions of the self-energy part of the electron propagator were then performed later on using the diagrammatic many body approach (28-30). At the same time, Purvis and Öhrn (31-33), using the superoperator formalism developed by Goscinski and co-workers (34,35), have developed and applied the electron propagator theory to calculate ionization energies and electron affinities of atoms and molecules. A related direct approach to calculate electron binding energies was developed by Simons and collaborators (36-40) based on the equations of motion (EOM) formalism of Rowe (41,42). Some deficiencies in this later development have been pointed out (31-33,43) with emphasis in the complete implementation of the electron propagator through third order. Further development of the electron propagator theory included the higher order decoupling with the inclusion of the five-fold operator set effects (44) and the nuclear motion effects (vibrational structure) in the ionization spectra (45). From the mathematical point of view, several points have been addressed during the early development of the electron propagator theory, such as, the Hermiticity of the superoperator Hamiltonian (46), the completeness of the operator manifolds (47,48), the resolution of the superoperator identity (49) and alternative decoupling methods (50,51). The N-representability problem has also been addressed in the case of the superoperator formulation of the electron propagator theory (52-55) and in order to solve this problem a truncated operator manifold has to satisfy the vacuum annihilation conditions (52-55) leading to the self-consistent propagators (56). These self-consistent propagators are obtained by rotations in the Fock space which are represented by an exponential ansatz for the exact ground state wave function, that is, the coupled-cluster method (57-61). It has then been shown that the use of either the extended coupled-cluster (62) or the normal coupled-cluster (63) leads to a consistent electron propagator which decouples the $(N+1)$- and $(N-1)$-electron Hilbert spaces. These formulations involve an effective (super)Hamiltonian obtained by a similarity transformation, which makes these propagators equivalent to the Fock-Space Coupled-Cluster (FSCC) (64-71) and equation-of-motion coupled-cluster (EOM-CC) (72-77) formalisms for the electron binding energy calculations. The use of an operator manifold which is complete in Fock space permits to express the electron propagator in a perturbation expansion (78) where there is no need to introduce a Manne-Dalgaard type of operator manifold (47,48), which is complete on a given N-particle Hilbert space, for inner projection (79-81). This approach also leads to a consistent electron propagator, but unlike the FSCC or EOM-CC formulation, it does not involve an effective Hamiltonian. More recently, Ortiz (82-84) has proposed an asymmetric metric for the operator manifold which effectively leads to a decoupling between the $(N+1)$- and $(N-1)$-electron Hilbert space for the normal electron propagator. It should be noted that, the similarities and equivalencies between the electron propagator theory, transition state, EOM, ΔE_{SCF}, and configuration interaction methods have been explored in the early days of the propagators (85-87).

Another important aspect of the electron propagator theory, started also at the early days, is the relative simplicity in using the available techniques as well as the development of new approaches to treat open shell and multiconfigurational systems. The first and most simple implementation of the electron propagator for treating open shells consisted of employing the unrestricted Hartree-Fock (UHF) single reference in the calculation of the electron binding energies of the triplet O_2 molecule (88). Unitary group approach (89-91) as well as spin adapted references (92) have also been implemented and applied to open shell atoms and molecules. A multiconfiguration reference state formulation of the electron propagator (MCEP) has also been developed (93,94), implemented and applied (94) for open shell systems which are degenerate and/or inherently multiconfigurational. These MCEP approaches attempt also to solve problems related to the proper choice of the operator manifold by including the transfer-type the $(N+1)$- and $(N-1)$-operators and where there is no obvious or unique choice of the Fock operator that reliably can represent the zeroth-order perturbation operator. In order to solve some of the difficulties of the MCEP formalism, namely, the complicated and non-economical matrix elements and its impractical extension to include larger manifolds involving strings of three or more adapted creation and annihilation operators, a new technique called multiconfigurational spin-tensor electron propagator (MCSTEP) has been developed, implemented and applied to atoms (95,96) and molecules, which significantly improves the results for electron binding energies (97,98).

It is worthwhile to notice that the one of the first applications of the electron propagator theory involved the calculation of properties other than electron binding energies, namely, ground state energy and natural orbitals (23,24). Since then, the electron propagator has been used to calculate ground state (99-108) and correlation energies (103-108) and one-particle density matrix, thus any single-particle operator expectation values (109-112), such as, dipole moment, kinetic and nuclear attraction energies, averages of z and $(x^2 + y^2)$, and electric field gradients (110). Also, elastic scattering (99,113), photoionization cross section (114-120), and from the dilated electron propagator, resonance energies and their associated lifetimes (121) can be obtained. A qualitative interpretation of the molecular electronic structure, including excited states (110), can be obtained from the Feynman-Dyson amplitudes and Dyson orbitals, allowing for an one-electron (orbital) picture (122,123). More recently, two major advances in the electron propagator theory have been performed, namely, the development and implementation of the coupled perturbed electron propagator (124-126) and the analytical gradients of the electron binding energy (111,112). These approaches allow the calculation and analysis of static linear and nonlinear response of a molecular system to an external applied field, yielding, for instance, dipole polarizabilities (124,125) and also the molecular geometry of ground and excited states of open shell (radical) molecules (111,112). It should be noted, that this analytical gradient approach is very general, in the sense that it can provide geometry of any excited state, independent of its symmetry.

2. Propagators

The double-time Green's function associated to operators $A(t)$ and $B(t')$ is defined as (21,127)

$$\langle\langle A(t); B(t')\rangle\rangle = -\mathrm{i}\theta(t-t')\langle 0 \mid A(t)B(t') \mid 0\rangle + \eta\mathrm{i}\theta(t'-t)\langle 0 \mid B(t')A(t) \mid 0\rangle \tag{2.1}$$

where $\theta(t)$ is the Heaviside function, $|0\rangle$ being the reference state of an N-particle system, and $\eta = +1$ for A and B fermion-like (non-number conserving) and $\eta = -1$ for boson-like (number conserving) operators, respectively.

For a pure state density operator, the Fourier transform of this double-time Green's function yields the spectral representation of the propagator (21)

$$\langle\langle A; B\rangle\rangle_E = \lim_{\varepsilon\to+0} \sum_m \left(\frac{\langle 0|A|m\rangle\langle m|B|0\rangle}{E - E_m + E_0 + \mathrm{i}\varepsilon} + \eta \frac{\langle 0|B|m\rangle\langle m|A|0\rangle}{E + E_m - E_0 - \mathrm{i}\varepsilon} \right) \tag{2.2}$$

with $H|m\rangle = E_m|m\rangle$ where H is the Hamiltonian of the system. Applying the identity, $E/(E-C) = 1 + C/(E-C)$, leads to the equations of motion of the propagator (118,128),

$$\langle\langle A; B\rangle\rangle_E = E^{-1}\langle 0|[A, B]_\eta|0\rangle + E^{-1}\langle\langle A; [H, B]_-\rangle\rangle_E \tag{2.3}$$

which can be iterated to yield de moment expansion of the propagator (34),

$$\langle\langle A;B\rangle\rangle_E = E^{-1}\langle 0|[A,B]_\eta|0\rangle + E^{-2}\langle 0|[A,[H,B]_-]_\eta|0\rangle + E^{-3}\langle 0|[A,[H,[H,B]_-]_-]_\eta|0\rangle + \cdots \tag{2.4}$$

It is possible to perform a systematic decoupling of this moment expansion using the superoperator formalism (34,35). An infinite dimensional operator vector space defined by a basis of field operators $\{X_i\}$ which supports the scalar product (or metric)

$$(X|Y) = \langle 0|[X^\dagger, Y]_\eta|0\rangle \tag{2.5}$$

for any two operators X and Y belonging to this linear space. The action of the identity and Hamiltonian superoperators are also defined

$$\hat{H}X = [X,H]_- \text{ and } \hat{I}X = X \tag{2.6}$$

The reason for the choice of this specific algebra can be justified by the correspondence between the Heisenberg and Schrödinger representations (129). Applying this algebra to the moment expansion yields the propagators in the resolvent form,

$$\langle\langle A;B\rangle\rangle_E = (A^\dagger \,|\, (E\hat{I} - \hat{H})^{-1} \,|\, B) \tag{2.7}$$

In this form, it is necessary to perform the inversion of a superoperator which is very cumbersome, so an inner projection technique (79-81) or the resolution of the superoperator identity (49),

$$\hat{I} = |\tilde{\mathbf{h}})(\mathbf{h}\,|\,\tilde{\mathbf{h}})^{-1}(\mathbf{h}| \tag{2.8}$$

can be used to express the resolvent in a matrix form

$$\langle\langle A;B\rangle\rangle_E = (A^\dagger \,|\, \tilde{\mathbf{h}})(\mathbf{h}\,|\,(E\hat{I} - \hat{H})\,|\,\tilde{\mathbf{h}})^{-1}(\mathbf{h}\,|\,B) \tag{2.9}$$

where $\mathbf{h}$ is a complete manifold of field operators arranged as a column vector and $\tilde{\mathbf{h}}$ is the transposed row vector. It has been shown that, at least for the electron propagator, the resolvent matrix form preserves the exact nature of its poles even for a non-exact reference function (49). As a result, the problem of inverting a superoperator is circumvented by a matrix inversion operation and it becomes the starting point for most of the approximate propagator methods. One of these approximate propagator methods (128) consists in treating the part of the manifold mostly responsible for the properties of interest better than the rest of the manifold by using the partitioning technique. The manifold $\mathbf{h}$ is divided into two sub-manifolds, $\mathbf{h}_a$ and $\mathbf{h}_b$, that is, $\mathbf{h} = \mathbf{h}_a \cup \mathbf{h}_b$, such that, if the manifold $\mathbf{h}$ is still complete, without loss of generality or exactness, the propagators can be expressed as

$$\langle\langle A;B\rangle\rangle_E = \begin{pmatrix}\mathbf{A}_a^\dagger & \mathbf{A}_b^\dagger\end{pmatrix}\begin{pmatrix}\mathbf{R}_{aa} & \mathbf{R}_{ab}\\ \mathbf{R}_{ba} & \mathbf{R}_{bb}\end{pmatrix}^{-1}\begin{pmatrix}\mathbf{B}_a\\ \mathbf{B}_b\end{pmatrix} \tag{2.10}$$

where,

$$\mathbf{R}_{ab} = (\mathbf{h}_a|(E\hat{I}-\hat{H})|\tilde{\mathbf{h}}_b)\,,\ \mathbf{A}_a^\dagger = (A^\dagger|\mathbf{h}_a)\,,\ \text{and}\ \mathbf{B}_a = (B|\mathbf{h}_a) \tag{2.11}$$

Solving the inverse matrix and assuming that $\mathbf{R}_{bb}^{-1}$ exists yields the partitioned propagators

$$\langle\langle A;B\rangle\rangle_E = (\mathbf{A}_a^\dagger - \mathbf{A}_b^\dagger\mathbf{R}_{bb}^{-1}\mathbf{R}_{ba})\overline{\mathbf{R}}_{aa}(\mathbf{B}_a - \mathbf{R}_{ab}\mathbf{R}_{bb}^{-1}\mathbf{B}_b) + \mathbf{A}_b^\dagger\mathbf{R}_{bb}^{-1}\mathbf{B}_b \tag{2.12}$$

with,

$$\overline{\mathbf{R}}_{aa} = (\mathbf{R}_{aa} - \mathbf{R}_{ab}\mathbf{R}_{bb}^{-1}\mathbf{R}_{ba})^{-1} \tag{2.13}$$

This form of the propagators reproduces exactly the same poles and residues as the original propagator as long as all the individual $\mathbf{R}_{ab}$ matrices are computed at the same level of accuracy (128). Assuming that the operator basis manifold $\mathbf{h}$ has been constructed to be orthonormal, that is,

$$\mathbf{A}_a^\dagger = \mathbf{B}_a = \mathbf{1}\,,\ \text{and}\ \mathbf{A}_b^\dagger = \mathbf{B}_b = \mathbf{0} \tag{2.14}$$

then the partitioned propagators in this specific manifold are written as

$$\langle\langle A;B\rangle\rangle_E = [E\mathbf{1} - \mathbf{H}_{aa} - \mathbf{H}_{ab}(E\mathbf{1} - \mathbf{H}_{bb})^{-1}\mathbf{H}_{ba}]^{-1} \tag{2.15}$$

where,

$$\mathbf{H}_{ab} = (\mathbf{h}_a|\hat{H}|\tilde{\mathbf{h}}_b) \tag{2.16}$$

This expression for the propagators is still exact, as long as, the principal sub-manifold $\mathbf{h}_a$ and its complement sub-manifold $\mathbf{h}_b$ are complete, and the characteristics of the propagator is reflected in the construction of these sub-manifolds (47,48). It should be noted that a different (asymmetric) metric for the superoperator space, Eq. (2.5), could be invoked so that another decoupling of the equations of motion is obtained (62,63,82-84). Such a metric will not be explored here, but it just shows the versatility of the propagator methods.

2.1. The Electron Propagator

When the operators A and B in Eq. (2.7) are simple creation and annihilation operators the resulting propagator is called electron propagator or one-particle Green's function, and $\eta = +1$. Collecting all these creation and annihilation operators in a row vector $\mathbf{a}$, the electron propagator can be expressed as,

$$\langle\langle \mathbf{a}^\dagger;\mathbf{a}\rangle\rangle_E = (\mathbf{a}|(E\hat{I}-\hat{H})^{-1}|\tilde{\mathbf{a}}) = \mathbf{G}(E) \tag{2.1.1}$$

where $\widetilde{\mathbf{a}}$ is the transposed column vector. As a result, for this propagator, the operator basis manifold $\mathbf{h}$ and the sub-manifolds, $\mathbf{h}_a$ and $\mathbf{h}_b$, are (47,48)

$$\mathbf{h} = \mathbf{h}_1 \cup \mathbf{h}_3 \cup \mathbf{h}_5 \cup \ldots = \{q, q^\dagger rs\ (r > s), p^\dagger q^\dagger rst\ (p > q, r > s > t), \ldots\} \quad (2.1.2)$$

$$\mathbf{h}_a = \mathbf{h}_1 = \mathbf{a} \text{ and } \mathbf{h}_b = \mathbf{h}_3 \cup \mathbf{h}_5 \cup \mathbf{h}_7 \cup \ldots \quad (2.1.3)$$

Thus, the inverse of the electron propagator can be written as,

$$\mathbf{G}^{-1}(E) = E\mathbf{1} - \mathbf{H}_{aa} - \mathbf{H}_{ab}(E\mathbf{1} - \mathbf{H}_{bb})^{-1}\mathbf{H}_{ba} \quad (2.1.4)$$

with $\mathbf{H}_{ab}$ given by Eq. (2.16).

2.2. Method of Solution

There are several approaches to solve the electron propagator equation, such as iterative pole search (31-33,130), CI-like solution (131,132), inversion expansion (109), etc. The diagonalization of the electron propagator matrix can yield some or all eigenvalues and eigenvectors (131,132) that can then be used to provide a spectral representation of the propagator. For a given value of E it is possible to construct the matrix (31),

$$\mathbf{W}(E) = E\mathbf{1}_{11} - \mathbf{G}^{-1}(E) = \mathbf{H}_{11} + \mathbf{H}_{1b}(E\mathbf{1}_{bb} - \mathbf{H}_{bb})^{-1}\mathbf{H}_{b1} \quad (2.2.1)$$

which allows for,

$$(E\mathbf{1}_{11} - \mathbf{W}(E))\mathbf{G}(E) = \mathbf{1}_{11} \quad (2.2.2)$$

In the case of a finite system described by a finite basis set the spectrum of $\mathbf{G}^{-1}(E)$ and $\mathbf{W}(E)$ are discrete and $\mathbf{G}^{-1}(E)$ has isolated real poles (31,99). As a result, the solution for the propagator consists in the diagonalization of the $\mathbf{W}(E)$ matrix

$$\mathbf{U}^\dagger(E)\mathbf{W}(E)\mathbf{U}(E) = \Lambda_{11}(E) \quad (2.2.3)$$

by the unitary matrix $\mathbf{U}(E)$. The k-th eigenvalue, $\lambda_k(E)$, can be obtained from the eigenvector, which is just the k-th column of $\mathbf{U}(E)$, by

$$\mathbf{u}_k^\dagger(E)\mathbf{W}(E)\mathbf{u}_k(E) = \lambda_k(E) \text{ with } \mathbf{u}_i^\dagger\mathbf{u}_j = \delta_{ij} \quad (2.2.4)$$

and $\mathbf{u}_k^\dagger(E)$ being the k-th row of $\mathbf{U}^\dagger(E)$.

The main goal of the procedure adopted to solve the propagator equation is to find the pole(s) of interested and the residues associated to it. Since $\mathbf{G}^{-1}(E)$ has single isolated poles then the residue theorem can be applied to yield

$$\text{Res}\, G_{pq}(E) = P_k u_{pk}(E_k) u^*_{kq}(E_k) \qquad (2.2.5)$$

the residue of $G_{pq}(E)$ associated with the k -th pole E_k, where P_k is it the pole strength

$$P_k = \lim_{E \to E_k} (E - E_k)/\lambda_k(E) \qquad (2.2.6)$$

which after application of the l'Hôpital theorem becomes

$$P_k^{-1} = \frac{\mathrm{d}}{\mathrm{d}E} \lambda_k(E)\bigg|_{E=E_k} = \mathbf{u}_k^\dagger(E_k)[\mathbf{1}_{11} + \mathbf{H}_{1b}(E_k \mathbf{1}_{bb} - \mathbf{H}_{bb})^{-2} \mathbf{H}_{b1}]\mathbf{u}_k(E_k) \qquad (2.2.7)$$

P_k can also be interpreted as the residue of $1/[E - \lambda_k(E)]$ taken at the E_k pole. Thus, the electron propagator matrix elements can be written as,

$$G_{pq}(E) = \sum_k \frac{P_k u_{pk}(E_k) u^*_{kq}(E_k)}{E - E_k} \qquad (2.2.8)$$

3. Hyperfine Splitting from the Electron Propagator

The complete Hamiltonian of the molecular system can be written as $H = H^0 + H^S$ or $\hat{H} = \hat{H}^0 + \hat{H}^S$ for the commutator being linear, where H^S is the Hamiltonian corresponding to the spin contribution(s) such as, Fermi contact term, dipolar term, spin-orbit coupling, etc. (5). As a result, H^0 would correspond to the spin free part of the Hamiltonian, which is usually employed in the electron propagator implementation. Accordingly, the k -th pole associated with the complete Hamiltonian H is E_k, so that E_k^0 is the k -th pole of the electron propagator for the spin free Hamiltonian H^0.

Expanding the k -th eigenvalue, Eq. (2.2.4), around E_k^0 yields,

$$\lambda_k(E) = \lambda_k(E_k^0) + \lambda_k'(E)\big|_{E=E_k^0}(E - E_k^0) + (2!)^{-1}\lambda_k''(E)\big|_{E=E_k^0}(E - E_k^0)^2 + \cdots \qquad (3.1)$$

At the k -th pole, $E \to E_k$, the k -th eigenvalue vanishes, $\lambda_k(E \to E_k) \to 0$, and since, $\delta_k = E_k - E_k^0$ is the hyperfine splitting associated with the k -th pole,

$$0 = \lambda_k(E_k^0) + \lambda_k'(E)\big|_{E=E_k^0}\delta_k + O(\delta_k^2) \qquad (3.2)$$

where,

$$\lambda_k(E_k^0) = \mathbf{u}_k^\dagger(E_k^0)[E_k^0 \mathbf{1}_{11} - \mathbf{H}_{11} - \mathbf{H}_{1b}(E_k^0 \mathbf{1}_{bb} - \mathbf{H}_{bb})^{-1}\mathbf{H}_{b1}]\mathbf{u}_k(E_k^0) \qquad (3.3)$$

and,

$$\lambda_k'(E)\big|_{E=E_k^0} = \mathbf{u}_k^\dagger(E_k^0)[\mathbf{1}_{11} - \mathbf{H}_{1b}(E_k^0 \mathbf{1}_{bb} - \mathbf{H}_{bb})^{-2}\mathbf{H}_{b1}]\mathbf{u}_k(E_k^0) \qquad (3.4)$$

As a result,

$$\begin{aligned}&\delta_k \mathbf{u}_k^\dagger(E_k^0)[\mathbf{1}_{11} + \mathbf{H}_{1b}(E_k^0\mathbf{1}_{bb} - \mathbf{H}_{bb})^{-2}\mathbf{H}_{b1}]\mathbf{u}_k(E_k^0)\\ &= -\mathbf{u}_k^\dagger(E_k^0)[E_k^0\mathbf{1}_{11} - \mathbf{H}_{11} - \mathbf{H}_{1b}(E_k^0\mathbf{1}_{bb} - \mathbf{H}_{bb})^{-1}\mathbf{H}_{b1}]\mathbf{u}_k(E_k^0)\end{aligned} \tag{3.5}$$

Expanding the matrices involving the complete Hamiltonian, $\mathbf{H}_{ab} = \mathbf{H}_{ab}^0 + \mathbf{H}_{ab}^S$, and using the identity $(\mathbf{A}-\mathbf{B})^{-1} = \mathbf{A}^{-1} + \mathbf{A}^{-1}\mathbf{B}(\mathbf{A}-\mathbf{B})^{-1}$ yields the intermediate,

$$\begin{aligned}\delta_k / P_k^0 &= -\mathbf{u}_k^\dagger(E_k^0)[E_k^0\mathbf{1}_{11} - \mathbf{H}_{11} - \mathbf{H}_{1b}(E_k^0\mathbf{1}_{bb} - \mathbf{H}_{bb})^{-1}\mathbf{H}_{b1}]\mathbf{u}_k(E_k^0)\\ &- \delta_k \mathbf{u}_k^\dagger(E_k^0)\{\mathbf{H}_{1b}^0[(E_k^0\mathbf{1}_{bb} - \mathbf{H}_{bb}^0)^{-1}\Omega_{\mathbf{H}_{bb}^S} + \Omega_{\mathbf{H}_{bb}^S}(E_k^0\mathbf{1}_{bb} - \mathbf{H}_{bb}^0)^{-1} + \Omega_{\mathbf{H}_{bb}^S}^2]\mathbf{H}_{b1}^0\\ &\qquad + \mathbf{H}_{1b}^0(E_k^0\mathbf{1}_{bb} - \mathbf{H}_{bb})^{-2}\mathbf{H}_{b1}^S + \mathbf{H}_{1b}^S(E_k^0\mathbf{1}_{bb} - \mathbf{H}_{bb})^{-2}\mathbf{H}_{b1}^0\\ &\qquad + \mathbf{H}_{1b}^S(E_k^0\mathbf{1}_{bb} - \mathbf{H}_{bb})^{-2}\mathbf{H}_{b1}^S\}\mathbf{u}_k(E_k^0)\end{aligned} \tag{3.6}$$

where,

$$\Omega_{\mathbf{H}_{bb}^S} = (E_k^0\mathbf{1}_{bb} - \mathbf{H}_{bb}^0)^{-1}\mathbf{H}_{bb}^S(E_k^0\mathbf{1}_{bb} - \mathbf{H}_{bb}^0 - \mathbf{H}_{bb}^S)^{-1} \tag{3.7}$$

and P_k^0 is the pole strength associated with the k-th pole, E_k^0, of the spin free Hamiltonian. Continuing the expansion in terms of $\mathbf{H}^0$ and realizing that

$$0 = \mathbf{u}_k^\dagger(E_k^0)[E_k^0\mathbf{1}_{11} - \mathbf{H}_{11}^0 - \mathbf{H}_{1b}^0(E_k^0\mathbf{1}_{bb} - \mathbf{H}_{bb}^0)^{-1}\mathbf{H}_{b1}^0]\mathbf{u}_k(E_k^0) \tag{3.8}$$

enables to express the exact hyperfine splitting, up to the quadratic contribution, Eq. (3.2), in terms of the spin free matrices only,

$$\begin{aligned}\delta_k &= P_k^0\mathbf{u}_k^\dagger(E_k^0)\{\mathbf{H}_{11}^S + \mathbf{H}_{1b}^S(E_k^0\mathbf{1}_{bb} - \mathbf{H}_{bb}^0)^{-1}\mathbf{H}_{1b}^0 + \mathbf{H}_{1b}^0(E_k^0\mathbf{1}_{bb} - \mathbf{H}_{bb}^0)^{-1}\mathbf{H}_{b1}^S\\ &\quad + \mathbf{H}_{1b}^S(E_k^0\mathbf{1}_{bb} - \mathbf{H}_{bb}^0)^{-1}\mathbf{H}_{b1}^S + (\mathbf{H}_{1b}^0 + \mathbf{H}_{1b}^S)\Omega_{\mathbf{H}_{bb}^S}(\mathbf{H}_{b1}^0 + \mathbf{H}_{b1}^S)\}\mathbf{u}_k(E_k^0)\\ &\quad - \delta_k P_k^0\mathbf{u}_k^\dagger(E_k^0)\{[(\mathbf{H}_{1b}^0 + \mathbf{H}_{1b}^S)][(E_k^0\mathbf{1}_{bb} - \mathbf{H}_{bb}^0)^{-1}\Omega_{\mathbf{H}_{bb}^S}\\ &\qquad + \Omega_{\mathbf{H}_{bb}^S}(E_k^0\mathbf{1}_{bb} - \mathbf{H}_{bb}^0)^{-1} + \Omega_{\mathbf{H}_{bb}^S}^2](\mathbf{H}_{b1}^0 + \mathbf{H}_{b1}^S)\\ &\qquad + \mathbf{H}_{1b}^0(E_k^0\mathbf{1}_{bb} - \mathbf{H}_{bb}^0)^{-2}\mathbf{H}_{b1}^S + \mathbf{H}_{1b}^S(E_k^0\mathbf{1}_{bb} - \mathbf{H}_{bb}^0)^{-2}\mathbf{H}_{b1}^0\\ &\qquad + \mathbf{H}_{1b}^S(E_k^0\mathbf{1}_{bb} - \mathbf{H}_{bb}^0)^{-2}\mathbf{H}_{b1}^S\}\mathbf{u}_k(E_k^0)\end{aligned} \tag{3.9}$$

The most important contributions to the spin Hamiltonian H^S can be expressed as one-electron operators, and it will be shown that the matrices $\mathbf{H}_{1b}^S$ and $\mathbf{H}_{b1}^S$ vanish, as long as the reference state is computed up to one order of perturbation smaller than these matrices. Thus,

$$\begin{aligned}\delta_k &= P_k^0\mathbf{u}_k^\dagger(E_k^0)\{\mathbf{H}_{11}^S + \mathbf{H}_{1b}^0\Omega_{\mathbf{H}_{bb}^S}\mathbf{H}_{b1}^0\\ &\qquad - \delta_k\mathbf{H}_{1b}^0[(E_k^0\mathbf{1}_{bb} - \mathbf{H}_{bb}^0)^{-1}\Omega_{\mathbf{H}_{bb}^S} + \Omega_{\mathbf{H}_{bb}^S}(E_k^0\mathbf{1}_{bb} - \mathbf{H}_{bb}^0)^{-1}\\ &\qquad + \Omega_{\mathbf{H}_{bb}^S}^2]\mathbf{H}_{b1}^0\}\mathbf{u}_k(E_k^0)\end{aligned} \tag{3.10}$$

which can be solved iteratively assuming that the matrices from the spin free propagator and the spin matrices $\mathbf{H}_{11}^S$ and $\mathbf{H}_{bb}^S$ are known.

4. The Hamiltonian and the Spin Hamiltonian H^S

The expression for the splitting in Eq. (3.9) is general and can be employed for all kinds of coupling Hamiltonians as long as the quadratic contribution can be neglected according to Eq. (3.2). Usually these coupling operators are obtained from the Dirac Hamiltonian (133-135), such as the spin-orbit coupling operator (H^{SO}) and the magnetic hyperfine Hamiltonian (h^{hf}), yielding the so called spin Hamiltonian H^S (5). For instance, the one-electron and two-electron terms of the spin-orbit coupling Hamiltonian, which is not going to be of any concern here, can be derived by applying the Breit-Pauli approximation to the Dirac Hamiltonian.

The hyperfine interaction Hamiltonian can be obtained (133-135) from the Dirac Hamiltonian by considering the effects of the magnetic field due to a nuclear magnetic dipole moment originated by a nuclear spin $\vec{I} \neq 0$. This Hamiltonian will be able to remove the degeneracy of an open shell specie and the splitting between the formerly degenerate state gives rise to the so called hyperfine splitting. The magnetic hyperfine Hamiltonian due to nuclei M can be written as

$$h_M^{\text{hf}} = \sum_{p,q} \langle p | (\vec{I}_M \cdot \overline{\overline{A}}_M \cdot \vec{s} + g_e \beta_e g_I \beta_n \vec{L} \cdot \vec{I}_M r_M^{-3}) | q \rangle p^\dagger q = \sum_{p,q} h_{pq,M}^{\text{hf}} p^\dagger q \tag{4.1}$$

where, $\vec{I}$ and $\vec{s}$ are the nuclear and electronic spin operators, g_e and g_I denote the electronic and nuclear g factors, with β_e and β_n being the Bohr and the nuclear magnetons, respectively. The hyperfine interaction tensor can be separated into an isotropic (spherically symmetric) and an anisotropic (dipolar) terms (5,133-135)

$$\overline{\overline{A}}_M = a_{\text{iso}}^{(M)} \overline{\overline{1}} + \overline{\overline{T}}^{(M)} \tag{4.2}$$

with

$$a_{\text{iso}}^{(M)} = g_e \beta_e g_I \beta_n \frac{8\pi}{3} \delta(\vec{r}_M) \tag{4.3}$$

and

$$\left(\overline{\overline{T}}^{(M)}\right)_{\mu,\nu} = g_e \beta_e g_I \beta_n r_M^{-5} (3 r_{M,\mu} r_{M,\nu} - r_M^2 \delta_{\mu\nu}) \tag{4.4}$$

where, $r_{M,\mu}$, with $\mu = (x, y, z)$ centered in $\vec{R}_M$, are the cartesian components of the vector $\vec{r}_M$, which is equal to $\vec{r} - \vec{R}_M$, with $\vec{R}_M$ being the coordinate of the nucleus M. As a result, the spin Hamiltonian will be approximated by the hyperfine Hamiltonian containing all nuclei contributions,

$$H^S \cong h^{\text{hf}} = \sum_M \sum_{pq} h_{pq,M}^{\text{hf}} p^\dagger q \tag{4.5}$$

The spin free electronic Hamiltonian of the system, H^0, is partitioned according to the usual Møller-Plesset form (129),

$$H^0 = h^0 + V = \sum_i^{occ} \varepsilon_i i^\dagger i + \frac{1}{4}\sum_{pqrs}\langle pq \| rs\rangle p^\dagger q^\dagger sr - \sum_{pq}\sum_i^{occ}\langle pi \| qi\rangle p^\dagger q \tag{4.6}$$

where the unperturbed Hamiltonian is written in terms of the occupied Hartree-Fock spin orbital energies ε_i and $\langle pq \| rs\rangle$ being the anti-symmetrized representation of two-electron integral. The following convention then applies, $a,b,c,d,\ldots$ denote unoccupied ("particles") and $i,j,k,l,\ldots$ occupied spin orbitals ("holes") in the reference state that diagonalize h^0, with $p,q,r,s,\ldots$ being either of the above.

Using the above partitioning into the Rayleigh-Schrödinger perturbation theory (RSPT) allows the perturbed reference function to be written as,

$$|0\rangle = |\text{HF}\rangle + |0^1\rangle + |0^2\rangle + \cdots \tag{4.7}$$

where, $|0^0\rangle = |\text{HF}\rangle$, the zero order, is the Hartree-Fock reference state, and $|0^n\rangle$ are the n-order RSPT perturbed wavefunction (129).

5. The Basis Manifold

Since the field operators satisfy the anticommutation relations (21,129)

$$[p,q]_+ = [p^\dagger,q^\dagger]_+ = 0 \text{ and } [p^\dagger,q]_+ = \delta_{pq} \tag{5.1}$$

then the principal manifold can be expressed as

$$\mathbf{h}_1 = \langle 0|0\rangle^{-1/2}\mathbf{a}_1 \text{ and } \mathbf{h}_1^\dagger = \langle 0|0\rangle^{-1/2}\mathbf{a}_1^\dagger \tag{5.2}$$

where the column and row vector are

$$\mathbf{a}_1 = \{i,a\} \equiv \{p\} \text{ and } \mathbf{a}_1^\dagger = \{j^\dagger,b^\dagger\} \equiv \{q^\dagger\} \tag{5.3}$$

and it can be assumed that the reference state is normalized, that is, $\langle 0|0\rangle = 1$.

The first element of the complementary manifold, $\mathbf{h}_3$, can be expressed as

$$\mathbf{h}_3 = (\mathbf{f}_3|\mathbf{f}_3)^{-1/2}\mathbf{f}_3 \text{ and } \mathbf{h}_3^\dagger = (\mathbf{f}_3|\mathbf{f}_3)^{-1/2}\mathbf{f}_3^\dagger \tag{5.4}$$

which has been Schmidt orthogonalized to $\mathbf{h}_1$ and symmetric normalized (136),

$$\mathbf{f}_3 = \mathbf{a}_3 - \mathbf{h}_1(\mathbf{h}_1|\mathbf{a}_3) \text{ and } \mathbf{f}_3^\dagger = \mathbf{a}_3^\dagger - (\mathbf{a}_3|\mathbf{h}_1)\mathbf{h}_1^\dagger \tag{5.5}$$

when there is no risk of linear dependencies, where

$$\mathbf{a}_3 = \{a^\dagger ij, i^\dagger ab\} \equiv \{p^\dagger qr\} \text{ and } \mathbf{a}_3^\dagger = \{j^\dagger i^\dagger a, a^\dagger b^\dagger i\} \equiv \{r^\dagger q^\dagger p\}, \quad q<r \tag{5.6}$$

and,

$$\mathbf{f}_3 = \{p^\dagger qr - \rho_{pq} r + \rho_{pr} q\} \text{ and } \mathbf{f}_3^\dagger = \{r^\dagger q^\dagger p - \rho_{qp} r^\dagger + \rho_{rp} q^\dagger\}, \quad q < r \tag{5.7}$$

with $\rho_{xy} = \langle 0 | x^\dagger y | 0 \rangle = \mathrm{Tr}\{\rho x^\dagger y\}$ being the one-particle reduced density matrix. It is straightforward to show that this construction of the inner projection basis manifold is orthonormal, that is,

$$(\mathbf{h}_1 \mid \mathbf{h}_1) = (\mathbf{h}_3 \mid \mathbf{h}_3) = \mathbf{1} \text{ and } (\mathbf{h}_1 \mid \mathbf{h}_3) = (\mathbf{h}_3 \mid \mathbf{h}_1) = \mathbf{0} \tag{5.8}$$

In a similar fashion, the second element of the complementary manifold $\mathbf{h}_5$ could be constructed to ensure normalization and orthogonality to $\mathbf{h}_1$ and $\mathbf{h}_3$.

6. Electron Propagator Spin Matrix Elements

Using the basis manifold defined before it can be shown that

$$\left([\mathbf{h}_1, H^S]_-\right)_s = \sum_u \mathbf{h}_1 h_{su}^{\mathrm{hf}} \tag{6.1}$$

that is, $\hat{H}^S$ maps the primary manifold $\mathbf{h}_1$ into itself. Thus,

$$\mathbf{H}_{b1}^S = \left(\mathbf{h}_b \mid [\mathbf{h}_1, H^S]_-\right) = \left(\mathbf{h}_b \mid \sum_u \mathbf{h}_1 h_{su}^{\mathrm{hf}}\right) = \sum_u h_{su}^{\mathrm{hf}} \left(\mathbf{h}_b \mid \mathbf{h}_1\right) = \mathbf{0} \tag{6.2}$$

since, $\mathbf{h}_b$ has constructed to be orthogonal to $\mathbf{h}_1$.

In the case of $\mathbf{h}_b = \mathbf{h}_1$, the resulting matrix elements are

$$\left(\mathbf{H}_{11}^S\right)_{pq} = \left(\mathbf{h}_1 \mid [\mathbf{h}_1, H^S]_-\right) = \sum_u h_{qu}^{\mathrm{hf}} \left(\left(\mathbf{h}_1 \mid \mathbf{h}_1\right)\right)_{pu} = \sum_u h_{qu}^{\mathrm{hf}} \delta_{pu} = h_{qp}^{\mathrm{hf}} \tag{6.3}$$

The matrix $\mathbf{H}_{1b}^S$ would be the transpose of $\mathbf{H}_{b1}^S$ if it were Hermitian. The Hermiticity of the superoperator Hamiltonian has been a concern since the beginnings of the electron propagator theory (46,129). For a Hermitian spin free Hamiltonian (H^0) the following relation can be written describing the Hermiticity problem,

$$\mathbf{H}_{ab}^0 - \mathbf{H}_{ba}^{0*} = \left\langle 0 \mid [[\mathbf{h}_a^\dagger, \mathbf{h}_b]_+, H^0]_- \mid 0 \right\rangle = \left\langle 0 \mid \hat{H}^0 [\mathbf{h}_a^\dagger, \mathbf{h}_b]_+ \mid 0 \right\rangle \tag{6.4}$$

which vanishes for an exact reference state, since $H^0 |0\rangle = E^0 |0\rangle$. When the reference state is expressed through a certain order n in the electron-electron interaction, then the Schrödinger equation is solved through the same order, that is, $H^0 |0\rangle^{(n)} = E_n^0 |0\rangle^{(n)} + O(n+1)$ where $|0\rangle^{(n)} = \sum_{i=0}^{n} |0^i\rangle$ and $E_n^0 = \sum_{i=0}^{n} E^{0(i)}$.

Then the non-Hermitian terms the will be of ($n+1$)-order, that is,

$$\mathbf{H}_{ab}^{0(n)} - \mathbf{H}_{ba}^{0(n)*} = \mathbf{0} + O(n+1) \tag{6.5}$$

The inclusion of the spin Hamiltonian H^S leads to a system involving a double perturbation (137), namely, the electron-electron interaction and H^S. Thus the reference state will be expressed as a double perturbed wavefunction $|0\rangle = |0^{(m,n)}\rangle$, where m and n denote the order in H^S and the electron-electron repulsion, respectively. The complete Hamiltonian is still Hermitian,

$$\mathbf{H}_{ab} - \mathbf{H}_{ba}^{*} = \langle 0 | [[\mathbf{h}_a^{\dagger}, \mathbf{h}_b]_+, H]_- | 0 \rangle = \langle 0 | \hat{H} [\mathbf{h}_a^{\dagger}, \mathbf{h}_b]_+ | 0 \rangle \tag{6.6}$$

and also it satisfy the Schrödinger equation through a given double order (m, n),

$H|0\rangle^{(n,m)} = E_{(n,m)}|0\rangle^{(n,m)} + O(n+1) + O(m+1)$ and $E_{(n,m)} = \sum_{i=0}^{n}\sum_{j=0}^{m} E^{(i,j)}$. So,

$$\mathbf{H}_{ab}^{(n,m)} - \mathbf{H}_{ba}^{(n,m)*} = \mathbf{0} + O(n+1) + O(m+1) \tag{6.7}$$

This means that the matrix $\mathbf{H}_{ab}$ is still Hermitian up to first order in H^S and $(n+1)$ order in the electron-electron interaction.

The matrix $\mathbf{H}_{33}^{S}$ can be obtained from the general expression for the super-operator Hamiltonian matrix $\mathbf{H}_{33}$ recently derived (126). For the non-diagonal one-electron Hamiltonian it can be written as

$$\left(\mathbf{H}_{33}^{S}\right)_{p'q'r',pqr} = \sum_{x,y>z} \left((\mathbf{f}_3 | \mathbf{f}_3)^{-1}\right)_{p'q'r',xyz} \left((\mathbf{f}_3 | \hat{H}^S | \mathbf{f}_3)\right)_{xyz,pqr} \tag{6.8}$$

where

$$\begin{aligned} &\left((\mathbf{f}_3 | \hat{H}^S | \mathbf{f}_3)\right)_{xyz,pqr} \\ &= \sum_s (1-P_{qr}) h_{rs}^{\mathrm{hf}} [\rho_{zy,qs}\delta_{xp} + (1-P_{yz})(\rho_{py,sx} + \rho_{ps}\rho_{yx})\delta_{zq}] \\ &- \sum_s (1-P_{yz}) h_{sp}^{\mathrm{hf}} [\rho_{sx}\delta_{zr}\delta_{qy} + (1-P_{qr})(\rho_{sy,rx} + \rho_{sr}\rho_{yx})\delta_{zq}] \\ &+ (1-P_{qr})(1-P_{yz}) h_{rz}^{\mathrm{hf}} (\rho_{px}\delta_{yq} - \rho_{py,qx} - \rho_{pq}\rho_{yx}) - h_{xp}^{\mathrm{hf}} \rho_{zy,qr} \end{aligned} \tag{6.9}$$

and

$$\begin{aligned} &\left((\mathbf{f}_3 | \mathbf{f}_3)\right)_{p'q'r',xyz} \\ &= \rho_{r'q',yz}\delta_{p'x} + (1-P_{yz})[\rho_{xp'}\delta_{q'y} - (1-P_{q'r'})(\rho_{xq',yp'} + \rho_{q'p'}\rho_{xy})]\delta_{r'z} \end{aligned} \tag{6.10}$$

with, $\rho_{pq,rs} = \langle 0 | p^{\dagger} q^{\dagger} rs | 0 \rangle$ being the two-particle density and P_{rs} an operator that permutes the rs indices. For a Hartree-Fock density this matrix become

$$\left(\mathbf{H}_{33}^{S}\right)_{a'i'j',aij}^{\mathrm{HF}} = \delta_{a'a} h_{i'i}^{\mathrm{hf}} \delta_{j'j} + \delta_{a'a}\delta_{i'i} h_{j'j}^{\mathrm{hf}} - h_{a'a}^{\mathrm{hf}} \delta_{i'i}\delta_{j'j} \tag{6.11a}$$

$$\left(\mathbf{H}_{33}^{S}\right)_{i'a'b',iab}^{\mathrm{HF}} = \delta_{i'i} h_{a'a}^{\mathrm{hf}} \delta_{b'b} + \delta_{i'i}\delta_{a'a} h_{b'b}^{\mathrm{hf}} - h_{i'i}^{\mathrm{hf}} \delta_{a'a}\delta_{b'b} \tag{6.11b}$$

since, $\rho_{pq}^{\mathrm{HF}} = \delta_{pq} n_p$ and $\rho_{pq,rs}^{\mathrm{HF}} = (1-P_{pq})\delta_{ps}\delta_{qr} n_p n_q$.

7. Order Analysis

It has been shown that independent of the truncation of the complementary manifold the matrix $\mathbf{H}_{b1}^{S}$ is null, and as long as its Hermiticity is maintained, so is $\mathbf{H}_{1b}^{S}$, yielding then Eq. (3.10) for the exact splitting δ_k for the state k. For practical implementations, the complementary manifold has to be truncated, so that for the smallest size of this manifold the splitting becomes

$$\begin{aligned}\delta_k &= P_k^0 \mathbf{u}_k^{\dagger}(E_k^0)\{\mathbf{H}_{11}^{S} + \mathbf{H}_{13}^{0}\Omega_{\mathbf{H}_{33}^{S}}\mathbf{H}_{31}^{0} \\ &- \delta_k \mathbf{H}_{13}^{0}[(E_k^0\mathbf{1}-\mathbf{H}_{33}^{0})^{-1}\Omega_{\mathbf{H}_{33}^{S}} + \Omega_{\mathbf{H}_{33}^{S}}(E_k^0\mathbf{1}-\mathbf{H}_{33}^{0})^{-1} + \Omega_{\mathbf{H}_{33}^{S}}^{2}]\mathbf{H}_{31}^{0}\}\mathbf{u}_k(E_k^0)\end{aligned} \tag{7.1}$$

with $\Omega_{\mathbf{H}_{33}^{S}}$ being defined according to Eq. (3.7). After α iterations the equation above becomes

$$\delta_{k(\alpha)} = \delta_{k(0)}\sum_{\beta=0}^{\alpha}(-P_k^0\Delta_k)^{\beta} \tag{7.2}$$

where,

$$\delta_{k(0)} = P_k^0\mathbf{u}_k^{\dagger}(E_k^0)\mathbf{H}_{11}^{S}\mathbf{u}_k(E_k^0) + P_k^0\mathbf{u}_k^{\dagger}(E_k^0)\mathbf{H}_{13}^{0}\Omega_{\mathbf{H}_{33}^{S}}\mathbf{H}_{13}^{0\dagger}\mathbf{u}_k(E_k^0) \tag{7.3}$$

and

$$\Delta_k = \mathbf{u}_k^{\dagger}(E_k^0)\{\mathbf{H}_{13}^{0}[(E_k^0\mathbf{1}-\mathbf{H}_{33}^{0})^{-1}\Omega_{\mathbf{H}_{33}^{S}} + \Omega_{\mathbf{H}_{33}^{S}}(E_k^0\mathbf{1}-\mathbf{H}_{33}^{0})^{-1} + \Omega_{\mathbf{H}_{33}^{S}}^{2}]\mathbf{H}_{13}^{0\dagger}\}\mathbf{u}_k(E_k^0) \tag{7.4}$$

This expression for δ_k could include up to infinite order correction in H^S by taking the limit $\alpha \to \infty$ since $\Omega_{\mathbf{H}_{33}^{S}}$ is at least of first order in H^S as well as via the iterative nature $\Omega_{\mathbf{H}_{33}^{S}}$ according of Eq. (3.7). However, H^S will, in general, give rise of very small effects, such as hyperfine splitting, so it can be expected that retaining all contributions up to second order in H^S will suffice for an accurate description.

7.1 First order in H^S and ESR

Since $\delta_{k(0)}$, Eq. (7.3), contains at least first order in H^S the summation in Eq. (7.2) disappears and the expression for splitting at first order simplifies to

$$\delta_k^{(1)} = \delta_{k(0)}^{(1)} = P_k^0\mathbf{u}_k^{\dagger}(E_k^0)\mathbf{H}_{11}^{S}\mathbf{u}_k(E_k^0) + P_k^0\mathbf{u}_k^{\dagger}(E_k^0)\mathbf{H}_{13}^{0}\Omega_{\mathbf{H}_{33}^{S}}^{(1)}\mathbf{H}_{13}^{0\dagger}\mathbf{u}_k(E_k^0) \tag{7.1.1}$$

where

$$\Omega_{\mathbf{H}_{33}^{S}}^{(1)} = (E_k^0\mathbf{1}-\mathbf{H}_{33}^{0})^{-1}\mathbf{H}_{33}^{S}(E_k^0\mathbf{1}-\mathbf{H}_{33}^{0})^{-1} \tag{7.1.2}$$

losses its iterative nature in H^S.

The $\mathbf{H}_{13}^{0}(E_k^0\mathbf{1}-\mathbf{H}_{33}^{0})^{-1}$ matrix is usually computed in the iterative pole search of the electron propagator (31,130). Thus, the implementation of the above expression for the splitting, Eq. (7.1.1), becomes simple, since it requires only the additional calculation of the $\mathbf{H}_{11}^{S}$ and $\mathbf{H}_{33}^{S}$ matrices.

It is still necessary to perform an order analysis of the correlation potential in the calculation of $\delta_k^{(1)}$. The usual implementation of the electron propagator is performed up to the third or partial fourth orders (31,32,129,130), which needs

$$\mathbf{H}_{13}^{0} \approx \mathbf{H}_{13}^{0(1)} + \mathbf{H}_{13}^{0(2)} \text{ and } \mathbf{H}_{33}^{0} \approx \mathbf{H}_{33}^{0(0)} + \mathbf{H}_{33}^{0(1)} \tag{7.1.3}$$

where the inverse matrix $(E_k^0\mathbf{1}-\mathbf{H}_{33}^{0})^{-1}$ has to be expanded (129), allowing for the following order analysis,

$$\delta_k^{(1)} \approx \delta_k^{(1,0)} + \delta_k^{(1,2)} + \delta_k^{(1,3)} \tag{7.1.4}$$

where,

$$\delta_k^{(1,0)} = P_k^0 \mathbf{u}_k^{\dagger}(E_k^0)\mathbf{H}_{11}^{S}\mathbf{u}_k(E_k^0) \tag{7.1.4a}$$

$$\delta_k^{(1,2)} = P_k^0 \mathbf{u}_k^{\dagger}(E_k^0)\mathbf{H}_{13}^{0(1)}(E_k^0\mathbf{1}-\mathbf{H}_{33}^{0(0)})^{-1}\mathbf{H}_{33}^{S}(E_k^0\mathbf{1}-\mathbf{H}_{33}^{0(0)})^{-1}\mathbf{H}_{13}^{0(1)\dagger}\mathbf{u}_k(E_k^0) \tag{7.1.4b}$$

and

$$\begin{aligned}\delta_k^{(1,3)} = P_k^0 \mathbf{u}_k^{\dagger}(E_k^0)\{&\mathbf{H}_{13}^{0(1)}(E_k^0\mathbf{1}-\mathbf{H}_{33}^{0(0)})^{-1}\mathbf{H}_{33}^{S}(E_k^0\mathbf{1}-\mathbf{H}_{33}^{0(0)})^{-1}\mathbf{H}_{13}^{0(2)\dagger}\\ &+\mathbf{H}_{13}^{0(1)}(E_k^0\mathbf{1}-\mathbf{H}_{33}^{0(0)})^{-1}\mathbf{H}_{33}^{S}(E_k^0\mathbf{1}-\mathbf{H}_{33}^{0(0)})^{-1}\mathbf{H}_{33}^{0(1)}(E_k^0\mathbf{1}-\mathbf{H}_{33}^{0(0)})^{-1}\mathbf{H}_{13}^{0(1)\dagger}\\ &+\mathbf{H}_{13}^{0(1)}(E_k^0\mathbf{1}-\mathbf{H}_{33}^{0(0)})^{-1}\mathbf{H}_{33}^{0(1)}(E_k^0\mathbf{1}-\mathbf{H}_{33}^{0(0)})^{-1}\mathbf{H}_{33}^{S}(E_k^0\mathbf{1}-\mathbf{H}_{33}^{0(0)})^{-1}\mathbf{H}_{13}^{0(1)\dagger}\\ &+\mathbf{H}_{13}^{0(2)}(E_k^0\mathbf{1}-\mathbf{H}_{33}^{0(0)})^{-1}\mathbf{H}_{33}^{S}(E_k^0\mathbf{1}-\mathbf{H}_{33}^{0(0)})^{-1}\mathbf{H}_{13}^{0(1)\dagger}\}\mathbf{u}_k(E_k^0)\end{aligned} \tag{7.1.4c}$$

Unlike the electron propagator expression for the electron binding energies (118,129) the expression for the splitting has a zero order contribution in the electronic correlation potential. It should be noted however, that even in this lowest level of correlation correction there are still contributions to the splitting due to the many body effects via P_k^0 and $\mathbf{u}_k^{\dagger}(E_k^0)$ terms. In fact, the diagonal or quasi-particle approximation, $\mathbf{u}_k^{\dagger}(E_k^0)=\mathbf{1}$, still yields these many body effects, namely $\delta_k^{(1,0)} \approx P_k^0\mathbf{h}_{kk}^{S}$. These many body effects can be quantified considering how much P_k^0 is less that unity (118,120). Since $\mathbf{h}_{kk}^{S}$ is related to the electronic density and δ_k describes the ionic state (open shell system), then the quantity $P_k^0\mathbf{h}_{kk}^{S}$ can be interpreted as an effective spin density of this open shell system. Similarly to the electron binding expression there is no first order contribution in the correlation potential, that is, $\delta_k^{(1,1)}=0$, so that $\delta_k^{(1,0)}$ is correct through second order. However, the second order correction in the electron correction for the splitting, $\delta_k^{(1,2)}$, is very important since it is simplest approximation that will take into account the spin-polarization effects (5,11) as well as the explicit effects of electron correlation. The latter can be taken even further into account by the third order expression $\delta_k^{(1,3)}$, Eq. (7.1.4c). A partial fourth order (P4) correction can also be performed with the matrices available for the third order implementation (130), namely,

$$\begin{aligned}\delta_k^{(1,P4)} &= P_k^0\mathbf{u}_k^\dagger(E_k^0)\\ &\times\{\mathbf{H}_{13}^{0(2)}(E_k^0\mathbf{1}-\mathbf{H}_{33}^{0(0)})^{-1}\mathbf{H}_{33}^S(E_k^0\mathbf{1}-\mathbf{H}_{33}^{0(0)})^{-1}\mathbf{H}_{33}^{0(1)}(E_k^0\mathbf{1}-\mathbf{H}_{33}^{0(0)})^{-1}\mathbf{H}_{13}^{0(1)\dagger}\\ &+\mathbf{H}_{13}^{0(1)}(E_k^0\mathbf{1}-\mathbf{H}_{33}^{0(0)})^{-1}\mathbf{H}_{33}^{0(1)}(E_k^0\mathbf{1}-\mathbf{H}_{33}^{0(0)})^{-1}\mathbf{H}_{33}^S(E_k^0\mathbf{1}-\mathbf{H}_{33}^{0(0)})^{-1}\mathbf{H}_{13}^{0(2)\dagger}\\ &+\mathbf{H}_{13}^{0(2)}(E_k^0\mathbf{1}-\mathbf{H}_{33}^{0(0)})^{-1}\mathbf{H}_{33}^S(E_k^0\mathbf{1}-\mathbf{H}_{33}^{0(0)})^{-1}\mathbf{H}_{13}^{0(2)\dagger}\}\mathbf{u}_k(E_k^0)\end{aligned} \tag{7.1.4d}$$

7.2 Second order in H^S and NMR

The contributions of the second order terms in H^S for the splitting in ESR is usually neglected since they are very small, and in fact they correspond to the NMR lines detected in some ESR experiments (5). However, the analysis of the second order $\delta_k^{(2)}$ expressions is important since it allows for the calculation of the indirect nuclear spin-spin couplings in NMR spectroscopy. These spin-spin couplings are usually calculated via a closed shell polarization propagator (138-140), so that, the approach described here would allow for the same calculations to be performed within the electron propagator theory for open shell systems. The general expression for the second order $\delta_k^{(2)}$ can be written as

$$\delta_k^{(2)} = \delta_{k(0)}^{(2)} - \delta_k^{(1)}P_k^0\Delta_k^{(1)} \tag{7.2.1}$$

where

$$\delta_{k(0)}^{(2)} = P_k^0\mathbf{u}_k^\dagger(E_k^0)\mathbf{H}_{13}^0\Omega_{\mathbf{H}_{33}^S}^{(2)}\mathbf{H}_{13}^{0\dagger}\mathbf{u}_k(E_k^0) \tag{7.2.1a}$$

$$\Omega_{\mathbf{H}_{33}^S}^{(2)} = (E_k^0\mathbf{1}-\mathbf{H}_{33}^0)^{-1}\mathbf{H}_{33}^S(E_k^0\mathbf{1}-\mathbf{H}_{33}^0)^{-1}\mathbf{H}_{33}^S(E_k^0\mathbf{1}-\mathbf{H}_{33}^0)^{-1} \tag{7.2.1b}$$

with

$$\Delta_k^{(1)} = \mathbf{u}_k^\dagger(E_k^0)\{\mathbf{H}_{13}^0(E_k^0\mathbf{1}-\mathbf{H}_{33}^0)^{-1}\Omega_{\mathbf{H}_{33}^S}^{(1)}\mathbf{H}_{31}^{0\dagger}+\text{c.c.}\}\mathbf{u}_k(E_k^0) \tag{7.2.1c}$$

and $\delta_k^{(1)}$ being the first order hyperfine splitting given in Eq. (7.1.1), and c.c. denotes the complex conjugate of the preceding term. Using the hyperfine interaction Hamiltonian, Eqs. (4.1) – (4.4), due to nuclei M and M' in the expressions above leads to the complete spin-spin coupling tensor containing the following terms, (SO) – paramagnetic spin-orbital due to $\vec{L}$ in Eq. (4.1), (SD) – spin-dipole due to $\overline{\overline{T}}^{(M)}$ in Eq. (4.4), and (FC) – Fermi contact due to $\delta(\vec{r}_M)$ in

Eq. (4.3), as well as the cross term SD/FC (139). This approach allows then for a unified picture of the ESR and NMR spectroscopy.

As before, the order analysis for the correlation effects can be performed and the lowest non-zero contributions arise from second order as,

$$\delta_k^{(2,2)} = \delta_{k(0)}^{(2,2)} - \delta_{k(0)}^{(1,0)} P_k^0 \Delta_k^{(1,2)} \tag{7.2.2}$$

where,

$$\delta_{k(0)}^{(2,2)} = P_k^0 \mathbf{u}_k^\dagger(E_k^0)\mathbf{H}_{13}^{0(1)}\Omega_{\mathbf{H}_{33}^S}^{(2)}\mathbf{H}_{13}^{0(1)\dagger}\mathbf{u}_k(E_k^0) \tag{7.2.2a}$$

$$\Delta_k^{(1,2)} = \mathbf{u}_k^\dagger(E_k^0)\{\mathbf{H}_{13}^{0(1)}(E_k^0\mathbf{1} - \mathbf{H}_{33}^0)^{-1}\Omega_{\mathbf{H}_{33}^S}^{(1)}\mathbf{H}_{31}^{0(1)\dagger} + \text{c.c.}\}\mathbf{u}_k(E_k^0) \tag{7.2.2b}$$

and $\delta_k^{(1,0)}$ is given by Eq. (7.1.4a). It should be noted that all matrices involved in such a calculation are available from an electron propagator-ESR implementation. The polarization propagator approach to spin-spin coupling allows for zero and first orders (FOPPA) expressions in the fluctuation potential (138-140). The implementation of higher order corrections are quite demanding in numerical and algebraic terms, see for instance the second order polarization propagator approximation (SOPPA) expressions (139).

The inclusion of the correlation corrections to the spin-spin coupling calculation via electron propagator is quite straightforward since at third order it can be written as

$$\delta_k^{(2,3)} = \delta_{k(0)}^{(2,3)} - \delta_{k(0)}^{(1,0)} P_k^0 \Delta_k^{(1,3)} \tag{7.2.3}$$

with

$$\delta_{k(0)}^{(2,3)} = P_k^0 \mathbf{u}_k^\dagger(E_k^0)\{\mathbf{H}_{13}^{0(1)}\Omega_{\mathbf{H}_{33}^S}^{(2)}\mathbf{H}_{13}^{0(2)\dagger} + \text{c.c.}\}\mathbf{u}_k(E_k^0) \tag{7.2.3a}$$

$$\begin{aligned}\Delta_k^{(1,3)} = \mathbf{u}_k^\dagger(E_k^0)\{&\mathbf{H}_{13}^{0(1)}(E_k^0\mathbf{1} - \mathbf{H}_{33}^0)^{-1}\Omega_{\mathbf{H}_{33}^S}^{(1)}\mathbf{H}_{31}^{0(2)\dagger} + \text{c.c.}\\ &+ \mathbf{H}_{13}^{0(1)}\Omega_{\mathbf{H}_{33}^S}^{(1)}(E_k^0\mathbf{1} - \mathbf{H}_{33}^0)^{-1}\mathbf{H}_{31}^{0(2)\dagger} + \text{c.c.}\}\mathbf{u}_k(E_k^0)\end{aligned} \tag{7.2.3b}$$

and at partial fourth order is expressed as

$$\delta_k^{(2,P4)} = \delta_{k(0)}^{(2,4)} - \delta_{k(0)}^{(1,0)} P_k^0 \Delta_k^{(1,P4)} - \delta_{k(0)}^{(1,2)} P_k^0 \Delta_k^{(1,2)} \tag{7.2.4}$$

where

$$\delta_{k(0)}^{(2,4)} = P_k^0 \mathbf{u}_k^\dagger(E_k^0)\mathbf{H}_{13}^{0(2)}\Omega_{\mathbf{H}_{33}^S}^{(2)}\mathbf{H}_{13}^{0(2)\dagger}\mathbf{u}_k(E_k^0) \tag{7.2.4a}$$

$$\Delta_k^{(1,P4)} = \mathbf{u}_k^\dagger(E_k^0)\{\mathbf{H}_{13}^{0(2)}(E_k^0\mathbf{1} - \mathbf{H}_{33}^0)^{-1}\Omega_{\mathbf{H}_{33}^S}^{(1)}\mathbf{H}_{31}^{0(2)\dagger} + \text{c.c.}\}\mathbf{u}_k(E_k^0) \tag{7.2.4b}$$

and

$$\delta_{k(0)}^{(1,2)} = P_k^0 \mathbf{u}_k^\dagger(E_k^0)\mathbf{H}_{13}^{0(2)}\Omega_{\mathbf{H}_{33}^S}^{(1)}\mathbf{H}_{13}^{0(2)\dagger}\mathbf{u}_k(E_k^0) \tag{7.2.4c}$$

which use only the matrices already available in the electron propagator-ESR implementation.

8. Concluding Remarks

It has been shown that the electron propagator theory allows for the direct calculation of hyperfine coupling constants (ESR) as well as spin-spin coupling (NMR) of open shell systems using a closed shell reference. In fact it provides a unified picture of presenting and calculating these couplings from the spin free electron propagator matrices. In addition, the approach developed here appears to be a very effective tool for introducing the effects of the electronic correlation in the calculations of such a properties.

The approach developed here for calculating the splitting is quite general, since it could be applied to other types of propagators such as, polarization propagator and two-electron propagator. The latter would allow access to other open shell systems than doublets, and the former for closed shell or number-conserving particle states. Thus, in conjunction with some recent developments such as, coupled perturbed electron propagator (124-136) and analytical gradients of electron binding energy (111,112), the present approach has aimed, and apparently succeeded, in extending the applicability and the unifying nature of the propagator theory to all kinds of properties and systems.

Acknowledgment

The author wishes to thank N. Yngve Öhrn firstly for introducing him to the fascinating areas of propagator methods and electron-nuclear dynamics, then for providing him with the best scientific formation expected, and most importantly for his unconditional friendship. Thus, the author is very thankful for this opportunity of honoring Yngve Öhrn in the occasion of his 65th birthday.

References

1. See, e.g., Elschenbroich, Ch., and Salzer, A., *Organometallics – A Concise Introduction* (VCH, New York, 1992).
2. See, e.g., Wertz, J. E., and Bolton, J. R., *Electron Spin Resonance* (McGraw-Hill, New York, 1992).
3. See, e.g., Goodman, B. A., and Raynor, J. B., *Adv. Inorg. Chem. Radiochem.*, **13**, 136 (1970).
4. See, e.g., So, H., and Pope, M. T. ***in*** *Electron and Proton Transfer in Chemistry and Biology*, Müller, A., Ratajczak, H., Junge, W., and Diemann, E., eds. (Elsevier, New York, 1992), p. 71.
5. See, e.g., Weltner, Jr., W., *Magnetic Atoms and Molecules* (Dover, New York, 1983).
6. See, e.g., In Lund, A., and Shiotani, M., *Radical Ionic Systems – Properties in Condensed Phases* (Kluwer, Dordrecht, 1991).

7. See, e.g., Knight, Jr., L. B., *Acc. Chem. Res.*, **19**, 313 (1986).
8. Hirota, E., *High Resolution Spectroscopy of Transient Molecules* (Springer, Berlin, 1985).
9. McConnell, H. M., *J. Chem. Phys.*, **28**, 1188 (1958).
10. Karplus, M., and Fraenkel, G. K., *J. Chem. Phys.*, **35**, 1312 (1961).
11. Chipman, D. M., *Theor. Chim. Acta*, **82**, 93 (1992).
12. Sekino, H., and Bartlett, R. J., *J. Chem. Phys.*, **82**, 4225 (1985).
13. Hamrick, Y. M., Van Zee, R. J., Godbout, J. T., Weltner, Jr., W., Lauderdale, W. J., Stanton, J. F., and Bartlett, R. J., *J. Phys. Chem.*, **95**, 2840 (1991).
14. Feller, D., and Davidson, E. R., *J. Chem. Phys.*, **80**, 1006 (1984).
15. Feller, D., and Davidson, E. R., *J. Chem. Phys.*, **88**, 7580 (1988).
16. Momose, T., Nakatsuji, H., and Shida, T., *J. Chem. Phys.*, **89**, 4185 (1988).
17. Momose, T., Yamagucchi, M., and Shida, T., *J. Chem. Phys.*, **93**, 7284 (1990).
18. Eriksson, L. A., Malkin, V. G., Malkina, O. L., and Salahub, D. R., *J. Chem. Phys.*, **99**, 9756 (1993).
19. Kong, J., Eriksson, L. A., and Boyd, R. J., *Chem. Phys. Lett.*, **217**, 24 (1994).
20. Linderberg, J., and Öhrn, Y., *Chem. Phys. Lett.*, **1**, 295 (1967), *J. Chem. Phys.*, **49**, 716 (1968).
21. Linderberg, J., and Öhrn, Y., *Propagators in Quantum Chemistry* (Academic Press, New York, 1973), pages 145-147.
22. Cederbaum, L. S., Matschke, F. E. P., and von Niessen, W., *Phys. Rev. A*, **12**, 6 (1975).
23. Reinhardt, W. P., and Doll, J. D., *J. Chem. Phys.*, **50**, 2767 (1969).
24. Doll, J. D., and Reinhardt, W. P., *J. Chem. Phys.*, **57**, 1169 (1972).
25. Cederbaum, L. S., Hohlneicher, G, and Peyerimhoff, *Chem. Phys. Lett.*, **11**, 421 (1971).
26. Cederbaum, L. S., *Chem. Phys. Lett.*, **25**, 562 (1974).
27. Cederbaum, L. S., Hohlneicher, G, and von Niessen, W., *Chem. Phys. Lett.*, **18**, 503 (1973).
28. Cederbaum, L. S., Hohlneicher, G, and von Niessen, W., *Mol. Phys.*, **26**, 1405 (1973).
29. Cederbaum, L. S., and von Niessen, W., *Chem. Phys. Lett.*, **24**, 263 (1974).
30. Cederbaum, L. S., *Mol. Phys.*, **28**, 479 (1974).
31. Purvis, G., and Öhrn, N. Y., *J. Chem. Phys.*, **60**, 4063 (1974).
32. Purvis, G., and Öhrn, N. Y., *J. Chem. Phys.*, **62**, 2045 (1975).
33. Purvis, G., and Öhrn, N. Y., *Chem. Phys. Lett.*, **33**, 396 (1975).
34. Goscinski, O., and Lukman, B., *Chem. Phys. Lett.*, **7**, 573 (1970).
35. Pickup, B. T., and Goscinski, O., *Mol. Phys.*, **26**, 1013 (1973).
36. Simons, J., and Smith, W. D., *J. Chem. Phys.*, **58**, 4899 (1973).
37. Chen, T. T., Smith, W. D., and Simons, J., *J. Chem. Phys.*, **61**, 2670 (1974).

38. Simons, J., *Chem. Phys. Lett.*, **25**, 122 (1974).
39. Smith, W. D., Chen, T. T., and Simons, J., *Chem. Phys. Lett.*, **26**, 296 (1974).
40. Griffing, K., and Simons, J., *J. Chem. Phys.*, **62**, 535 (1975).
41. Rowe, D. J., *Rev. Mod. Phys.*, **40**, 153 (1968).
42. Rowe, D. J., *Phys. Rev.*, **175**, 1283 (1968).
43. Jørgensen, P., and Simons, J., *J. Chem. Phys.*, **63**, 5302 (1975).
44. Redmon, L. T., Purvis, G., and Öhrn, Y., *J. Chem. Phys.*, **63**, 5011 (1975).
45. Cederbaum, L. S., and Domcke, W., *Adv. Chem. Phys.*, **36**, 205 (1977).
46. Nehrkorn, C., Purvis, G. D., and Öhrn, Y., *J. Chem. Phys.*, **64**, 1752 (1976).
47. Manne, R., *Chem. Phys. Lett.*, **45**, 470 (1977).
48. Dalgaard, E., *Int. J. Quantum Chem.*, **15**, 169 (1979).
49. Simons, J., *J. Chem. Phys.*, **64**, 4541 (1976).
50. Born, G., and Öhrn, Y., *Int. J. Quantum Chem., Quantum Chem. Symp.*, **12**, 143 (1978).
51. Born, G., and Öhrn, Y., *Phys. Scr.*, **21**, 378 (1980).
52. Goscinski, O., and Weiner, B., *Phys. Scr.*, **21**, 385 (1980).
53. Weiner, B., and Goscinski, O., *Phys. Rev. A*, **22**, 2374 (1980).
54. Weiner, B., and Goscinski, O., *Int. J. Quantum Chem.*, **18**, 1105 (1980).
55. Weiner, B., and Goscinski, O., *Int. J. Quantum Chem.*, **21**, 269 (1982).
56. Prasad, M. D., Pal, S., and Mukherjee, D., *Phys. Rev. A*, **31**, 1287 (1985).
57. Coester, F., *Nucl. Phys.*, **1**, 421 (1958).
58. Coester, F., and Kummel, H., *Nucl. Phys.*, **17**, 477 (1960).
59. Cizek, J., *J. Chem. Phys.*, **45**, 4256 (1966).
60. Cizek, J., *Adv. Chem. Phys.*, **14**, 35 (1969).
61. Cizek, J., and Paldus, J., *Phys. Scr.*, **21**, 251 (1980).
62. Datta, B., Mukhopadhyay, D., and Mukherjee, D., *Phys. Rev. A*, **47**, 3632 (1993).
63. Meissner, L., and Bartlett, R. J., *Int. J. Quantum Chem., Quantum Chem. Symp.*, **27**, 67 (1993).
64. I. Lindgren, *J. Phys. B*, **7**, 2441 (1974).
65. I. Lindgren, *Int. J. Quantum Chem., Quantum Chem. Symp.*, **12**, 33 (1978).
66. Haque, A., and Mukherjee, D., *J. Chem. Phys.*, **80**, 5058 (1984).
67. Stolarczyk, L, and Monkhorst, H., *Phys. Rev. A*, **32**, 725, 743 (1985).
68. Pal, S., Rittby, M., Bartlett, R. J., Sinha, D, and Mukherjee, D., *Chem. Phys. Lett.*, **137**, 273 (1987).
69. Pal, S., Rittby, M., Bartlett, R. J., Sinha, D, and Mukherjee, D., *J. Chem. Phys.*, **88**, 4357 (1988).
70. Sinha, D, Mukhopadhyay, S. K., Chaudhuri, R., and Mukherjee, D., *Chem. Phys. Lett.*, **154**, 544 (1989).
71. Mukhopadhyay, D., Mukhopadhyay, S. K., Chaudhuri, R., and Mukherjee, D., *Theor. Chim. Acta*, **80**, 483 (1991).

72. Monkhorst, H., *Int. J. Quantum Chem., Quantum Chem. Symp.*, **11**, 421 (1977).
73. Emrich, K., *Nucl. Phys. A*, **351**, 392 (1981).
74. Sekino, H., and Bartlett, R. J., *Int. J. Quantum Chem., Quantum Chem. Symp.*, **18**, 255 (1984).
75. Nooijen, M., and Bartlett, R. J., *J. Chem. Phys.*, **102**, 3629 (1995).
76. Nooijen, M., and Bartlett, R. J., *J. Chem. Phys.*, **106**, 6441, 6449 (1997).
77. Nooijen, M., and Bartlett, R. J., *J. Chem. Phys.*, **107**, 6812 (1997).
78. Kutzelnigg, W., and Mukherjee, D., *J. Chem. Phys.*, **90**, 5578 (1989).
79. Löwdin, P. O., *Phys. Rev.*, **139**, 357 (1965).
80. Löwdin, P. O., *Int. J. Quantum Chem.*, **2**, 867 (1968).
81. Löwdin, P. O., *Int. J. Quantum Chem., Quantum Chem. Symp.*, **16**, 485 (1982).
82. Ortiz, J. V., *J. Chem. Phys.*, **104**, 7599 (1996).
83. Ortiz, J. V., *Int. J. Quantum Chem.*, **63**, 291 (1997).
84. Ortiz, J. V., and Zakrzewski, V. G., *J. Chem. Phys.*, **105**, 2762 (1996).
85. Purvis, G., and Öhrn, N. Y., *J. Chem. Phys.*, **65**, 917 (1976).
86. Born, G., Kurtz, H. A., and Öhrn, N. Y., *J. Chem. Phys.*, **68**, 74 (1978).
87. Freed, K. F., Herman, M. F., and Yeager, D. L., *Phys. Scr.*, **21**, 242 (1980).
88. Purvis, G., and Öhrn, N. Y., *J. Chem. Phys.*, **62**, 2045 (1975).
89. Born, G., and Shavitt, I., *J. Chem. Phys.*, **76**, 558 (1982).
90. Born, G., *Int. J. Quantum Chem., Quantum Chem. Symp.*, **16**, 633 (1982).
91. Born, G., *Int. J. Quantum Chem.*, **28**, 335 (1985).
92. Albertsen, P., and Jørgensen, P., *J. Chem. Phys.*, **70**, 3254 (1979).
93. Banerjee, A., Shepard, R., and Simons, J., *Int. J. Quantum Chem., Quantum Chem. Symp.*, **12**, 389 (1978).
94. Nichols, J. A., Yeager, D. L., and Jørgensen, P., *J. Chem. Phys.*, **80**, 293 (1984).
95. Golab, J. T., and Yeager, D. L., *J. Chem. Phys.*, **87**, 2925 (1987).
96. Graham, R. L., Yeager, D. L., and Rizzo, A., *J. Chem. Phys.*, **91**, 5451 (1989).
97. Yeager, D. L. ***in*** *Applied Many-Body Methods in Spectroscopy and Electronic Structure*, Mukherjee, D., ed. (Plenum Press, New York, 1992), pp. 133-161.
98. Zakrzewski, V. G., Ortiz, J. V., Nichols, J. A., Heryadi, D., Yeager, D. L., and Golab, J. T., *Int. J. Quantum Chem.*, **60**, 29 (1996).
99. Csanak, G., Taylor, H. S., and Yaris, R., *Adv. At. Mol. Phys.*, **7**, 287 (1971).
100. Langhoff, P. W., and Hernández, A. J., *Chem. Phys. Lett.*, **49**, 361 (1977).
101. Carravetta, V., and Moccia, R., *Mol. Phys.*, **35**, 129 (1978).
102. Cacelli, I., Moccia, R., and Carravetta, V., *Chem. Phys. Lett.*, **70**, 569 (1980).
103. Holleboom, L. J., Snijders, J. G., Baerends, E. J., and Buijse, M. A., *J. Chem. Phys.*, **89**, 3638 (1988).

104. Holleboom, L. J., Snijders, J. G., and, Baerends, E. J., *Int. J. Quantum Chem.*, **34**, 289 (1988).
105. Holleboom, L. J., and Snijders, J. G., *J. Chem. Phys.*, **93**, 5826 (1990).
106. Holleboom, L. J., and Snijders, J. G., *Int. J. Quantum Chem.*, **46**, 259 (1992).
107. Ortiz, J. V., *Int. J. Quantum Chem., Quantum Chem. Symp.*, **29**, 331 (1995).
108. Ortiz, J. V., *J. Chem. Phys.*, **103**, 5630 (1995).
109. Schirmer, J., and Angonoa, G., *J. Chem. Phys.*, **91**, 1754 (1989).
110. Ortiz, J. V., *J. Am. Chem. Soc.*, **113**, 1102(1992); **113**, 3593 (1992).
111. Cioslowski, J., and Ortiz, J. V., *J. Chem. Phys.*, **96**, 8379 (1992).
112. Ortiz, J. V., *Int. J. Quantum Chem., Quantum Chem. Symp.*, **26**, 1 (1992).
113. Berman, M., Walter, O., and Cederbaum, L. S., *Phys. Rev. Lett.*, **50**, 1979 (1983).
114. Pickup, B. T., *Chem. Phys.*, **19**, 193 (1977).
115. Deleuze, M., Pickup, B. T., and Delhalle, J., *Mol. Phys.*, **83**, 655 (1994).
116. Mishra, M., and Öhrn, Y., *Int. J. Quantum Chem., Quantum Chem. Symp.*, **14**, 335 (1980).
117. Mishra, M., and Öhrn, Y., *Chem. Phys. Lett.*, **71**, 549 (1980).
118. Öhrn, Y., and Born, G., *Adv. Quantum Chem.*, **13**, 1 (1981).
119. von Niessen, W., Schirmer, J., and Cederbaum, L. S., *Comput. Phys. Rep.*, **1**, 57 (1987).
120. Cederbaum, L. S., Domcke, W., Schirmer, J., and von Niessen, W., *Adv. Chem. Phys.*, **65**, 115 (1986).
121. Mishra, M., and Öhrn, Y., *Chem. Phys. Lett.*, **81**, 339 (1981).
122. Ortiz, J. V. ***in*** *Computational Chemistry: Reviews of Current Trends*, vol. 2, Leszczynski, J., ed. (World Scientific, Singapore, 1997), pp. 1-61.
123. Ortiz, J. V., Zakrzewski, V. G., and Dolgounitcheva, O. ***in*** *Conceptual Perspectives in Quantum Chemistry*, Calais, J.-L., and Kryachko, E., eds. (Kluwer Academic Publishers, Dordrecht, 1997), pp. 465-517.
124. Pickup, B. T., *Int. J. Quantum Chem., Quantum Chem. Symp.*, **26**, 13 (1992).
125. Pickup, B. T., *Philos. Mag. B*, **69**, 799 (1994).
126. Deleuze, M. S., and Pickup, B. T., *Int. J. Quantum Chem.*, **63**, 483 (1997).
127. Zubarev, D. N., *Sov. Phys. Usp.*, **3**, 320 (1960).
128. Oddershede, J., *Adv. Chem. Phys.*, **69**, 201 (1987).
129. Jørgensen, P., and Simons, J., *Second Quantization-Based Methods in Quantum Chemistry* (Academic Press, New York, 1981).
130. Ortiz, J. V., *Int. J. Quantum Chem., Quantum Chem. Symp.*, **23**, 321 (1989).
131. Baker, J., and Pickup, B. T., *Chem. Phys. Lett.*, **76**, 537 (1980).
132. Baker, J., and Pickup, B. T., *Mol. Phys.*, **49**, 651 (1983).
133. Frosch, R. A., and Foley, H. M., *Phys. Rev.*, **88**, 1337 (1952).
134. Dousmanis, G. C., *Phys. Rev.*, **97**, 967 (1955).

135. Kristiansen, P., and Veseth, L., *J. Chem. Phys.*, **84**, 2711 (1986).
136. Löwdin, P. O., *Adv. Phys.*, **5**, 1 (1956).
137. Wilson, S. ***in*** *Methods in Computational Chemistry*, vol. 5, Wilson, S., ed. (Plenum Press, New York, 1992), p.267.
138. Oddershede, J., *Adv. Quantum Chem.*, **11**, 275 (1978).
139. Geertsen, J., and Oddershede, J., *Chem. Phys.*, **90**, 301 (1984).
140. Diz, A. C., Ruiz de Azua, M. C., Giribet, C. G., and Contreras, R. H., *Int. J. Quantum Chem.*, **37**, 663 (1990).

The New Challenges of the Theory of Ionization for Polymers and Solids

M. S. Deleuze

Theoretische Scheikunde, Departement SBG, Limburgs Universitair Centrum, Universitaire Campus, B-3590 Diepenbeek, Belgium

L.S. Cederbaum

Lehrstuhl für Theoretische Chemie, Institut für Physikalische Chemie, Universität Heidelberg, D69120 Heidelberg, Germany

TABLE OF CONTENTS

1. INTRODUCTION

Photoelectron spectroscopy (PES) has been extensively used since the seventies to probe the electronic structure of polymers and assess band structure data (first ionization potential, effective electron mass, ...) of importance in deducing the electrical, optical and chemical properties of advanced materials. Valence photoionization spectra can also yield invaluable information on the molecular architecture prevailing at polymer surfaces (or regions near to it), provided one can set up safe relationships between ionization bands and the chemical bonding characteristics. These include not only the overall atomic connectivity (primary structure or configuration) of chains (1), but also the possibility of tracing more subtle changes of the molecular conformation (or secondary structure) (2-6). Owing to the short escape depth of electrons in matter, photoionization spectroscopy can then be regarded as a very promising alternative to other surface sensitive

ADVANCES IN QUANTUM CHEMISTRY, VOLUME 35

0065-3276/99 $30.00

techniques such as STM or AFM, the reliablity of which is often impeded by a necessary preconditioning of organic samples.

Except for a very few studies (4-12), the photoionization spectra of polymers and large molecules have most oftenly been interpreted using the one-particle picture (or equivalently the orbital picture) of ionization, which is based on the assumption of a one-to-one correspondence between the recorded lines and canonical states computed at various methodological levels (13). Among others, these comprise semi-empirical or ab initio methods such as extended Hückel (1), CNDO, MNDO, INDO ..., Hartree-Fock Roothaan (HFR) (2-3), Valence effective Hamiltonian (VeH) (14), ... More recently, simple Green's function schemes have been implemented in a quasi-particle version (4-6,15), to account for the leading correlation and relaxation effects on the ionization energy of a single electron level, by means of a self-energy or Perturbation Theory correction to the HF eigenvalues.

The orbital picture has apparently always been relevant for reproducing the main features found within the valence photoionization spectra of large compounds, clusters and solids. Quite naturally therefore, and since molecular orbitals provide a direct insight into bonding characteristics, this picture has been for decades the ultimate reference for most polymer scientists (and by extension solid state physicists in general). As a matter of fact, an analysis of the nodal structure and delocalization character of orbitals lies always behind any structural assignment (see e.g. 2,4d,5,6b,6g,6h) of organic molecules and polymers from their photoemission spectra. However, the physics behind the recorded signals can differ markedly from a single-particle process.

An increasing amount of experimental (16) and theoretical (17) studies on molecules of rather limited size points indeed to a complete failure of orbital models for the description of photoemission. Besides a shift of electron binding energies by several eV compared to the results of HF calculations, many-body effects (electronic correlation and relaxation) often manisfest themselves (18,19) through the fragmentation of inner valence lines into dense sets of shake-up (or satellite) lines of low intensity. This phenomenom is the result of strong interactions between the primary (one-hole ; 1h) and excited configurations of the cation. It is most oftenly referred to as a '*complete breakdown of the molecular orbital picture of ionization*' (19) , when it is no longer possible to distinguish a clearly dominant 1h contribution among the lines deriving from the ionization of a given electron level.

In this contribution, in order to illustrate the importance of shake-up bands for extended systems, we simulate and compare on correlated grounds the ionization spectra of polyethylene and polyacetylene, the most simplest systems one can consider to represent insulating or semi-conducting polymers. Conclusions for the infinite stereoregular chains are drawn by extrapolation of the trends observed with the first terms of the related *n*-alkane or acene series, C_nH_{2n+2} and C_nH_{n+2}, respectively, with n=2, 4, 6 and 8. Our simulations are also compared to X-ray photoionization spectra (7) recorded on gas phase samples of ethylene, butadiene and hexatriene, which provide a clear experimental manisfestation of the construction of correlation bands (8-12).

Calculations are carried out using one of the most efficient and reliable tool for the description of the photoemission process : the one-particle Green's function (GF) method (20) (also referred to as the one-particle propagator (21) approach), in the so-called third-order Algebraic Diagrammatic Construction (ADC[3]) scheme (22). By virtue of this scheme, these calculations account for electronic correlation both in the initial and final states of the ionization process and for multistate interactions within the cation. As such, they provide certainly one of the most thorough investigation available to date on the ionization spectra of large saturated (11) and unsaturated (12) hydrocarbons. In this contribution, we restrict ourselves to a basic presentation of the one-particle Green's function and its relationships with electron binding energies and photoionization intensities. More fundamental or technical details on its expansion and evaluation can be found elsewhere (10-12,17,19-25). For a general and thorough introduction to propagator methods, we also would like to draw the readers attention to a text book and lecture notes (21) by Prof. Y. Öhrn.

2. OUTLINE OF THEORY AND METHODOLOGY

For a correlated N-electron system with a non-degenerate ground state $|\Psi_0^N>$, the one-particle Green's function has the spectral representation (20,21) :

$$G_{ij}(\omega) = \sum_m \frac{<\Psi_0^N|a_i^+|\Psi_m^{N-1}><\Psi_m^{N-1}|a_j|\Psi_0^N>}{\omega-(E_0^N-E_m^{N-1})-i0^+} + \sum_p \frac{<\Psi_0^N|a_j|\Psi_p^{N+1}><\Psi_p^{N+1}|a_i^+|\Psi_0^N>}{\omega-(E_p^{N+1}-E_0^N)+i0^+} \tag{1}$$

This expression is derived as the Fourier transform of a time-dependent one-particle autocorrelation function (26) (i.e. *propagator*), and cast in matrix form **G**(ω) over a suitable molecular orbital (e.g. HF) basis, by means of the related set of one-electron creation (a_i^+) and annihilation (a_i) operators. In this equation, the sums over m and p run over all the states of the (N-1)- and (N+1)-electron system, $|\Psi_m^{N-1}>$ and $|\Psi_p^{N+1}>$, respectively. E_0^N and $E_n^{N\pm1}$ represent the energy of the N-particle system in its ground state and of the N±1-particle states.

The importance of the one-particle Green's function for the calculation of ionization and electron affinity spectra can already be appreciated from Eq. (1) : regardless of sign, ionization energies and electron affinities relate to the poles of its first and second components, respectively. The associated residues correspond to products of one-electron transition amplitudes. These are defined as partial overlaps between the neutral ground state and the ionized states :

$$x_j^{(n)} = \begin{cases} <\Psi_n^{N-1}|a_j|\Psi_0^N> & \forall n \in \{N-1\} \\ <\Psi_0^N|a_j|\Psi_n^{N+1}> & \forall n \in \{N+1\} \end{cases} \qquad (2)$$

and are therefore closely related to intensities. Regardless of scattering effects in the continuum, relative intensities are indeed consistently estimated (19,20) from the probability of a given ionic state $|\Psi_n^{N\pm1}>$, defining (12) its *pole strength* :

$$\Gamma_n = \sum_i \|x_i^{(n)}\|^2 \qquad (3),$$

The ionization process can be described as a one-particle event, when only one term dominates in summation (3), yielding a pole strength close unity. In this case, 1 - Γ_n gives an estimate of the fraction of photoemission intensity dispersed in many-body effects. On the contrary, small pole strengths are indicative (18-20) of a breakdown of the one-particle picture of ionization.

When molecular orbital cross sections σ_i are available, photoionization intensities are more conveniently obtained as (11,12)

$$I^{(n)} = \sum_i \gamma_{in}\, \sigma_i \qquad (4)$$

together with

$$I^{(n)} = \sum_i \gamma_{in}\, \sigma_i \qquad (5)$$

provided the photoionization process occurs at high (e.g. X-ray) photon energies.

Nowadays, the evaluation of electron binding energies and pole strengths from **G**(ω) can be efficiently carried out at relatively high orders in electronic correlation, by reformulating the well-known (27) iterative Dyson equation into a matrix eigenvalue problem (20) and by using a block extension (25,28) of the Lanczos algorithm (29) in a two-step diagonalization procedure (22).

In comparison with the more standard Configuration Interaction (CI) method, the one-particle Green's function approach offers the essential advantages, in the outlook of numerical applications on extended systems, of a stronger and *systematic compactness* (30) of the configuration spaces in high order approximations and of energy separability (5,31) in the *dissociation limit* (*size-consistency*). The latter is a necessary prerequisite (*) for a correct (i.e. *size-*

intensive (32)) scaling of transition moments and energies in the *thermodynamic limit* of an infinite system.

In principle, one can extract from **G**(ω) the complete series of the primary (one-hole, 1h) and excited (shake-up) states of the cation. In practice, one usually restricts the portion of shake-up space to be spanned to the 2h-1p (two-hole, one-particle) states defined by a single-electron transition, neglecting therefore excitations of higher rank (3h-2p, 4h-3p ...) in the ionized system. In the so-called ADC[3] scheme (22), electronic correlation effects in the reference ground state are included through third-order. In this scheme, multistate 2h-1p/2h-1p configuration interactions are also accounted for to first-order, whereas the couplings of the 1h and 2h-1p excitation manifolds are of second-order in electronic correlation.

These calculations have been conducted on the basis of RHF optimized geometries, considering the 6-31G basis set for the *n*-alkane compounds (11), and the 6-31G* basis set for the polyacene series (12). In both cases, the basis set contention has been checked by comparison with more thorough investigations on small compounds, such as ADC[3] calculations (11a) on *n*-butane based on the 6-311G, 6-31G* and 6-31G** basis, or the MRSDCI ionization spectrum of ethylene as obtained by Murray and Davidson (33) using a 196-CGTO basis set.

The self-consistent field computations have been carried out using the GAMESS series of programs (34). The requested convergence on each of the elements of the density matrix and the integral cutoff were fixed to 10^{-5} and 10^{-9} hartree, respectively. ADC results have been obtained using a slightly modified version of the one-electron Green's function packages (17,22,23,25) developed in Heidelberg, which ensures (*,24) the correct scaling of the static self-energy.

A threshold of 0.005 on pole strengths was imposed on the extraction of the poles of **G**(ω), in the final diagonalization step (22), which at present is carried out using a block version (35) of the conventional Davidson procedure. The spatial symmetry has been exploited to the extent of the largest molecular abelian group. The computed ionization spectra of the selected compounds are presented as convoluted densities of states (DOS) in Figures 1 and 2, together with the ADC[3] values obtained for binding energies and pole strengths, which are displayed as spike spectra. To visualize directly the relative importance of shake-up bands within the global result, we provide separately in these figures the contribution arising

---

(*) As mathematically discussed and numerically illustrated in (24,31), great care must be exercised with size-dependence questions. Third- and higher-order components of the self-energy **Σ**(ω) required to expand **G**(ω) by means of the Dyson equation *diverge logarithmically* with increasing system size. This difficulty relates to the long-range character (1/r) of the Coulomb force and to the *N-representability* of the correlated electron density ρ associated to **G**(ω) via a Coulson contour in the complex ω-space. *Size-intensivity* can only be guaranteed for **Σ**(ω) via a suitable cancellation (24) of logarithmic divergencies (31) in *antigraph sets* (20), or by a rescaling (24) of ρ, ensuring the preservation of the exact particle number (N).

from the lines with a pole strength smaller than 0.40 . In order to enable a quantitative comparison with the experimenal records (7) by M.P. Keane *et al* on ethylene, butadiene, and hexatriene, X-ray photoionization intensities for an AlKα monochromatized photon source (hν=1486.6 eV) have also (Figure 3) been computed (12) by means of Eq. (4), using molecular orbital cross sections evaluated from STO-3G calculations based on the parametric model of Gelius (36). The relative atomic photoionization cross sections used in this case for the C_{2s}, C_{2p} and H_{1s} atomic functions are 100, 7.69 and 0.00, respectively. In spite of its simplicity, this model has often brought invaluable insight into the valence XPS spectra of large organic compounds (2,6). Lacking more suitable tools to account for vibrational broadening and experimental resolution, ionization spectra have been convoluted using as spread function a linear combination of one lorentzian and one gaussian curve of equal weight and width (FWHM=1.1 eV), in analogy with a previous investigation (6g) of XPS records obtained from gas phase samples of saturated hydrocarbons.

3. RESULTS

3.1 Ionization Spectra of Alkane Chains

As shown by the intensity distributions (Figure 1), and as inferred from large pole strength values, close to unity, the idea of a one-particle process of ionization can be retained nearly over the whole valence region of *n*-alkane compounds. The orbital picture is completely preserved for ethane (18e), a molecule with much too large excitation energies to yield a significant dispersion of intensity into many-body effects. For larger saturated hydrocarbons, the innermost ionization lines tend to split into many satellites of weak and comparable intensity (11), as a result of strong configuration interactions in the cation. Shake-up structures remain nonetheless confined above a general threshold of 23 eV on electron binding energies, an observation which can be related (10) to a zeroth-order (i.e. HF) estimate of 27.69 eV for the shake-up energy threshold of polyethylene.

The preservation of the orbital picture of ionization in the upper part of the inner valence region is an essential result in regards with many structural investigations (3,5,6) of gas phase samples, polymer surfaces or thin coated organic fims by means of photoemission spectroscopy. It justifies indeed the assignment (5,6b) of conformational signatures for saturated hydrocarbons to alterations of their long-range methylenic hyperconjugation pattern, yielding a mixing of the C_{2s} and C_{2p} + H_{1s} character of one-electron states at the border of the inner and outer valence regions, in a binding energy interval comprised between 14 and 22 eV.

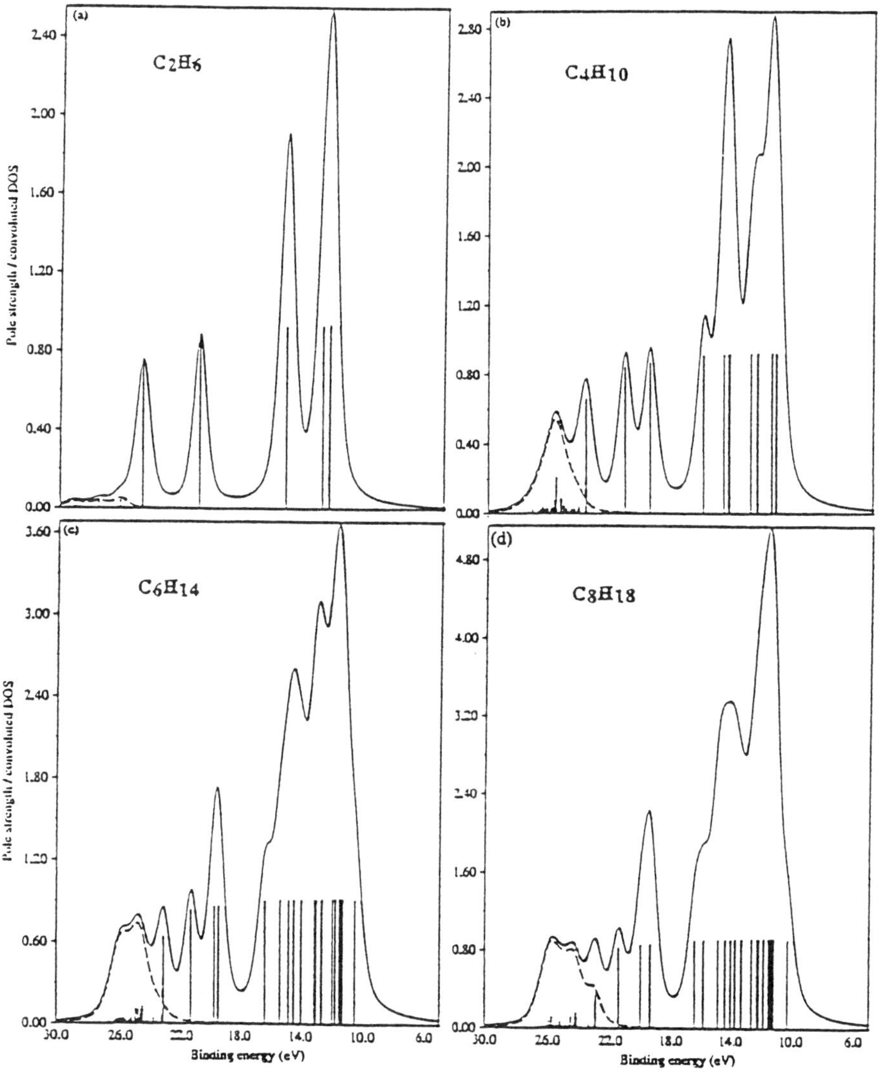

Figure 1 : Valence ADC[3] ionization spectra (full lines) of the (a) ethane, (b) *n*-butane, (c) *n*-hexane and (d) *n*-octane compounds in their all-staggered conformation (6-31G results). The dotted curves represent the partial contribution to the convoluted density of states arising from satellites with $\Gamma_n < 0.40$.

3.2 Ionization Spectra of Polyenes

Very different is the behaviour observed in the polyacene series (Figure 2). The smaller compound of this series, ethylene, gives rise already (Figure 2a) to a partial breakdown of the orbital picture of ionization for (18e) the $2a_g$ one-electron level, yielding a line with a slightly dominant 1h character at 23.8 eV (Γ_n=0.450), and two 2h-1p satellites of appreciable intensity, at 24.9 eV (Γ_n=0.153) and 27.6 eV (Γ_n=0.201). Shake-up structures spread rapidly (Figures 2b, 2c and 2d) towards lower binding energies when considering larger conjugated systems. As with *n*-butane, the breakdown of the molecular orbital picture is already virtually complete for the innermost ($3a_g$) C_{2s} level of 1,3-trans butadiene.

As contrasted with its saturated counterpart, the butadiene compound undergoes (Figure 2b) also a strong fragmentation of the other C_{2s} levels into satellites (18e), though the orbital picture of ionization can still be partially retained in the inner valence region, in the form of three relatively intense 1h lines emerging distinctively from the shake-up background at 18.24 eV ($4b_u$ level, Γ_n=0.67), 19.48 eV ($4a_g$ level, Γ_n=0.421) and 23.39 eV ($3b_u$ level, Γ_n=0.550). The idea of a one-electron ionization process becomes completely inadequate in the C_{2s} region of 1,3,5-trans hexatriene (Figure 2c) and 1,3,5,7-trans octatetraene (Figure 2d). For these compounds, and by extrapolation for polyacetylene, the inner valence structures are found to relate exclusively to satellites.

The ADC[3] results obtained for polyenes (Figures 2b, 2c and 2d) indicate furthermore a partial contamination of the outer valence region by a few satellites of appreciable strength (Table 1), falling among and borrowing their intensity to the outermost 1h lines. One can also notice a significant drop of the pole strengths of 1h lines at the bottom of that region, indicating a rather important dispersion (10-20 %) of the outer valence intensity at higher binding energies. These observations typically relate to the closure of the fundamental gap with increasing system size, and are consistent (10) with a zeroth-order estimate of 10.86 eV for the shake-up energy threshold of polyacetylene.

Our results fit also with a previous investigation (9) on polyenes based on a version of the 2h-1p CI scheme restricted to the virtual one-electron states generated by a minimal basis. In our case, however, the fragmentation of lines into satellites is much more pronounced. The reason lies in the size-consistency of the ADC[3] approach (as contrasted with the size-inconsistency of any truncated form of CI (27d), in the full handling of the virtual space, and (10) in the inclusion of correlation corrections to the reference ground state, leading to (37) a net reduction of the quasi-particle band gap of conjugated polymers.

The pole strength profiles obtained in the outer valence region of the 1,3-trans butadiene, 1,3,5-trans hexatriene and 1,3,5,7-trans octatetraene compounds relate also typically to that found (10) with low-gap hydrogen chains. They nicely reflect the competition for intensity between the main and the correlation (*i.e.* satellite) bands in that region. In both cases, the less energetic (HOMO^{-2}LUMO^{+1} (10,12)

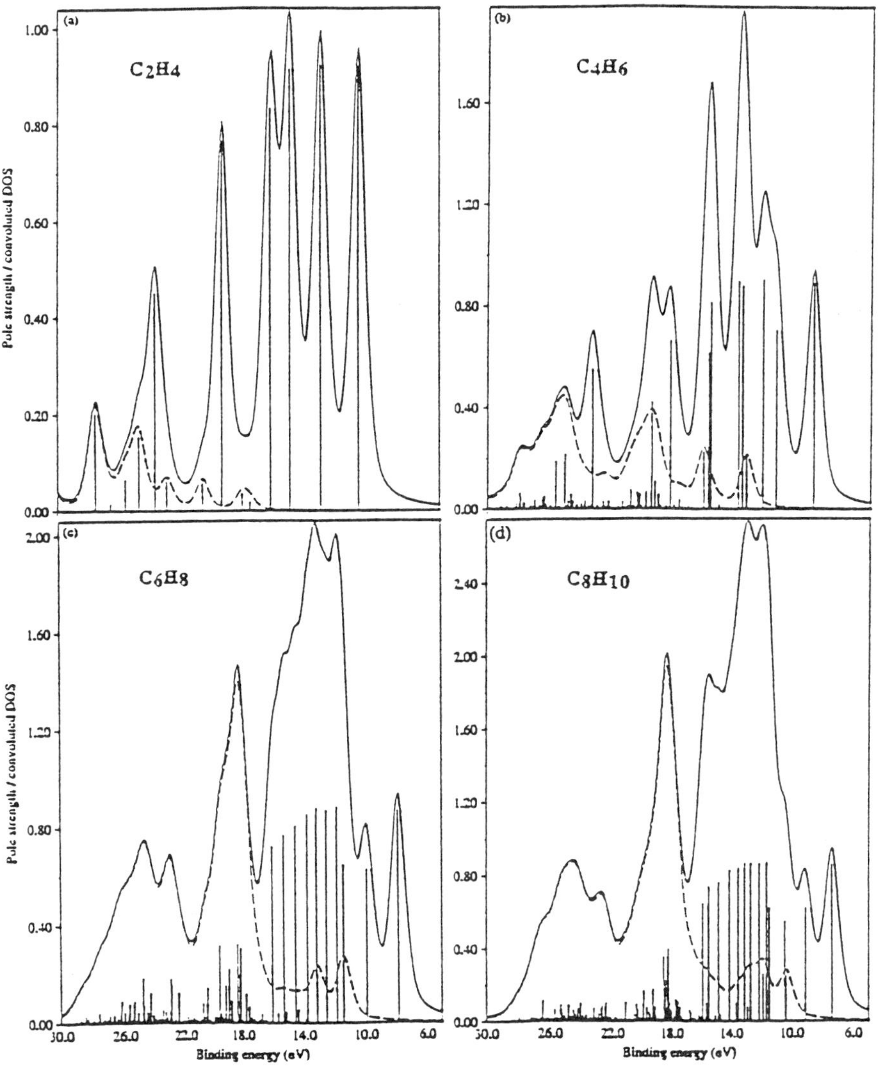

Figure 2 : Valence ADC[3] ionization spectra (full lines) of the (a) ethylene, (b) butadiene, (c) hexatriene and (d) octatetraene compounds in their all-trans configuration (6-31G* results). The dotted curves represent the partial contribution to the convoluted density of states arising from satellites with $\Gamma_n < 0.40$.

shake-up state roughly coincides with the most severe depletion of intensity among the 1h outer-valence lines, and generates correspondingly a rather intense line at the top of correlation bands. In all cases also, the outermost satellites are found to systematically borrow their intensity from the occupied orbital next to the HOMO (denoted as HOMO-1), exhibiting the right symmetry for an interaction with the $HOMO^{-2}LUMO^{+1}$ shake-up state. More details on these couplings can be found in Table 1.

Table 1 : Detailed assignment of the outermost satellite of polyenes, and of the main line from which it originates.

Compound	$HOMO^{-2}LUMO^{+1}$ line	$(HOMO - 1)^{-1}$ line
C_4H_6	$1b_g^{-2}2a_u^{+1}$ $IP_n = 13.10$ eV $\Gamma_n = 0.207$	$1a_u^{-1}$ $IP_n = 11.16$ $\Gamma_n = 0.706$
C_6H_8	$2a_u^{-2}2b_g^{+1}$ $IP_n = 11.39$ eV $\Gamma_n = 0.258$	$1b_g^{-1}$ $IP_n = 9.91$ eV $\Gamma_n = 0.630$
C_8H_{10}	$2b_g^{-2}3a_u^{+1}$ $IP_n = 10.34$ eV $\Gamma_n = 0.245$	$2a_u^{-1}$ $IP_n = 9.09$ eV $\Gamma_n = 0.620$

An excellent agreement with the X-ray photoionization spectra of ethylene, butadiene and hexatriene (7) is obtained (12) (Figure 3) when including in our calculations the Gelius (36) photoionization cross sections for an Alkα photon beam, by means of Eqs. (4) and (5). Such a direct comparison is impossible for octatetraene, a compound for which there is no available XPS data.

Besides a drastic reduction of the intensity obtained in outer valence region as compared with the convoluted DOS curves of Figure 2, cross section effects yield also, in fair agreement with experiment, a significant drop of relative intensities at the top of the inner valence region of 1,3-trans butadiene and 1,3,5-trans hexatriene. This relates to the fact that the orbitals bordering the C_{2s} band incorporate a substantial proportion of C_{2p} atomic contributions. Thus, although the one-electron picture of ionization severely breaks down in that energy region (18 eV), the composition of orbitals does still show through X-ray photoionization spectra.

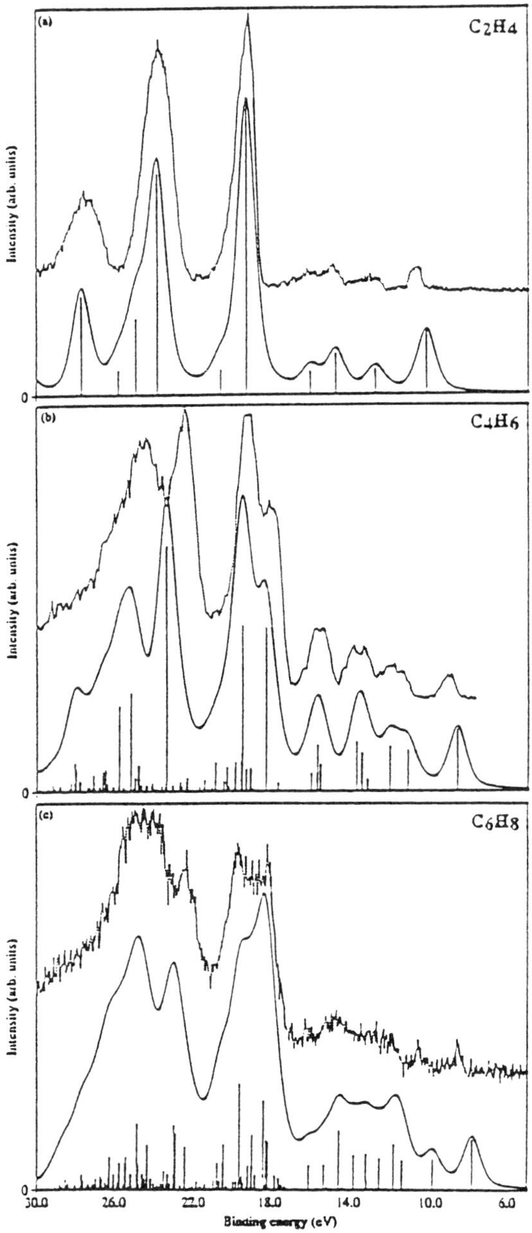

Figure 3 : The simulated ADC[3] X-ray (AlKα) photoionization spectra (full lines) of (a) ethylene, (b) butadiene and (c) hexatriene versus the corresponding experimental records (dotted lines) by M.P. Keane *et al* [7].

3.3 Size-Dependence Effects within Correlation Bands

As in preliminary investigations on model hydrogen chains (10), our calculations indicate (Figures 1 and 2) that shake-up lines tend to dissolve into a quasi-continuum of solutions with vanishingly small intensities, when going to the polymer limit. This phenomenom is at the root of a major deficiency of the ADC[3] scheme as it is currently implemented : by imposing a minimal threshold on pole strengths in the final extraction step, one may indeed exclude a susbtantial amount of the photoionization intensity to be found in correlation bands. Taking the *n*-nonane compound as the most extreme example observed so far (11), one may estimate that about 30 % of the C_{2s} intensity is lost into shake-up lines with $\Gamma_n < 0.005$, which are excluded from the calculations presented here.

The size-dependence of the intensity of single shake-up lines is dictated by the squares of the coupling amplitudes between the 1h and 2h-1p manifolds, which by definition (22) scale like bielectron integrals. Upon a development based on Bloch functions ($\phi_n(k)$), a LCAO expansion over atomic primitives (γ) and lattice summations over cell indices (μ), these, in the limit of a stereoregular polymer chain consisting of a large number (N_0) of cells of length a_0, take the form (31) :

$$< \phi_l(k_l)\phi_m(k_m)||\phi_n(k_n)\phi_o(k_o) >= (N_0)^{-1}\delta_{(k_n+k_o,k_l+k_m)}\tau\begin{pmatrix} k_l & k_m \\ l & m \end{pmatrix} \Bigg|\Bigg| \begin{matrix} k_n & k_o \\ n & o \end{pmatrix} \tag{6}$$

with

$$\tau\begin{pmatrix} k_l & k_m \\ l & m \end{pmatrix} \Bigg|\Bigg| \begin{matrix} k_n & k_o \\ n & o \end{pmatrix} = \sum_{pqrs}^{M_0} C^*_{pl}(k_l)C_{qn}(k_n)C^*_{rm}(k_m)C_{so}(k_o)\times \sum_{\mu\mu'\mu''}^{N_0}\left[e^{i(k_n\mu+k_o\mu''-k_m\mu')a_0} < \gamma_p^0\gamma_r^{\mu'}||\gamma_q^{\mu}\gamma_s^{\mu''} >\right] \tag{7}$$

Neglecting for simplicity the long-range character of the Coulomb force, the above summations yield (31) a bounded result (τ) when extended to infinity. Bielectron integrals can thus be regarded as scaling like N_0^{-1}, either in the thermodynamic limit ($N_0 \to \infty$), or (31) in the dissociation limit ($a_0 \to \infty$).

For the correlation bands obtained by a convolution of many shake-up lines, the size-consistency and size-intensivity requirements imply a convergence towards some asymptotic profile, when going to the polymer limit. This must ideally be achieved through a *balance* of the scaling properties of the individual shake-up lines, and the dispersion of the intensity of 1h lines over a rapidly increasing number of excited states with increasing system size.

From considerations on translational symmetry in the limit of a stereoregular polymer, which are more conveniently analyzed in terms of conservation constraints on momenta at interaction vertices and within self-energy diagrams (31), each 1h line can be easily shown (see e.g. Figure 4 for a second-order process) to generate a number of 2h-1p satellites proportional to N_0^{+2}. This multiplicity factor is exactly balanced by the N_0^{-2} scaling factor on shake-up intensities arising from the square of the 1h/2h-1p coupling amplitudes. This ensures the correct (i.e. *size-consistent*) distribution of the 1h intensity over satellites.

Very similarly, higher-order processes can be shown to yield a size-consistent redistribution of the intensity of shake-up states among themselves, via multiple 2h-1p/2h-1p interactions. Any restriction on this balance will therefore yield a *size-inconsistent* description of correlation bands, which will tend to vanish with increasing system size (11). A nice example is provided here, with the necessary introduction of a lower limit on pole strengths in the block-Davidson diagonalization procedure.

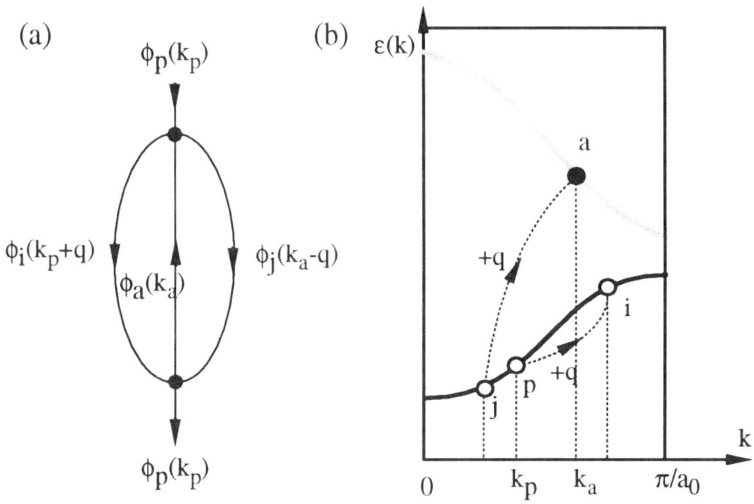

Figure 4 : (a) Scattering processes behind the coupling of a 1h primary ionized (a_p) state and a 2h-1p shake-up state ($a_a^+a_ia_j$), at second-order in electronic correlation (i,j,p : occupied band indices; a : virtual band index); (b) corresponding excitations in the asymmetric part of the first Brillouin zone ($-\pi / a_0 \leq k \leq \pi / a_0$) of a 2-band stereoregular polymer chain. Under the translational symmetry and periodic boundary (Born - Von Karman) constraints, this consists of a quasi-continuous but discrete series of N_0 points in k-space, if the polymer chain contains N_0 unit cells. Scanning the independent momenta present in the second-order $\Sigma_{pp}(k_p)$ self-energy diagram 4(a) is therefore equivalent to exploring a grid of N_0^2 points.

4. CONCLUSIONS AND OUTLOOK FOR THE FUTURE

The ionization spectra displayed in this study illustrate the importance of satellite bands for extended systems, an aspect of band-structure theory which has been largely overlooked until now. For these systems, these bands are, by far, much more difficult to investigate than one-particle bands, as the intensity of shake-up lines tend to individually vanish when the size of chains or clusters extend to infinity. An ideal approach to overcome this difficulty lies on the use of the block-Lanczos algorithm, which is known to preserve the total moments of the spectrum and is thus exactly suitable for exploring dense spectra. Work is in progress along these lines (38), at the university of Heidelberg. In the long term, one should also clearly consider the development and implementation of a crystal orbital version of the ADC/GF scheme.

In this contribution, evidences have been given for a breakdown of the orbital picture of ionization of large saturated hydrocarbons related to polyethylene, above a binding energy threshold of about 23 eV. The one-particle picture is on the other hand well-preserved below that threshold. At this stage, however, it is rather unclear yet whether the extent of shake-up contamination will remain unchanged for polyethylene. The correlation bands in polyethylene may reach higher binding energies, as a substantial proportion of correlation states of very low intensity has not been determined for the longest chains.

As contrasted with n-alkanes, a severe fragmentation of lines into satellites is observed throughout the inner valence region of conjugated oligomers converging to polyacetylene. Our results indicate also a partial contamination of the outer valence region by satellites, which is consistent with an HF estimate of 10.86 eV for the minimal energy required for a 2h-1p shake-up transition in this system. By straightforward extrapolation to higher excitations, it is quite obvious that the breakdown of the one-particle picture of ionization will in reality be even more severe than computed here, because 2h-1p states may themselves strongly interact with 3h-2p shake-up states above 16 eV. At this stage, it should be noted that GF/ADC calculations being exact through fourth- and fifth-orders would be required to describe these states at a consistent (first-order) level in electronic correlation !

Thus, the main conclusion is that: although the photoionization spectra of stereoregular polymers (and by extension crystalline solids) have been extensively investigated during more than three decades, a sound and correct description of the physics behind the recorded signals still remains a very challenging request for theoreticians. Many studies, in particular those carried out on semi-conducting systems, should be reconsidered under the light of the results presented here.

ACKNOWLEDGEMENTS

M. Deleuze is particularly grateful to Prof. Y. Öhrn for his teaching of propagator theory, for his interest in this work and for his encouragements. The authors gratefully acknowledge Professor S. Svensson (Uppsala University, Sweden) for providing the experimental data reproduced in Figs 3(a-c). M. Deleuze would like to thank the *FNRS*, the Fund for Scientific Research for the French Community of Belgium, for his 1994-1996 position as a Senior Research Assistant of Prof. J. Delhalle, at the former *laboratoire de chimie théorique appliquée* of the *FUNDP-Namur;* and the *FWO_Vlaanderen*, the Flemish Fund for Scientific Research, for his new position as a Senior Research Assistant of Prof. J.-P. Francois at the *Limburgs Universitair Centrum*. Financial support by the *Deutsche Forschungsgemeinschaft* is gratefully acknowledged by Lorenz S. Cederbaum.

REFERENCES

(1) (a) Delhalle, J.; *Chem. Phys.*, **5**, 306 (1974); (b) Pireaux, J.J.; Riga, J.; Caudano, R.; Verbist, J.J.; André, J.-M.; Delhalle, J.; Delhalle, S. ; *J. Elect. Spectr. Rel. Phenom* **1974**, *5*, 531; (c) Pireaux, J.J.; Riga, J.; Caudano, R.; Verbist, J.J.; Delhalle, J.; Delhalle, S.; André, J.-M.; Gobillon, Y.; *Phys. Script.* **1977**, *16*, 329; (d) Delhalle, J.; Montigny, R.; Demanet, Ch., André, J.-M.; *Theor. Chim. Acta* **1979**, *50*, 343.

(2) (a) Boulanger, P.; Lazzaroni, R.; Verbist, J.; Delhalle, J.; *Chem. Phys. Lett.* **1986**, *129*, 275; (b) Boulanger, P.; Riga, J.; Verbist, J.; Delhalle, J.; *Macromolecules* **1989**, *21*, 173; (c) Hennico, G.; Delhalle, J.; Boiziau, G.; Lecayon, G.; *J. Chem. Soc. Faraday Trans.* **1990**, *86*, 1025; (d) Boulanger, P.; Magermans, C.; Verbist, J.; Delhalle, J.; Urch, D.S.; *Macromolecules* **1991**, *24*, 2757.

(3) (a) Delhalle, J.; Delhalle, S.; Riga, J.; *J. Chem. Soc., Faraday Trans. 2* **1987**, *83*, 503; (b) Delhalle, J.; Deleuze, M.; *J. Mol. Struct. (Theochem)* **1992**, *261*, 187; (c) Flamant, I.; Mosley, D.H.; Deleuze, M.; André, J.M.; Delhalle, J.; *Int. J. Quantum Chem.* **1994**, *S28*, 469.

(4) (a) Ortiz, J.V.; *J. Am. Chem. Soc.* **1988**, *110*, 4522; (b) Ortiz, J.V.; *Macromolecules* **1988**, *21*, 1189; (c) Ortiz, J.V.; *J. Chem. Phys.* **1991**, *94*, 6064; (d) Ortiz, J.V.; *Macromolecules* **1993**, *26*, 7282.

(5) Deleuze, M.; "Etude de la Structure Electronique de Chaînes Modèles par l'Approche en Propagateurs à une Particule", *PhD thesis* (FUNDP-Namur, Belgium, **1993**).

(6) (a) Deleuze, M.; Horeczky, P.; Delhalle, J.; Pickup, B.T.; *Int. J. Quantum. Chem.* **1992**, *S26*, 31; (b) Deleuze, M.; Denis, J.P.; Delhalle, J.; Pickup, B.T.; *J. Phys. Chem.* **1993**, *97*, 5115; (c) Delhalle, J.; Denis, J.P.; Deleuze, M.; Riga, J.; Dosière, M.; *Chem. Phys. Lett.* **1993**, *210*, 21; (d) Deleuze, M.; Delhalle, J.; Pickup, B.T.; *Chem. Phys.* **1993**, *175*, 427; (e) Deleuze, M.; Delhalle, J.; Pickup, B.T.; *J. Phys. Chem.* **1994**, *98*, 2382; (f) Riga, J.; Delhalle, J.; Deleuze, M.; Pireaux, J.J.; Verbist, J.; *Surf. Int. Anal.* **1994**, *22*, 507; (g) Deleuze, M.; Delhalle, J.; Pickup, B.T., Svensson, S.; *J. Am. Chem. Soc.* **1994**, *116*, 10715; (h) Deleuze, M.; Delhalle, J.; Mosley, D.H.; André, J.-M. *Phys. Script.* **1995**, *51*, 111; (i) Deleuze. M.; Delhalle, J.; *Int. J. Quantum Chem.* **1996**, *S30*, 1505; (j) Duwez, A.S.; Di Paolo, S.; Ghijsens, J.; Riga, J.; Deleuze M., Delhalle, J.; *J. Phys. Chem. B* **1997**, *101*, 884.
(7) Keane, M.P.; Naves de Brito, A.; Correia, N.; Svensson, S.; Karlsson, L.; Wannberg, B.; Gelius, U.; Lunell, S.; Salaneck, W.R.; Lödlung, M.; Swanson, D.B.; Mac Diarmid, A.G.; *Phys. Rev. B* **1992**, *45*, 6390.
(8) Liegener, C.-M.; *Phys. Rev. B* **1993**, *47*, 1607.
(9) Fronzoni, G.; Decleva, P.; Lisini, A.; De Alti, G. *J. Electron. Spectr.* **1994**, *69*, 207.
(10) Deleuze, M.; Cederbaum, L.S.; *Phys. Rev. B* **1996**, *53*, 13326.
(11) Deleuze, M. ; Cederbaum, L.S.; *J. Chem. Phys.* **1996**, *105*, 7583.
(12) Deleuze, M.; Cederbaum, L.S.; *Int. J. Quantum Chem.* **1997**, *63*, 465.
(13) For a review, see e.g. André, J.-M.; *Adv. Quantum Chem.* **1980**, *12*, 65.
(14) (a) Brédas, J.-L.; *Chem. Phys. Lett.* **1985**, *115*, 119; (b) Salaneck, W.R.; Wu, C.R.; Brédas, J.-L.; Ritsko, J.J. *Chem. Phys. Lett.* **1986**, *127*, 88; (c) Brédas, J.-L.; Salaneck, W.R.; *J. Chem. Phys.* **1986**, *85*, 2219; (d) Lazzaroni, R.; Riga, J.; Verbist, J.; Brédas, J.-L.; Delhalle, J.; Wudl, F.; *J. Chem. Phys.* **1988**, *88*, 4257; (e) Salaneck, W.R.; Inganäs, O.; Thémans, B.; Nilsson, J.O.; Sjögren, B.; Österholm, J.E.; Brédas, J.-L.; Svensson, S.; *J. Chem. Phys.* **1988**, *88*, 4613; (f) Orti, E. ; Brédas, J.L.; *J. Chem. Phys.* **1988**, *89*, 1009.
(15) (a) Liegener, C.M.; *Chem. Phys. Lett.* **1990**, *167*, 555; (b) Sun J.-Q.; Bartlett, R.J.; *Phys. Rev. Lett.* **1996**, *77*, 3669.
(16) (a) Salaneck, W.R.; Thomas, H.R.; Bigelow, R.W.; Duke, C.B.; Plummer, E.W.; Heeger, J.J.; A.G. MacDarmid, A.G.; *J. Chem. Phys.* **1980**, *72*, 3674; (b) Schultz, R.; Schweig, A.; Zittlau, A.; *J. Am. Chem. Soc.*, *105*, 2980 (**1983**); (c) Keane, M.P.; Svensson, S.; Naves de Brito, A.; Correia, N.; Lunell, S.; Sjögren, B.; Inganes, O.; Salaneck, W.R. *J. Chem. Phys.* **1990**; *93*, 6357; (d) Sjögren, B.; Naves de Brito, S. Lunell, B. Wannberg, A.; Gelius, U.; Svensson, S. *J. Electron Spectr.* **1992**, *59*, 16; (e) Sjögren, B.; Svensson, S.; Naves de Brito, A.; Correia, N.; Keane, M.P.; Enkvist, C.; Lunell, S.; *J. Chem. Phys.* **1992**, *96*, 6389; (f) Lisini, A.; Keane, M.P.; Lunell, S.; Correia, N.; Naves de Brito, A.; Svensson, S. *Chem. Phys.* **1993**, *169*, 379.
(17) See e.g. von Niessen, W.; Schirmer, J.; Cederbaum, L.S.; *Comput. Phys. Rep.* **1984**, *1*, 57 and references therein.

(18) (a) Cederbaum, L.S.; Hohlneicher, G.; Von Niessen, W.; *Chem. Phys. Lett.* **1973**, *18*, 503; (b) Cederbaum, L.S.; *Chem. Phys. Lett.* **1974**, *25*, 562; (c) Cederbaum, L.S.; *Mol. Phys.* **1974**, *28*, 479; (d) Cederbaum, L.S.; Schirmer, J.; Domcke, W.; von Niessen, W.; *Int. J. Quantum Chem.* **1978**, *14*, 593; (e) Cederbaum, L.S.; Domcke, W.; Schirmer, J.; von Niessen, W.; Diercksen, G.H.F.; Kraemer, W.P.; *J. Chem. Phys.* **1978**, *69*, 1591; (f) Schirmer, J.; Domcke, W.; Cederbaum, L.S.; von Niessen, W.; *J. Phys. B* **1978**, *11*, 1901; (g) Domcke, W.; Cederbaum, L.S.; Schirmer, J.; Von Niessen, W.; *Chem. Phys.* **1979**, *40*, 171; (h) Schirmer, J.; Domcke, W.; Cederbaum, L.S.; von Niessen, W.; Åsbrink, L.; *Chem. Phys. Lett.* **1979**, *61*, 30; (i) von Niessen, W.; Diercksen, G.H.F.; *J. Electron. Spectr. Rel. Phenom* **1980**, *20*, 95.

(19) Cederbaum, L.S.; Schirmer, J.; Domcke, W.; von Niessen, W.; *Adv. Chem. Phys.* **1986**, *65*, 115.

(20) Cederbaum, L.S.; Domcke, W.; *Adv. Chem. Phys.* **1977**, *36*, 205.

(21) (a) Linderberg, J.; Y. Öhrn, "Propagators in Quantum Chemistry'; Academic Press: London, 1973; (b) Öhrn, Y.; in "NATO Advanced Institute Series", 1978, 317-328; (c) Öhrn, Y.; Born, G.; *Adv. Quantum Chem.* **1981**, *13*, 1; (d) Öhrn, Y.; in "Lecture Notes in Chemistry", Mukherjee, D.; Ed.; 1988, Vol 50, 185-206.

(22) Schirmer, J.; Cederbaum, L.S. ; Walter, O.; *Phys. Rev. A* **1983**, *28*, 1237.

(23) Schirmer, J.; Angonoa, G.; *J. Chem. Phys.* **1989**, *91*, 1754.

(24) Deleuze, M.; Scheller, M.K.; Cederbaum, L.S.; *J. Chem. Phys.* **1995**, *103*, 3578.

(25) Weikert, H.-G.; Meyer, H.-D.; Cederbaum, L.S.; Tarantelli, F.; *J. Chem. Phys.* **1996**, *104*, 7122.

(26) Zubarev, D.N.; *Usp. Fiz. Nauk.* **1960**, *71*, 71; *Engl. Trans., Sov. Phys. Usp.* **1960**, *3*, 320.

(27) (a) March, N.H.; Young, W.H.; Sampatar, S.; "The Many-Body Problem in Quantum Mechanics", University Press: Cambridge, 1967; (b) Mattuck, A.D.; "A Guide to Feynman Diagrams in the Many-Body Problem", Mc Graw Hill: New-York, 1967; (c) Fetter,A.L. ; Walecka, J.D.; "Quantum Theory of Many-Particle Systems"; Mc Graw Hill: New-York, 1971; (d) Szabo, A.; Ostlund, N.S.; "Modern Quantum Chemistry", Mc Millian Publishing Co: New-York, 1982.

(28) (a) Ruhe, A.; *Math. Comp.* **1979**, *33*, 680; (b) Meyer, H.-D.; Pal, S.; *J. Chem. Phys.* **1989**, *91*, 6195.

(29) (a) Lanczos, G.; *J. Res. Natl. Bur. Stand.* **1950**, *45*, 255; (b) Parlett, B.N.; "The Symmetric Eigenvalue Problem"; Prentice-Hall: Englewood Cliffs, N.J., 1980; (c) Cullum, J.K.; Wiloughby, R.A.; "Lanczos Algorithms for Large Symmetric Eigenvalue Computations"; Birkhäuser: Boston, 1985.

(30) (a) Schirmer, J.; *Phys. Rev. A* **1991**, 43, 4647; (b) Mertins, F.; Schirmer, J.; *Phys. Rev. A* **1996**, *53*, 2140.

(31) Deleuze, M.; Delhalle, J.; Pickup, B.T.; Calais, J.-L.; *Adv. Quantum Chem.* **1995**, *26*, 35. Besides the final proof for a logarithmic scaling of some static self-energy forms and the cancellation of divergencies in antigraph sets (24), this PhD work [chapter V of (5)] stimulated other investigations of the convergence properties of lattice summations in the context of correlated methods, such as: (b) CPHF-CPEP (Deleuze, M.; Pickup, B.T.; *J. Chem. Phys.,* **1995**, *102*, 8967, (c) RPA (Champagne, B.; Fripiat, J.G.; Mosley, D.H.; André, J.-M.; *Int J. Quantum Chem.* **1995**, *S29*, 429; Champagne, B.; Mosley, D.H.; Fripiat, J.G.; André J.M.; and Öhrn, Y.; *J. Chem. Phys.* **1996**, *104*, 1166), MBPT (Sun, J.-Q.; Bartlett, R.J.; *J. Chem. Phys.* **1997**, *106*, 5058, 5554; *Phys. Rev. Lett.* **1998**, *80*, 349) and Coupled-Clusters (Nooijen, M; Bartlett, R.D.; *Int. J. Quantum Chem.* **1997**, *63*, 601).

(32) Deleuze, M.; Delhalle, J.; André, J.-M.; *Int. J. Quantum Chem.* **1992**, *41*, 243; (b) Deleuze, M.; Delhalle, J.; Pickup, B.T.; Calais, J.-L.; *Phys. Rev. B* **1992**, *46*, 15668.

(33) Murray, C.W.; Davidson, E.R.; *Chem. Phys. Lett.* **1992**, *190*, 231; (b) Desjardin, S.J.; Bawagan, A.D.O; Liu, Z.F.; Tan, K.H.; Wang, Y.; Davidson, E.R.; *J. Chem. Phys.* **1995**, *102*, 6385.

(34) Schmidt, M.W.; *et al., QCPE Bulletin* **1990**, *10*, 52.

(35) F. Tarantelli, unpublished.

(36) Gelius, U.; *J. Electron Spectr. Relar. Phenom.* **1974**, *5*, 985.

(37) (a) S. Suhai, *Chem. Phys. Lett.* **1983**, *96*, 619; (b) Suhai, S.; *Int. J. Quantum Chem.* **1983**, *23*, 1239; (c) Suhai, S.; *Phys. Rev. B* **1983**, *27*, 3506.

(38) (a) Golod, A.; Cederbaum, L.S.; unpublished results; (b) Golod, A. *Diploma Thesis* (University of Heidelberg, Germany, 1997).

Towards the Calculations of Polarizabilities of Stereoregular Polymers ♠

Benoît CHAMPAGNE[#], David H. MOSLEY, Joseph G. FRIPIAT, and Jean-Marie ANDRÉ

Laboratoire de Chimie Théorique Appliquée, Facultés Universitaires Notre-Dame de la Paix, rue de Bruxelles, 61, B-5000 Namur (BELGIUM).

Table of contents

1. INTRODUCTION

The linear polarizability, α, describes the first-order response of the dipole moment with respect to external electric fields. The polarizability of a solute can be related to the dielectric constant of the solution through Debye's equation and molar refractivity through the Clausius-Mosotti equation [1]. Together with the dipole moment, α dominates the intermolecular forces such as the van der Waals interactions, while its variations upon vibration determine the Raman activities. Although α corresponds to the linear response of the dipole moment, it is the first quantity of interest in nonlinear optics (NLO) and particularly for the deduction of structure-property relationships and for the design of new

---

♠ dedicated to Professor Yngve ÖHRN at the occasion of his 65th birthday.

[#] Research Associate of the National Fund for Scientific Research (Belgium).

ADVANCES IN QUANTUM CHEMISTRY, VOLUME 35

0065-3276/99 $30.00

NLO chromophores. Indeed, a close relation often exists between the linear and nonlinear responses of π-conjugated organic compounds which are known to present substantial NLO responses [2-4].

It is well-known that in saturated compounds, α presents an additive behaviour. The polarizability as well as the molar refractivity of molecules are well estimated from adding the contributions of its constitutive bonds or groups. However, for π-conjugated systems, α increases supralinearly with its size due to the important electron delocalization. In particular, for one-dimensional extended systems such as polayacetylene (PA) chains, the chain length dependence of the longitudinal tensor component of the polarizability per unit cell, $\Delta\alpha_L(N) = \alpha_L(N) - \alpha_L(N-1)$, first shows an increase with the number of unit cells (N), then a saturation towards an asymptotic value which characterizes the polymer. The $\Delta\alpha_L(N)$ representation for the polarizability per unit cell is often preferred to the $\alpha_L(N)/N$ function as it converges faster with increasing chain length. The evolutions of the polarizabilities per unit cell with increasing chain length are compared for saturated (polyethylene) and unsaturated (polyacetylene) chains in Figure 1.

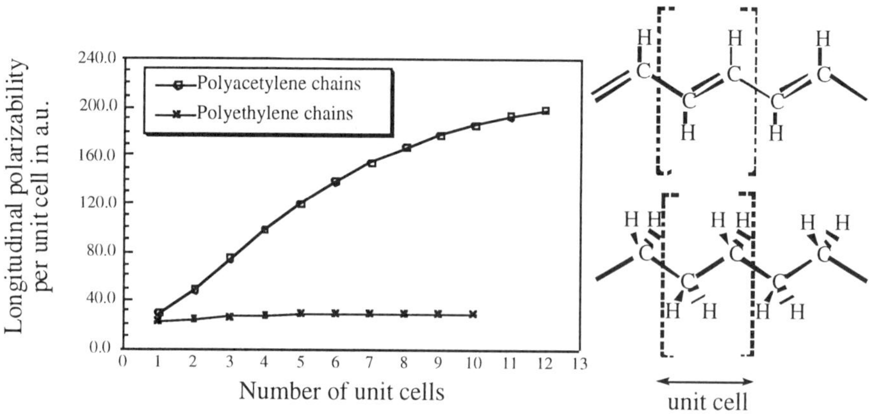

Figure 1 : Evolution as a function of the number of unit cells (N) of the longitudinal polarizability per unit cell, $\Delta\alpha_L(N) = \alpha_L(N) - \alpha_L(N-1)$ for prototypes of saturated (polyethylene) and unsaturated (polyacetylene) polymers.

One way of determining the asymptotic value, $\Delta\alpha_L(\infty)$, consists of computing the properties in larger and larger oligomers and then of extrapolating the results to the infinite chain limit. This has the advantage that no new methodological development is required: it is sufficient to employ well-defined techniques to characterize the properties of increasingly large systems. However, the analytical form of the fitting function is unknown and therefore, when the oligomeric results are far from the plateau, it can lead to significant imprecisions. Several strategies combining various forms of fitting functions as well as different sets of data points have been designed to provide asymptotic values within a given stability criteria [5-6]. The recent work of Dalskov *et al.* [7] demonstrates the difficulties related to these fitting procedures and addresses the evolution of the imprecision as a function of the size of the oligomers and the nature of the polynomial. For exact polynomial fits, *i.e.* when the degree of the polynomial is the number of points minus one, numerical errors in the evaluation of the oligomeric properties lead to variations in the polymeric values which could be larger than the variations of $\Delta\alpha_L(N)$ between $N = N_{max}$ and $N = N_{max} - 1$ where N_{max} is the number of unit cells of the largest oligomer.

The second procedure, several aspects of which are reviewed in this paper, consists of directly computing the asymptotic value by employing newly-developed polymeric techniques which take advantage of the one-dimensional periodicity of these systems. Since the polarizability is either the linear response of the dipole moment to the field or the negative of the second-order term in the perturbation expansion of the energy as a power series in the field, several schemes can be proposed for its evaluation. Section 3 points out that several of these schemes are inconsistent with band theory summarized in Section 2. In Section 4, we present the main points of the polymeric polarization propagator approaches we have developed, and in Section 5, we describe some of their characteristics in applications to prototype systems.

2. BAND STRUCTURE OF PERIODIC INFINITE SYSTEMS

The translational periodicity of the potential is the necessary and sufficient condition for describing the wavefunction as a linear combination of Bloch functions

$$\theta_p(k,\vec{r}) = \frac{1}{\sqrt{2N+1}} \sum_{j=-N}^{N} e^{ikja} \chi_p(\vec{r} - \vec{R}_p - ja\vec{e}_z) \qquad (1)$$

where (2N+1) is the (odd) number of unit cells ($N \to \infty$) that corresponds also to the number of k-states in one band or to the periodicity [(2N+1)a] of the crystalline orbitals imposed by the Born-Karman cyclic boundary conditions. These conditions assume that, for large N, the wavefunction in the 0th and $2N+1^{th}$ unit cells is the same. k is the wavevector or quasimomentum associated with the particle, the normalization factor is $1/\sqrt{2N+1}$, and the translational unit cell length is a. $\chi_p(\vec{r} - \vec{R}_p - ja\vec{e}_z)$ is the pth atomic orbital centered in the jth unit cell which, in practice, is chosen to be a contraction of Gaussian functions. $\vec{e}_z$ is the unit vector in the periodicity direction. In restricted Hartree-Fock theory, the many-electron wavefunctions of closed-shell polymers are approximated by Slater determinants which are constructed from doubly-occupied crystalline orbitals given as linear combinations of Bloch functions,

$$\phi_n(k,\vec{r}) = \sum_{p=1}^{\omega} C_{pn}(k)\theta_p(k,\vec{r}) \tag{2}$$

The polymeric LCAO coefficients, $C_{pn}(k)$ and their associated energies $\varepsilon_n(k)$ are, respectively, the eigenvectors and eigenvalues of the matrix equation:

$$\mathbf{F}(k)\mathbf{C}(k)=\mathbf{S}(k)\mathbf{C}(k)\varepsilon(k) \tag{3}$$

where $\mathbf{F}(k)$ and $\mathbf{S}(k)$ are the k-dependent Fock and overlap matrices. The standard theory of band structure calculations is reviewed in two books [8-9] to which we refer the reader for more details. Improved description of the electron distribution, energies and properties can be obtained after considering the electron correlation effects both size-consistently and by performing the different lattice summations until convergence [10-11].

3. UNBOUNDED NATURE OF THE DIPOLE MOMENT OPERATOR

When an external electric field is applied along the periodicity axis of the polymer, the potential becomes non periodic (Fig. 2), Bloch's theorem is no longer applicable and the monoelectronic wavefunctions can not be represented under the form of crystalline orbitals. In the simple case of the free electron in a one-dimensional box with an external electric field, the solutions of the Schrödinger equation are given as combinations of the first- and second-species Airy functions and do not show any periodicity [12-16].

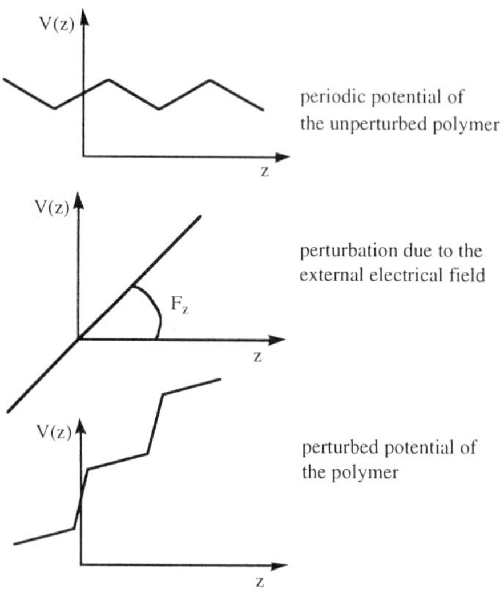

Figure 2 : Schematic representation of the effect of an external electric field applied along the longitudinal axis of a stereoregular polymer.

As a consequence, *finite field* methods, which consist of computing the energy or dipole moment of the system for external electric field of different amplitudes and then evaluating their first, second derivatives with respect to the field amplitude numerically, cannot be applied. Similarly, procedures such as the coupled-perturbed Hartree-Fock (CPHF) or time-dependent Hartree-Fock (TDHF) approaches which determine the first-order response of the density matrix with respect to the perturbation cannot be applied due to the breakdown of periodicity.

In order to solve this problem of unboundedness, Argyres and Sfiat [17] decomposed the dipole moment operator into a periodic sawtooth function and its non periodic stair-case complement. The stair-case component is responsible of the localization of the electronic wavefunction whereas the sawtooth potential is associated with the periodic character of the polarization. Otto and Ladik [18-19] have proposed an alternative decomposition of the dipole moment operator.

These treatments of periodic parts of the dipole moment operator are supported by several studies which show that, for large oligomeric chains, the perturbed electronic density exhibits a periodic potential in the middle of the chain whereas the chain end effects are related to the charge transfer through the chain [20-21]. Obviously, approaches based on truncated dipole moment operators still need to demonstrate that the global polarization effects are accounted for. In other words, one has to ensure that the polymeric value corresponds to the asymptotic limit of the oligomeric results obtained with the full operator.

4. POLARIZATION PROPAGATOR FOR COMPUTING THE POLARIZABILITY PER UNIT CELL OF POLYMERS

4.1 Coupled Hartree-Fock values within the Random Phase approximation

The strategy we have adopted consists of computing the polarizability without determining the field-dependent energies and wavefunctions by adopting perturbation-like approaches. This is accomplished within the framework of the polarization propagator methods at the level of the random phase approximation (RPA). Although the vibrational contributions to the polarizability have been shown to be important [22], only the electronic contribution has so far been considered at this level of theory when taking the translational periodicity into account. This approach provides coupled Hartree-Fock (CHF) polarizability values which are equivalent to the CPHF and TDHF schemes or to the Hartree-Fock finite field technique, including electron reorganizational effects self-consistently in terms of the Coulomb and Pauli fields.

The main quantity providing the dynamic longitudinal polarizability of closed-shell infinite periodic systems is the polarization propagator which at the RPA level takes the form [23-25]:

$$\alpha_{zz}(\text{RPA},\omega) = 2\left(\Omega^*, \Omega\right)\begin{pmatrix} \mathbf{A}+\omega\mathbf{1} & \mathbf{B} \\ \mathbf{B}^* & \mathbf{A}^*-\omega\mathbf{1} \end{pmatrix}^{-1}\begin{pmatrix} \Omega \\ \Omega^* \end{pmatrix} \tag{4}$$

Since the dipole moment operator is spin and number conserving, the poles of the polarization propagator correspond to the singlet single excitation energies. The corresponding residues are the transition moments. The static and dynamic longitudinal polarizabilities per unit cell are then given by the same integration over the first Brillouin zone and summations over all the occupied (i) and unoccupied (a) bands [26-29]:

$$\frac{\alpha_{zz}(\text{RPA},\omega)}{(2N+1)} = \sum_{i}^{occ}\left\{\int_{-\frac{\pi}{a}}^{\frac{\pi}{a}}\left[\frac{a}{\pi}\sum_{a}^{unocc}\Omega^*_{ai}(k)X_{ai}(k,\omega)+\Omega_{ai}(k)Y_{ai}(k,\omega)\right]dk\right\} \tag{5}$$

where the frequency-dependent $X_{ai}(k,\omega)$ and $Y_{ai}(k,\omega)$ terms are obtained from solving the following linear equation system:

$$\begin{cases} (\mathbf{A}+\omega\mathbf{1})\mathbf{X}+\mathbf{B}\mathbf{Y}=\boldsymbol{\Omega} \\ \mathbf{B}^*\mathbf{X}+\left(\mathbf{A}^*-\omega\mathbf{1}\right)\mathbf{Y}=\boldsymbol{\Omega}^* \end{cases} \tag{6}$$

The matrices **A**, **B**, Ω are of infinite dimension since there are an infinite number (2N+1, $N \to \infty$) of k-values and thus an infinite number of k-states in each band. Moreover, there is an equation for each triplet formed by a k-value and two band indices. This triplet represents a particle-hole excitation that is vertical in order to preserve the momentum. As is the case in many polymeric techniques, the infinite sum over k is transformed into an integration in the first Brillouin zone :

$$\begin{cases} \frac{(2N+1)a}{2\pi}\sum_{j}^{occ}\sum_{b}^{unocc}\int_{-\frac{\pi}{a}}^{\frac{\pi}{a}}\left[\left(A(k,k')\right)_{ai,bj}+\omega\delta_{ij}\delta_{ab}\delta_{kk'}\right]X_{bj}(k') \\ \qquad\qquad +\left(B(k,k')\right)_{ai,bj}Y_{bj}(k')dk' = \Omega_{ai}(k) \\ \frac{(2N+1)a}{2\pi}\sum_{j}^{occ}\sum_{b}^{unocc}\int_{-\frac{\pi}{a}}^{\frac{\pi}{a}}\left(B^*(k,k')\right)_{ai,bj}X_{bj}(k') \\ \qquad\qquad +\left[\left(A^*(k,k')\right)_{ai,bj}-\omega\delta_{ij}\delta_{ab}\delta_{kk'}\right]Y_{bj}(k')dk' = \Omega^*_{ai}(k) \end{cases} \tag{7}$$

where

$$\begin{aligned}\left(A(k,k')\right)_{ai,bj} &= \left(\varepsilon_a(k)-\varepsilon_i(k)\right)\delta_{ab}\delta_{ij}\delta_{kk'} \\ &\quad +2\left(\phi_a(k)\phi_i(k)\middle|\phi_j(k')\phi_b(k')\right)-\left(\phi_a(k)\phi_b(k')\middle|\phi_j(k')\phi_i(k)\right)\end{aligned} \tag{8}$$

$$\left(B(k,k')\right)_{ai,bj} = \left(\phi_i(k)\phi_b(k')\middle|\phi_j(k')\phi_a(k)\right)-2\left(\phi_i(k)\phi_a(k)\middle|\phi_j(k')\phi_b(k')\right) \tag{9}$$

The subscript labels a, b,... (i, j,...) correspond to unoccupied (occupied) bands. The Mulliken notation has been chosen to define the two-electron integrals between crystalline orbitals. Two recent studies demonstrate the nice converging behaviour of the different direct lattice sums involved in the evaluation of these two-electron integrals between crystalline orbitals [30]. According to Blount's procedure [31], the z-dipole matrix elements are defined by the following integration which is only non zero for k=k' :

$$\left(\phi_a(k')|z|\phi_i(k)\right) = i\delta_{kk'}\left(u_a(k)\middle|\frac{\partial}{\partial k}u_i(k)\right) = i\delta_{kk'}\Omega_{ai}(k) \tag{10}$$

The properties and computational procedures of the dipole transition strength elements, $\Omega_{ai}(k)$, are described in several papers [32-33] to which we refer the reader for further details. By partitioning $\mathbf{A}(k,k')$ into $\mathbf{A_0}(k,k')$ and $\mathbf{A_1}(k,k')$ according to:

$$\left(A_0(k,k')\right)_{ai,bj} = \left(\varepsilon_a(k)-\varepsilon_i(k)\right)\delta_{ab}\delta_{ij}\delta_{kk'} \tag{11}$$

$$\left(A_1(k,k')\right)_{ai,bj} = 2\left(\phi_a(k)\phi_i(k)\middle|\phi_j(k')\phi_b(k')\right)-\left(\phi_a(k)\phi_b(k')\middle|\phi_j(k')\phi_i(k)\right) \tag{12}$$

and by taking advantage of the relation between the Kroneker delta and the Dirac delta function,

$$\delta_{kk'} \leftrightarrow \frac{2\pi}{(2N+1)a}\delta(k-k'), \tag{13}$$

the linear equation problem takes the following form :

$$\begin{cases}\dfrac{(2N+1)a}{2\pi}\displaystyle\sum_j^{occ}\sum_b^{unocc}\int_{-\frac{\pi}{a}}^{\frac{\pi}{a}}\left(A_1(k,k')\right)_{ai,bj}X_{bj}(k')+\left(B(k,k')\right)_{ai,bj}Y_{bj}(k')dk' \\ \qquad\qquad +\left(\varepsilon_a(k)-\varepsilon_i(k)+\omega\right)X_{ai}(k) = \Omega_{ai}(k) \\ \dfrac{(2N+1)a}{2\pi}\displaystyle\sum_j^{occ}\sum_b^{unocc}\int_{-\frac{\pi}{a}}^{\frac{\pi}{a}}\left(B^*(k,k')\right)_{ai,bj}X_{bj}(k')+\left(A_1^*(k,k')\right)_{ai,bj}Y_{bj}(k')dk' \\ \qquad\qquad +\left(\varepsilon_a(k)-\varepsilon_i(k)-\omega\right)Y_{ai}(k) = \Omega_{ai}^*(k)\end{cases} \tag{14}$$

At this stage, trapezoïdal quadrature is used to obtain **X** and **Y**.

4.2 Uncoupled Hartree-Fock approximation

By making the approximation of setting matrix **B** to zero and **A** to $\mathbf{A_0}$, the resolvent matrix becomes diagonal, every coupling between the crystal orbitals disappears and the polarizability reads:

$$\frac{\alpha_{zz}(\mathrm{UCHF},\omega)}{(2N+1)} = \sum_{i}^{occ}\left\{\int_{-\frac{\pi}{a}}^{\frac{\pi}{a}}\left[\frac{2a}{\pi}\sum_{a}^{unocc}\frac{\left|\Omega_{ai}(k')\right|^2(\varepsilon_a(k')-\varepsilon_i(k'))}{\left[(\varepsilon_a(k')-\varepsilon_i(k'))^2-\omega^2\right]}\right]dk'\right\} \quad (15)$$

This is the expression of the uncoupled Hartree Fock polarizability per unit cell [33]. Since in this case, there is no coupling between the different vertical transitions, the coupling, and thus the matrices $\mathbf{A_1}$ and **B**, are responsible for the field-induced electron reorganizational effects.

In the following Section we present results of the application of the method to two model prototype systems, namely molecular hydrogen chains and all-trans polyacetylene.

5. APPLICATIONS TO PROTOTYPE SYSTEMS

5.1 Molecular hydrogen chains

For model chains of H_2 (the intermolecular distance is fixed at 3.0 a.u., the intramolecular distance is 2.0 a.u.) the longitudinal polarizability per unit cell has been computed at the RPA/3-21G level of approximation as a function of chain length by using the SUPERMOLECULE program [34]. The results given in Table I for different values of the optical frequency show the expected converging behaviour. Then, the polymeric results, given at the bottom of Table I, have been obtained with the PLH [35] and PLHRPA [36] programs. A particular feature of the PLH program consists of the inclusion of the long-range coulombic interactions which are accounted for via a multipole expansion technique including all the monopole-quadrupole and dipole-dipole interactions. The numbers of unit cells in the short and intermediate range regions are 21 and 41, respectively, in order to ensure properly converged results. In the band structure calculation, we set the threshold for the two-electron integrals at 10^{-7} a.u. . The criterion for convergence on the density matrix elements is fixed at 10^{-5}. In the evaluation of the two-electron integrals between crystalline orbitals (Eqs. 8 and 9), we have restricted the short- and intermediate-range regions to 41 and 81 unit cells. The number of k-points in the first Brillouin zone has been fixed at 81. The values obtained with the oligomers of increasing size clearly tend smoothly towards the polymeric results; showing the nice extrapolation property of our polymeric procedure as well as the correspondence between the oligomeric and polymeric approaches.

Figure 3 shows the frequency-dispersion curves of the polymeric H_2 systems by using both the coupled and uncoupled Hartree-Fock polymeric procedures. As a consequence of including the field-induced electron reorganizational effects, the coupled Hartree-Fock values are larger than the uncoupled values. Several studies [26-28] have pointed out that this UCHF underestimation increases when the alternating character of the chain decreases, *i.e.* when the system becomes more conjugated. In addition, increasing the frequency enhances this phenomenon because the estimated excitation energies are much larger by the uncoupled technique.

Table I : Longitudinal polarizability per unit cell as a function of the number of H_2 units cells, N in comparison with the polymeric value.

N	ω=0.0	ω=0.1	ω=0.2
4	25.79	29.57	51.63
5	26.89	31.12	57.98
6	27.44	31.95	62.11
7	27.75	32.41	64.79
8	27.93	32.68	66.55
9	28.04	32.84	67.71
10	28.11	32.96	68.50
11	28.16	33.03	69.05
12	28.19	33.09	69.43
13	28.22	33.12	69.71
14	28.24	33.15	69.90
15	28.25	33.17	70.06
16	28.26	33.18	70.18
17	28.27	33.20	70.26
18	28.28	33.21	70.33
19	28.28	33.21	70.39
20	28.29	33.23	70.43
$\Delta\alpha_{zz}(\infty)$	28.33	33.28	70.68

The results (in a.u.) obtained at the RPA/3-21G level of approximation correspond to electric field circular frequencies of 0.0, 0.1, and 0.2 a.u.

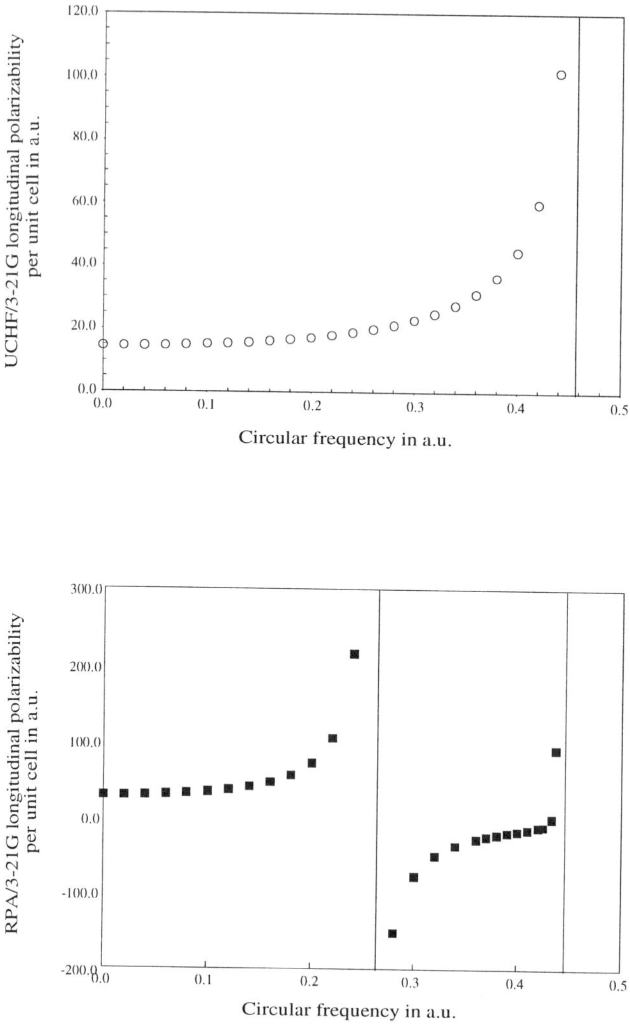

Figure 3 : Frequency-dispersion curves of the longitudinal polarizability per unit cell of infinite periodic chains of hydrogen molecules according to the method used (RPA (bottom) and UCHF (top)). All the values are in a.u. . The position of the first excitation energies which corresponds to the poles is indicated by vertical bars.

By scanning the polarizability as a function of the frequency, the dipole-allowed singlet excitation energies can be determined. Figures 3 displays the position of the first excitation energies as obtained from the UCHF and RPA procedure. At the uncoupled Hartree-Fock level, the poles of the propagator correspond to the crystalline orbital energy differences and form a band. By assigning the zero energy reference value to the top of the valence band, these excited levels also define the conduction band levels (Figure 4). The bottom of the conduction band characterizes thus the lowest singlet excitation energy. Using the 3-21G atomic basis set, this lowest excitation energy amounts to 12.41eV whereas the conduction bandwidth is 18.18 eV [29]. In that approximation, the excited states are completely delocalized over the entire system.

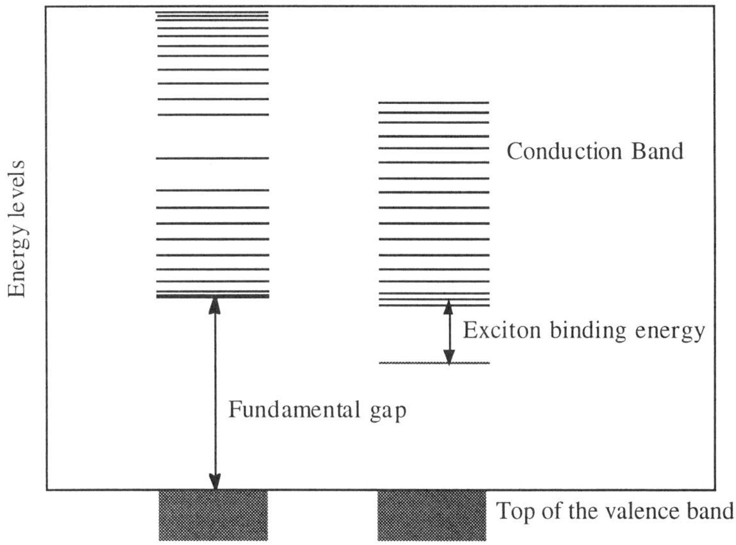

Figure 4 : Conduction band levels and excitation levels of infinite periodic hydrogen chains by using different approximations of the polarization propagator. The left part refers to the crystalline orbital energy differences, namely, the Hartree-Fock excitation energies; the right part refers to the random phase approximation results obtained by using 41 k-points in half the first Brillouin zone.

Within the RPA technique, due to an explicit account of the electron-hole pair interactions, the possible localization of the excited states, which is driven by energy stabilization, can be observed and reproduced. Such a method is suitable to detect possible excitons. In general, the search of the polarization propagator poles gives two features : a band formed by all but a few levels which are lower than the bottom of that band. Such isolated levels are called excitons and result from the localization of the excited state. In our study of the hydrogen chain models, one excitonic level is substantially stabilized and can be distinguished. However, there could be a second exciton with smaller binding energy. Only analysis of the excited state wavefunction could answer this question. The energy gap between that single level and the bottom of the conduction band is called the exciton binding energy, E_b. At the RPA/3-21G level of approximation, we find that E_b = 4.9 eV, the bottom of the excitation band is located 11.90 eV higher than the ground state [29]. This energy value associated with the lowest energy transition to the conduction band is called the fundamental gap. Compared to the Hartree-Fock results, the conduction band has been slightly shifted downward (0.50 eV) by taking care of the electron-hole interaction.

Since the excited levels are determined by searching the poles of the polarization propagator, the poles corresponding to very small dipole transition strengths are difficult to identify. This is particularly the case for the highest energy poles and, consequently, it is not possible to determine by such procedure the top of the conduction band.

5.2 All-trans polyacetylene

By adopting the RPA level of approximation several polymeric calculations of $\Delta\alpha_{zz}(\infty)$ have been carried out for polyethylene [37,39], polysilane [37,39], polyacetylene [28, 37-38], polyyne [38], polydiacetylene [38], and polybutatriene [38] in the static limit. For the purpose of illustrating the use of the polymeric RPA technique to real systems, we chose PA because refractive index measurements for the direction parallel to the periodicity axis, $n_{//}$, are available. Indeed, it has been measured for highly-oriented crystalline trans-polyacetylene by Fink and Leising [40-41] who determined $n_{//}$ in the low-frequency limit, *i.e.* the vibrational contribution is negligibly small. Combining $n_{//}$ = 3.25 with a density of 3.25 g/cm^3 leads to an estimate of the polarizability per unit length, α_{zz}/L, of 43.0 a.u. which corresponds to a α_{zz}/N value of 201 a.u. in the infinite chain length limit.

The geometry adopted for all-trans polyacetylene has been determined by Suhai [42] in an MP2 geometry optimization using an extended polarized basis set. The corresponding bond length alternation (BLA) attains 0.082Å whereas experimental values range between 0.06 and 0.08Å [43-44]. This choice is important because the BLA influences substantially the polarizability as well as the hyperpolarizabilities of π-conjugated molecules. The RPA/3-21G $\Delta\alpha(\infty)$ value attains 217 a.u. [38] and is in close agreement with the experimental data. However, several other aspects deserve to be considered for determining $\Delta\alpha_{zz}(\infty)$ and/or $n_{//}$ with an improved accuracy. Electron correlation effects [45] and the role of the medium [46] are expected to have a non negligible influence. On the other hand, the rather small 3-21G basis set is expected to provide satisfactory results for properties along the chain axis compared to basis sets containing polarization and diffuse functions. This fact is well-known and is related to the cooperative effects of the basis functions which describe also the electron density distribution on the neighboring atoms [47].

As can be seen from Eqs. 5 and 15, this approach allows the polarizability to be decomposed in terms of its most contributing excited states. In unsaturated polymers, the π-symmetry bands and particularly their highest-energy states present the largest contribution to $\Delta\alpha_{zz}(\infty)$. This fact has been highlighted several times [28,37-38].

6. OUTLOOK

The random phase approximation applied to periodic infinite systems is an efficient tool for determining their polarizabilities per unit cell as well as for addressing their excitation spectra. The neglect of electron correlation effects of this CHF method can however be alleviated in future steps based on higher-order treatments of the polarization propagator [48-49]. The surrounding medium typically associated with van der Waals or hydrogen bonding interactions presents a double effect : it modulates the field amplitudes which are experienced by the reference system and it modifies the geometry and electronic structure of the molecule/polymer from what it is in the isolated state. Some of these effects have recently been tackled for 3D periodic systems by Causa and coworkers at the UCHF level [50]. Its RPA and higher-order treatments would constitute another major step towards the description of real polymeric systems. Returning to one of the motivations of these investigations, following studies should concentrate on the extension of these polymeric methods to determine the hyperpolarizabilities of stereoregular polymers. Although such developments are going to be more involved than for the linear polarizability, they should be of crucial importance to address rapidly and efficiently the NLO properties of conjugated polymers for which the chain length dependence is important and the saturation with chain length very slow. Initial steps have already been accomplished along these lines [51] and they should be continued by including also the vibrational contributions .

ACKNOWLEDGMENTS

This work is dedicated to Professor Yngve ÖHRN at the occasion of his 65th birthday. His advice, experience and friendship have been both appreciated and important for the achievement of our study which has found its origin in the *Propagators in Quantum Chemistry* book he wrote with Professor Jan LINDERBERG.

This work has benefited from enlightening discussions with Prof. B.T. Pickup, Prof. J. Delhalle, and Dr. C. Barbier. B.C. would like to take this opportunity to thank the Belgian National Fund for Scientific Research for his successive Research Assistant, Senior Research Assistant and Research Associate positions.

This work would have been impossible without the continuous financial supports of the Belgian National Fund for Scientific Research, the 'Loterie Nationale' for the conventions N° 9.4593.92 and N° 2.4519.97, the Belgian National Interuniversity Research Programs PAI/IUAP N° P3/49 and P4/10. The calculations reported here were performed at the Namur Scientific Computing Facility (Namur-SCF).

REFERENCES

[1] P.W. Atkins, *Physical Chemistry* (Third Edition, Oxford University Press, 1987), Chap. 24.

[2] D.R. Kanis, M.A. Ratner, and T.J. Marks, Chem. Rev. **94**, 195 (1994).

[3] J.L. Brédas, C. Adant, P. Tackx, A. Persoons, and B.M. Pierce, Chem. Rev. **94**, 243 (1994).

[4] B. Kirtman and B. Champagne, Int. Rev. Phys. Chem. **16**, 389 (1997).

[5] B. Kirtman, J.L. Toto, K.A. Robins and M. Hasan, J. Chem. Phys. **102**, 5350 (1995).

[6] B. Champagne, D. Jacquemin, J.M. André and B. Kirtman, J. Phys. Chem. A **101**, 3158 (1997).

[7] E.K. Dalskov, J. Oddershede, and D.M. Bishop, J. Chem. Phys. **108**, 2152 (1998).

[8] J. Ladik, *Quantum Theory of Polymers as Solids*, (Plenum Press, New York, 1988).

[9] J.M. André, J. Delhalle, and J.L. Brédas, *Quantum Chemistry Aided Design of Organic Polymers for Molecular Electronics* (World Scientific Publishing Company, London, 1991).

[10] J.Q. Sun and R.J. Bartlett, J. Chem. Phys. **104**, 8553 (1996).

[11] J.Q. Sun and R.J. Bartlett, Phys. Rev. Lett. **80**, 349 (1998).

[12] J. Zak, Phys. Rev. **168**, 686 (1968).

[13] J.E. Churchill and F.E. Arntz, Am. J. Phys. **37**, 693 (1969).

[14] A. Rabinovitch and J. Zak, Phys. Rev. B **4**, 2358 (1971).

[15] J.E. Churchill and F.E. Holmström, Am. J. Phys. **50**, 848 (1982).

[16] J.E. Churchill and F.E. Holmström, Physica B. **123**, 1 (1983).

[17] P.N. Argyres and S. Sfiat, J. Phys. Cond. Matter **2**, 7089 (1990).

[18] P. Otto, Phys. Rev. B **45**, 10876 (1992).

[19] J. Ladik and P. Otto, Int. J. Quantum Chem. **S27**, 111 (1993).

[20] J.M. André in *Large Finite Systems*, edited by J. Jortner *et al.* (D. Reidel, Dordrecht, 1987) 277.

[21] B. Kirtman, Int. J. Quantum Chem. **43**, 147 (1992).

[22] B. Champagne, E.A. Perpète and J.-M. André, J. Chem. Phys. **101**, 10796 (1994). B. Champagne, E.A. Perpète, J.-M. André and B. Kirtman, J. Chem. Soc. Faraday Trans. **91**, 1641 (1995).

[23] J. Linderberg and Y. Öhrn, *Propagators in Quantum Chemistry* (Academic, New York, 1973).

[24] J. Oddershede, Advan. Quantum Chem. **11**, 257 (1978).

[25] J. Oddershede, P. Jørgensen, and D.L. Yeager, Comput. Phys. Rep. **2**, 33 (1984).

[26] B. Champagne, J.G. Fripiat, and J.M. André, J. Chem. Phys. **96**, 8330 (1992).

[27] B. Champagne, D.H. Mosley, J.G. Fripiat, and J.M. André, Int. J. Quantum Chem. **46**, 1 (1993).

[28] B. Champagne, D.H. Mosley, and J.M. André, Int. J. Quantum Chem. **S27**, 667 (1993).

[29] B. Champagne, J.M. André, and Y. Öhrn, Int. J. Quantum Chem. **57**, 811 (1996).

[30] B. Champagne, D.H. Mosley, and J.M. André, Chem. Phys. Letters **210**, 232 (1993). B. Champagne, J.G. Fripiat, D.H. Mosley, and J.M. André, Int. J. Quantum Chem. **S29**, 429 (1995).

[31] E.I. Blount in *Solid State Physics*, F. Seitz and D. Turnbull, Eds. **13**, 305 (Academic Press, New York, 1962).

[32] C. Barbier, J. Delhalle, and J.M. André, J. Molec. Struct. (Theochem) **188**, 299 (1989).

[33] B. Champagne and J.M. André, Int. J. Quantum Chem. **42**, 1009 (1992).

[34] J. Almlöf and M. Feyereisen, SUPERMOLECULE, a program for ab initio electronic structure calculations; M. Feyereisen, J. Nichols, J. Oddershede, and J. Simons, J. Chem. Phys. **96**, 2978 (1992).

[35] J.M. André, D.H. Mosley, B. Champagne, J. Delhalle, J.G. Fripiat, J.L. Brédas, D.J. Vanderveken, and D.P. Vercauteren, in METECC-94, *Methods and Techniques in Computational Chemistry*, edited by E. Clementi (STEF, Cagliari, 1993), Vol B, Chap. 10, 423.

[36] B. Champagne, PhD. Thesis, Namur (Belgium) (1992).

[37] B. Champagne, D.H. Mosley, J.G. Fripiat, and J.M. André, SPIE Proceedings, the International Society for Optical Engineering, *Nonlinear Optical Properties of Organic Materials V* **1775**, 237 (1992).

[38] B. Champagne and Y. Öhrn, Chem. Phys. Letters **217**, 551 (1994).

[39] B. Champagne and J.M. André, Nonlinear Optics **9**, 25 (1995).

[40] J. Fink and G. Leising, Phys. Rev. B**34**, 5320 (1986).

[41] G. Leising, Phys. Rev. B **38**, 10313 (1988).

[42] S. Suhai, Int. J. Quantum Chem. **42**, 193 (1992).

[43] C.R. Fincher, C.E. Chen, A.J. Heeger, A.G. MacDiarmid, and J.B. Hastings, Phys. Rev. Lett. **48**, 100 (1982).

[44] C.S. Yannoni and T.C. Clarke, Phys. Rev. Lett. **51**, 1191 (1983).

[45] B. Kirtman, J.L. Toto, K.A. Robins, and M. Hasan, J. Chem. Phys. **102**, 5350 (1995); B. Champagne, D.H. Mosley, M. Vračko, and J.M. André, Phys. Rev. A **52**, 1037 (1995).

[46] B. Kirtman, in *Theoretical and Computational Modeling of NLO and Electronic Materials*, edited by S.P. Karna and A.T. Yeates (ACS, Washington DC, 1996) ACS Symposium Series **628**, 58.

[47] G.J.B. Hurst, M. Dupuis, and E. Clementi, J. Chem. Phys. **89**, 385 (1988).

[48] P. Norman, D. Jonsson, O. Vahtras and H. Ågren, Chem. Phys. Lett. **242**, 7 (1995).

[49] M.J. Packer, E.K. Dalskov, T. Enevoldsen, H.J. Aa. Jensen, and J. Oddershede, J. Chem. Phys. **105**, 5886 (1996).

[50] D. Ayma, J.P. Campillo, M. Rérat, and M. Causa, J. Comp. Chem. **18**, 1253 (1997).

[51] B. Champagne, D. Jacquemin, and J.M. André, SPIE proceedings **2527**, 71 (1995).

Dispersion coefficients for second hyperpolarizabilities using coupled cluster cubic response theory

Christof Hättig and Poul Jørgensen

Department of Chemistry, University of Aarhus,
DK-8000 Aarhus C, Denmark

Contents

ADVANCES IN QUANTUM CHEMISTRY, VOLUME 35

0065-3276/99 $30.00

1 Introduction

Linear response function approaches were introduced into the chemistry literature about thirty years ago Ref. [1,2]. At that time they were referred to as Green functions or propagator approaches. Soon after the introduction it became apparent that they offered a viable and attractive alternative to the state specific approaches for obtaining molecular properties as excitation energies, transition moments and second order molecular properties.

The response function approaches have since developed into a useful tool, not only for calculating molecular properties related to the linear response but also for calculations of non-linear response molecular properties. For the most common electronic structure wave function models linear and non-linear response functions have been developed and have served as a useful tool for the calibration and interpretation of molecular properties. To give a specific example consider the history of the electric field induced second harmonic generation (ESHG) experiments of the second hyperpolarizability of the Ne atom. Based on a sequence of ESHG measurements using lasers of different frequencies Shelton found that the dispersion of the ESHG hyperpolarizability of Ne was anomalous [3,4], in contrast to the positive dispersion found in a previous experiment [5] and in an early multiconfiguration self consistent field (MCSCF) calculation [6]. Based on an empirical calculation Bishop could demonstrate [7] that the anomalous dispersion obtained in the ESHG experiment [3,4] can not be reproduced by theory. Accurate large-scale MCSCF response function calculations were then carried out and showed that the dispersion was not anomalous [8]. Subsequently this was confirmed by new experimental ESHG data [9]. Recently accurate coupled cluster singles and doubles (CCSD) response function calculations have been carried out and have challenged the accuracy of one of the new ESHG data points [10]. An experimental reinvestigation has not yet been performed for this point. The above example demonstrates clearly the enourmous development that has taken place since response functions were introduced. It also shows how important it is that the development in theory and experiment goes hand in hand.

In this paper we discuss how more direct approaches may be developed for calculating directly the dispersion of cubic response functions using coupled cluster theory. Power series expansions of frequency-dependent hyperpolarizabilities have in recent years become a widely-used tool for comparison of experimentally obtained and *ab initio* calculated hyperpolarizabilities for different optical proccesses, in particular for second hyperpolarizabilities where for many molecules experimental data is available for more than one optical process and often covers a wide range of frequencies [4, 11–14]. The most commonly used ansatz to expand the frequency-dependence of hyperpolarizabilities in the normal dispersion region [4, 9, 13–17] are even power series in the frequency arguments. The usefulness of this ansatz is emphasized by the fact that for the isotropic average of the second hyperpolarizability $\gamma_{||}$ a single, process-independent coefficient is sufficient to express the frequeny-dependence up to second-order in the frequencies [18, 19]:

$$\gamma_{||}(\omega_0; \omega_1, \omega_3, \omega_3) = \gamma_{||}(0)\Big(1 + A(\omega_0^2 + \omega_1^2 + \omega_2^2 + \omega_3^2) + \mathcal{O}(\omega_i^4)\Big). \quad (1)$$

A sum-over-states expression for the coefficient A for the expansion of the diagonal components $\gamma_{\alpha\alpha\alpha\alpha}$ was derived by Bishop and De Kee [20] and calculations were reported for the atoms H and He. However, the usual approach to calculate dispersion coefficients for many-electron systems by means of *ab initio* response methods is still to extract these coefficients from a polynomial fit to pointwise calculated frequency-dependent hyperpolarizabilities. Despite the inefficiency and the numerical difficulties of such an approach [16, 21], no *ab initio* implementation has yet been reported for analytic dispersion coefficients for frequency-dependent second hyperpolarizabilities which is applicable to many-electron systems.

In a recent publication [22] we reported the implementation of dispersion coefficients for first hyperpolarizabilities based on the coupled cluster quadratic response approach. In the present publication we extend the work of Refs. [22–24] to the analytic calculation of dispersion coefficients for cubic response properties, i.e. second hyperpolarizabilities. We define the dispersion coefficients by a Taylor expansion of the cubic response function in its frequency arguments. Hence, this approach is applicable to all hyperpolarizability components and the coefficients are independent of specific nonlinear optical proccesses.

The generality of a simple power series ansatz and an open-ended formulation of the dispersion formulas facilitate an alternative approach to the calculation of dispersion curves for hyperpolarizabilities complementary to the point-wise calculation of the frequency-dependent property. In particular, if dispersion curves are needed over a wide range of frequencies and for several optical proccesses, the calculation of the dispersion coefficients can provide a cost-efficient alternative to repeated calculations for different optical proccesses and different frequencies. The open-ended formulation allows to investigate the convergence of the dispersion expansion and to reduce the truncation error to what is considered tolerable.

In the next section we derive the Taylor expansion of the coupled cluster cubic response function in its frequency arguments and the equations for the required expansions of the cluster amplitude and Lagrangian multiplier responses. For the experimentally important isotropic averages $\gamma_{||}$, $\gamma_{\perp}$ and γ_K we give explicit expressions for the A and higher-order coefficients in terms of the coefficients of the Taylor series. In Sec. 4 we present an application of the developed approach to the second hyperpolarizability of the methane molecule. We test the convergence of the hyperpolarizabilities with respect to the order of the expansion and investigate the sensitivity of the coefficients to basis sets and correlation treatment. The results are compared with dispersion coefficients derived by least square fits to experimental hyperpolarizability data or to pointwise calculated hyperpolarizabilities of other *ab inito* studies.

2 Coupled Cluster response theory

In the derivation of response functions one considers a molecule or an atom described by the time-independent Hamiltonian $\hat{H}^{(0)}$ which is perturbed by an external one-electron perturbation $\hat{V}(t, \epsilon)$.

$$\hat{H}(t, \epsilon) = \hat{H}^{(0)} + \hat{V}(t, \epsilon) \qquad (2)$$

We assume that the perturbation operator $\hat{V}(t,\epsilon)$ can be expanded in a sum over Fourier components as

$$\hat{V}(t,\epsilon) = \sum_j \hat{H}^{X_j}\epsilon_j(\omega_j)e^{(-i\omega_j t)} \quad , \tag{3}$$

where $\hat{H}^{X_j}$ are hermitian time- and field-independent one-electron operators and $\epsilon_j(\omega_j)$ are associated field strengths for frequencies ω_j. The letter ϵ without index is in the following used as abbreviation for the complete set $\{\epsilon_j\}$ of field strengths included in $\hat{V}$. The time-dependent ground-state coupled cluster wavefunction for such a system is conveniently parametrized in a form, where the oscillating phase factor caused by the so-called level-shift or time-dependent quasienergy $\mathcal{W}(t,\epsilon)$ is explicitly isolated [25–28]:

$$\left|CC(t,\epsilon)\right\rangle = \exp\left(-i\int_{t_0}^{t}\mathcal{W}(t',\epsilon)dt'\right)\exp\left(\hat{T}(t,\epsilon)\right)\left|\text{HF}\right\rangle \tag{4}$$

The field- and time-dependent cluster operator is defined as $\hat{T}(t,\epsilon) = \sum_\mu t_\mu(t,\epsilon)\hat{\tau}_\mu$ and $|\text{HF}\rangle$ is the SCF wavefunction of the unperturbed molecule. By keeping the Hartree-Fock reference fixed in the presence of the external perturbation, a two step approach, which would introduce into the coupled cluster wavefunction an artificial pole structure form the response of the Hartree Fock orbitals, is circumvented. The quasienergy $\mathcal{W}$ and the time-dependent coupled cluster equations are determined by projecting the time-dependent Schrödinger equation onto the Hartree-Fock reference and onto the bra states $\langle\text{HF}|\hat{\tau}_\mu^\dagger\exp(-\hat{T})$:

$$\mathcal{W}(\epsilon,t) = \left\langle\text{HF}\middle|\hat{H}(\epsilon,t)\exp(\hat{T})\middle|\text{HF}\right\rangle. \tag{5}$$

$$\left\langle\text{HF}\middle|\hat{\tau}_\mu^\dagger\exp\left(-\hat{T}(t,\epsilon)\right)\hat{H}(t,\epsilon)\exp\left(\hat{T}(t,\epsilon)\right)\middle|\text{HF}\right\rangle - i\frac{d}{dt}t_\mu(t,\epsilon) = 0 \tag{6}$$

The latter equation may be written in the shorthand form

$$e_\mu(t,\epsilon) - i\tfrac{d}{dt}t_\mu(t,\epsilon) = 0 \quad . \tag{7}$$

Frequency-dependent higher-order properties can now by obtained as derivatives of the time-average of the quasienergy $\{\mathcal{W}\}_T$ with respect to the field strengths of the external perturbations. To derive computational efficient expressions for the derivatives of the coupled cluster quasienergy we use a variational formulation which is obtained by combining the quasienergy and the time-dependent coupled cluster equations to a quasienergy Lagrangian:

$$L(t,\epsilon) = \mathcal{W}(t,\epsilon) + \sum_{\mu} \bar{t}_{\mu} \left(e_{\mu}(t,\epsilon) - i\tfrac{d}{dt}t_{\mu} \right) \tag{8}$$

The time-average of the quasienergy Lagrangian

$$\{L(t,\epsilon)\}_T = \lim_{t_0\to\infty} \frac{1}{2t_0} \int_{-t_0}^{+t_0} L(t,\epsilon)dt \ , \tag{9}$$

may now be required to be stationary with respect to the Fourier components of the cluster amplitudes and the Lagrangian multipliers [23–25,28].

$$\delta\{L(t,\epsilon)\}_T = 0 \tag{10}$$

Inserting the perturbation and Fourier expansion of the cluster amplitudes and the Lagrangian multipliers,

$$t_{\mu}(t,\epsilon) = t_{\mu}^{(0)} + \sum_{n=1}^{\infty} \frac{1}{n!} \sum_{j_1} \cdots \sum_{j_n} t^{X_{j_1}\dots X_{j_n}}(\omega_{j_1},\dots,\omega_{j_n}) \times \prod_{m=1}^{n} \epsilon_{j_m}(\omega_{j_m}) e^{-i\omega_{j_m} t} \tag{11}$$

$$\bar{t}_{\mu}(t,\epsilon) = \bar{t}_{\mu}^{(0)} + \sum_{n=1}^{\infty} \frac{1}{n!} \sum_{j_1} \cdots \sum_{j_n} \bar{t}^{X_{j_1}\dots X_{j_n}}(\omega_{j_1},\dots,\omega_{j_n}) \times \prod_{m=1}^{n} \epsilon_{j_m}(\omega_{j_m}) e^{-i\omega_{j_m} t} \tag{12}$$

the variational condition for the time-averaged quasienergy (10) becomes a set of response equations for the individual expansion coefficients of $t_{\mu}(t,\epsilon)$ and $\bar{t}_{\mu}(t,\epsilon)$. In zeroth-order one obtains the usual coupled cluster equations, $e^{(0)} = 0$, and the equations for the zeroth-order Lagrangian multipliers,

$$\eta^{(0)} + \bar{t}^{(0)}\mathbf{A} = 0 \ , \tag{13}$$

where the vector $\eta^{(0)}$ and the Jacobian matrix $\mathbf{A}$ are defined as partial derivatives of the quasienergy and the Lagrangian with respect to cluster amplitudes and Lagrangian multipliers:

$$\eta_\nu^{(0)} = \left(\frac{\partial \mathcal{W}}{\partial t_\nu}\right)_0 , \tag{14}$$

$$A_{\mu\nu} = \left(\frac{\partial^2 L}{\partial \bar{t}_\mu \partial t_\nu}\right)_0 . \tag{15}$$

The index 0 indicates that the derivatives are taken for zero perturbation strengths. The equations for the first-order amplitudes and multipliers are obtained as:

$$\Big(\mathbf{A} - \omega\mathbf{1}\Big)t^A(\omega) = -\xi^A , \tag{16}$$

$$\bar{t}^A(\omega)\Big(\mathbf{A} + \omega\mathbf{1}\Big) = -\Big(\eta^A + \mathbf{F}t^A(\omega)\Big) , \tag{17}$$

with ξ^A, η^A, and $\mathbf{F}$ defined as:

$$\xi_\mu^A = \left(\frac{\partial^2 L}{\partial \epsilon_A(0)\partial \bar{t}_\mu}\right)_0 , \tag{18}$$

$$\eta_\mu^A = \left(\frac{\partial^2 L}{\partial \epsilon_A(0)\partial t_\mu}\right)_0 , \tag{19}$$

$$F_{\mu\nu} = \left(\frac{\partial^2 L}{\partial t_\mu \partial t_\nu}\right)_0 . \tag{20}$$

The second-order response equations for the cluster amplitudes and the Lagrangian multipliers are:

$$\Big(\mathbf{A} - (\omega_A + \omega_B)\mathbf{1}\Big)t^{AB}(\omega_A, \omega_B) = -\xi^{AB}(\omega_A, \omega_B) , \tag{21}$$

$$\bar{t}^{AB}(\omega_A, \omega_B)\Big(\mathbf{A} + (\omega_A + \omega_B)\mathbf{1}\Big) = -\bar{\xi}^{AB}(\omega_A, \omega_B) , \tag{22}$$

with the right hand side vectors defined as:

$$\xi^{AB}(\omega_A, \omega_B) = \hat{P}^{AB}\Big(\tfrac{1}{2}\mathbf{B}t^A(\omega_A)t^B(\omega_B) + \mathbf{A}^B t^A(\omega_A)\Big) , \tag{23}$$

$$\xi^{AB}(\omega_A,\omega_B) = \mathbf{F}t^{AB}(\omega_A,\omega_B) + \hat{P}^{AB}\Big(\tfrac{1}{2}\mathbf{G}t^A(\omega_A)t^B(\omega_B) + \mathbf{F}^A t^B(\omega_B) + \bar{T}^B(\omega_B)\mathbf{B}t^A(\omega_A) + \bar{T}^B(\omega_B)\mathbf{A}^A\Big) \quad (24)$$

The matrices $\mathbf{B}$, $\mathbf{G}$ and $\mathbf{F}^X$ are defined as partial third derivatives of the Lagrangian:

$$G_{\mu\nu\gamma} = \left(\frac{\partial^3 L}{\partial t_\mu \partial t_\nu \partial t_\gamma}\right)_0 . \quad (25)$$

$$B_{\gamma\mu\nu} = \left(\frac{\partial^3 L}{\partial \bar{t}_\gamma \partial t_\mu \partial t_\nu}\right)_0 . \quad (26)$$

$$F^X_{\mu\nu} = \left(\frac{\partial^2 L}{\partial t_\mu \partial t_\nu \partial \epsilon_X(0)}\right)_0 . \quad (27)$$

$$A^X_{\mu\nu} = \left(\frac{\partial^2 L}{\partial \bar{t}_\mu \partial t_\nu \partial \epsilon_X(0)}\right)_0 . \quad (28)$$

The response functions are obtained as derivatives of the real part of the time-averaged quasienergy Lagrangian:

$$\langle\langle X_1; X_2 \ldots, X_n \rangle\rangle_{\omega_2,\ldots,\omega_n} = \left(\frac{d^n \left\{\frac{1}{2}L(t,\epsilon) + \frac{1}{2}L(t,\epsilon)^*\right\}_T}{d\epsilon_1(\omega_1)\ldots d\epsilon_n(\omega_n)}\right)_0 \quad (29)$$

As a consequence of the time-averaging of the quasienergy Lagrangian, the derivative in the last equation gives only a nonvanishing result if the frequencies of the external fields fulfill the matching condition $\sum_i \omega_i = 0$. In fourth order Eq. (29) gives the cubic response function:

$$\begin{aligned}\langle\langle A; B, C, D\rangle\rangle_{\omega_B,\omega_C,\omega_D} = & \qquad (30)\\ & \tfrac{1}{2}\hat{C}^{\pm\omega}\hat{P}^{ABCD}\Big\{\tfrac{1}{24}\,\mathbf{H}\,t^A(\omega_A)\,t^B(\omega_B)\,t^C(\omega_C)\,t^D(\omega_D)\\ & + \tfrac{1}{4}\Big(\eta^{CD}(\omega_C,\omega_D) + \tfrac{1}{2}\mathbf{F}t^{CD}(\omega_C,\omega_D)\Big)t^{AB}(\omega_A,\omega_B)\\ & + \tfrac{1}{6}\,\bar{t}^A(\omega_A)\,\mathbf{C}\,t^B(\omega_B)\,t^C(\omega_C)\,t^D(\omega_D)\\ & + \tfrac{1}{2}\bar{t}^A(\omega_A)\,\mathbf{B}^B\,t^C(\omega_C)\,t^D(\omega_D)\Big\},\end{aligned}$$

with $\omega_A = -(\omega_B + \omega_C + \omega_D)$ and the vector η^{CD} defined as

$$\eta^{CD}(\omega_C, \omega_D) = \bar{\xi}^{CD}(\omega_C, \omega_D) - \mathbf{F}t^{CD}(\omega_C, \omega_D) \ . \tag{31}$$

The operator $\hat{P}^{ABCD}$ symmetrizes with respect to permutations of the perturbation indices A, B, C and D and the accompanied frequencies and $\hat{C}^{\pm\omega}$ symmetrizes with respect to an inversion of the signs of all frequencies and simultaneous complex conjugation. The three matrices $\mathbf{H}$, $\mathbf{C}$ and $\mathbf{B}^X$ are defined as:

$$H_{\mu\nu\gamma\delta} = \left(\frac{\partial^3 L}{\partial t_\mu\, \partial t_\nu\, \partial t_\gamma\, \partial t_\delta}\right)_0 , \tag{32}$$

$$B^X_{\mu\nu\gamma} = \left(\frac{\partial^3 L}{\partial \bar{t}_\mu\, \partial t_\nu\, \partial t_\gamma\, \partial \epsilon_X(0)}\right)_0 , \tag{33}$$

$$C_{\mu\nu\gamma\delta} = \left(\frac{\partial^3 L}{\partial \bar{t}_\mu\, \partial t_\nu\, \partial t_\gamma\, \partial t_\delta}\right)_0 . \tag{34}$$

An implementation of the cubic response function Eq. (30) for the coupled cluster model hierachy CCS, CC2 and CCSD was reported in Ref. [24].

3 Dispersion coefficients for dynamic second hyperpolarizabilities

3.1 Expansion of general hyperpolarizabilities components

In the normal dispersion region below the first pole, response functions can be expanded in power series in their frequency arguments. The four frequencies, associated with the operator arguments of the cubic response function are related by the matching condition $\omega_A + \omega_B + \omega_C + \omega_D = 0$. Thus second hyperpolarizabilities or in general cubic response properties are functions of only three independent frequency variables, which may be chosen as ω_B, ω_C and ω_D:

$$\gamma_{ABCD}(-\omega_B - \omega_C - \omega_D; \omega_B, \omega_C, \omega_D) = \sum_{l,m,n=0}^{\infty} \omega_B^l \omega_C^m \omega_D^n D_{ABCD}(l, m, n) \tag{35}$$

with the coefficients $D_{ABCD}(l,m,n)$ defined as:

$$D_{ABCD}(l,m,n) = \frac{l!m!n!}{(l+m+n)!}\left(\frac{d^{l+m+n}\gamma_{ABCD}}{d\omega_B^l\, d\omega_C^m\, d\omega_D^n}\right)_{\omega_i=0} \tag{36}$$

To derive working expressions for the dispersion coefficients D_{ABCD} we need the power series expansion of the first-order and second-order responses of the cluster amplitudes and the Lagrangian multipliers in their frequency arguments. In Refs. [22, 29] we have introduced the coupled cluster Cauchy vectors:

$$t^A(\omega_A) = \sum_{n=0}^{\infty} \omega_A^n C^A(n) \ , \tag{37}$$

$$C^A(n) = \frac{1}{n!}\left(\frac{d^n t^A(\omega_A)}{d\omega_A^n}\right)_{\omega_A=0} = -\,\mathbf{A}^{-n-1}\xi^A \ , \tag{38}$$

and the corresponding expansion for the first-order Lagrange multipliers:

$$\bar{t}^A(\omega_A) = \sum_{n=0}^{\infty} \omega_A^n \bar{C}^A(n) \ . \tag{39}$$

$$\bar{C}^A(n) = \frac{1}{n!}\left(\frac{d^n \bar{t}^A(\omega_A)}{d\omega_A^n}\right)_{\omega_A=0} \ , \tag{40}$$

$$= \sum_{m=0}^{n} (-1)^{m+1}\Big(\mathbf{F}C^A(n-m) + \delta_{m,n}\eta^A\Big)\mathbf{A}^{-m-1} \ , \tag{41}$$

The Cauchy vectors for $n \geq 0$ are obtained by solution of the recursive set of linear equations,

$$\mathbf{A}C^A(n) = C^A(n-1) \ , \tag{42}$$

starting with the zeroth-order Cauchy vector which is equal to the static limit of the first-order cluster amplitude response: $C^A(0) = t^A(0)$. The $\bar{C}^A(n)$ vectors for $n \geq 0$ are obtained from the recurrence relation:

$$\bar{C}^A(n)\mathbf{A} = -\Big[\mathbf{F}C^X(n) + \bar{C}^A(n-1)\Big] \ , \tag{43}$$

starting with $\bar{C}^A(0) = \bar{t}^A(0)$. The second-order response of the cluster amplitudes can be expanded in their two frequency arguments as:

$$t^{AB}(\omega_A, \omega_B) = \sum_{m,n=0}^{\infty} \omega_A^m \omega_B^n C^{AB}(m,n) \tag{44}$$

$$\begin{aligned} C^{AB}(m,n) &= \frac{1}{m!n!} \left(\frac{d^{m+n} t^{AB}(\omega_A, \omega_B)}{d\omega_A^m \, d\omega_B^n} \right)_{\omega_A=\omega_B=0} \\ &= -\sum_{p=0}^{m} \sum_{q=0}^{n} \binom{m-p+n-q}{m-p} \mathbf{A}^{-m+p-n+q-1} \, {}^c\xi^{AB}(p,q) \ , \end{aligned} \tag{45}$$

with

$$\begin{aligned} {}^c\xi^{AB}(m,n) &= \frac{1}{m!n!} \left(\frac{d^{m+n} \xi^{AB}(\omega_A, \omega_B)}{d\omega_A^m \, d\omega_B^n} \right)_{\omega_A=\omega_B=0} & (46) \\ &= \hat{P}^{AB}_{mn} \left(\tfrac{1}{2} \mathbf{B} C^A(m) C^B(n) + \delta_{n,0} \mathbf{A}^B C^A(m) \right) & (47) \end{aligned}$$

The operator $\hat{P}^{AB}_{mn}$ symmetrizes with respect to permutation of the operators A and B together with the accompanied "Cauchy" orders m, n; δ is the Kronecker symbol. Similar as for the first-order Cauchy vectors the second-order Cauchy vectors are obtained as solutions of a recursive set of equations:

$$\begin{aligned} \mathbf{A} C^{AB}(m,0) &= -\left({}^c\xi^{AB}(m,0) - C^{AB}(m-1,0) \right) \ , & (48) \\ \mathbf{A} C^{AB}(0,n) &= -\left({}^c\xi^{AB}(0,n) - C^{AB}(0,n-1) \right) \ , & (49) \\ \mathbf{A} C^{AB}(m,n) &= -\left({}^c\xi^{AB}(m,n) - C^{AB}(m-1,n) \right. & (50) \\ & \qquad \left. - C^{AB}(m,n-1) \right) \ , & \end{aligned}$$

starting with $C^{AB}(0,0) = t^{AB}(0,0)$. For the second-order response of the Larangian multipliers one finds the following expansion in the frequency argmuments:

$$\bar{t}^{AB}(\omega_A, \omega_B) = \sum_{m,n=0}^{\infty} \omega_A^m \omega_B^n \bar{C}^{AB}(m,n) \tag{51}$$

$$\bar{C}^{AB}(m,n) = \frac{1}{m!n!}\left(\frac{d^{m+n}\bar{t}^{AB}(\omega_A,\omega_B)}{d\omega_A^m\, d\omega_B^n}\right)_{\omega_A=\omega_B=0} \tag{52}$$

$$= -\sum_{p=0}^{m}\sum_{q=0}^{n}(-1)^{m-p+n-q}\binom{m-p+n-q}{m-p} \times {}^{c}\bar{\xi}^{AB}(p,q)\mathbf{A}^{-m+p-n+q-1} \tag{53}$$

with

$${}^{c}\bar{\xi}^{AB}(m,n) = \frac{1}{m!n!}\left(\frac{d^{m+n}\bar{\xi}^{AB}(\omega_A,\omega_B)}{d\omega_A^m\, d\omega_B^n}\right)_{\omega_A=\omega_B=0} \tag{54}$$

$$= \mathbf{F}C^{AB}(m,n) + \hat{P}^{AB}_{mn}\Big(\tfrac{1}{2}\mathbf{G}C^{A}(m)C^{B}(n) + \delta_{m,0}\mathbf{F}^{A}C^{B}(n) + \bar{C}^{B}(n)\mathbf{B}C^{A}(m) + \delta_{m,0}\bar{C}^{B}(n)\mathbf{A}^{A}\Big) \; .$$

The $\bar{C}^{AB}(m,n)$ vectors are calculated using the equations:

$$\bar{C}^{AB}(m,0)\mathbf{A} = -\Big({}^{c}\bar{\xi}^{AB}(m,0) + \bar{C}^{AB}(m-1,0)\Big) \; , \tag{55}$$

$$\bar{C}^{AB}(0,n)\mathbf{A} = -\Big({}^{c}\bar{\xi}^{AB}(0,n) + \bar{C}^{AB}(0,n-1)\Big) \; , \tag{56}$$

$$\bar{C}^{AB}(m,n)\mathbf{A} = -\Big({}^{c}\bar{\xi}^{AB}(m,n) + \bar{C}^{AB}(m-1,n) + \bar{C}^{AB}(m,n-1)\Big) \; , \tag{57}$$

The matrices $\mathbf{F}$, $\mathbf{G}$, $\mathbf{F}^X$, $\mathbf{H}$, $\mathbf{A}^X$, $\mathbf{B}$, and $\mathbf{C}$ which appear in the expression for the second hyperpolarizability in Eq. (30) are defined as partial derivatives of the quasienergy Lagrangian taken at zero field strengths and hence are frequency-independent.

To find the power series expansion of Eq. (30) in ω_B, ω_C, ω_D we can thus replace the first-order responses of the cluster amplitudes and Lagrangian multipliers and the second-order responses of the cluster amplitudes by the expansions in Eqs. (37), (39) and (44) and express ω_A as $-\omega_B - \omega_C - \omega_D$. However, doing so starting from Eq. (30) leads to expressions which involve an unneccessary large number of second-order Cauchy vectors $C^{AB}(m,n)$. To keep the number of second-order equations for the intermediates $C^{AB}(m,n)$ and $\bar{C}^{AB}(m,n)$ minimal, we use Eqs. (22) – (24) and (31) to rewrite the first term in Eq. (30) in the alternative forms:

$$\hat{P}^{ABCD}\,\tfrac{1}{4}\Big(\eta^{CD}(\omega_C,\omega_D)+\tfrac{1}{2}\mathbf{F}t^{CD}(\omega_C,\omega_D)\Big)t^{AB}(\omega_A,\omega_B) \tag{58}$$

$$=\hat{C}^{BCD}\Big\{\eta^{AB}(\omega_A,\omega_B)t^{CD}(\omega_C,\omega_D)+\bar{\xi}^{CD}(\omega_C,\omega_D)t^{AB}(\omega_A,\omega_B)\Big\}$$

$$=\hat{C}^{BCD}\Big\{\eta^{AB}(\omega_A,\omega_B)t^{CD}(\omega_C,\omega_D)+\bar{t}^{CD}(\omega_C,\omega_D)\xi^{AB}(\omega_A,\omega_B)\Big\}$$

where $\hat{C}^{BCD}$ is a permutation operator which generates the three cyclic permuations of B, C, and D: $\hat{C}^{BCD}(BCD)=(BCD)+(CDB)+(DBC)$. To obtain an efficient expression for the power series expansion of the cubic response function in the frequency arguments, we introduce the intermediates $P_{AB,CD}(i,j;k,l)$ defined for $i+j>k+l$ as

$$P_{AB,CD}(i,j;k,l) \;=\; {}^{c}\eta^{AB}(i,j)\,C^{CD}(k,l)+\bar{C}^{CD}(k,l)\,{}^{c}\xi^{AB}(i,j) \tag{59}$$

for $i+j<k+l$ as

$$P_{AB,CD}(i,j;k,l) \;=\; {}^{c}\eta^{CD}(k,l)\,C^{AB}(i,j)+\bar{C}^{AB}(i,j)\,{}^{c}\xi^{CD}(k,l) \tag{60}$$

and for $i+j=k+l$ as

$$\begin{aligned}P_{AB,CD}(i,j;k,l) \;=\;& \mathbf{F}C^{AB}(i,j)C^{CD}(k,l)\\ &+\Big({}^{c}\eta^{AB}(i,j)+\bar{C}^{AB}(i-1,j)+\bar{C}^{AB}(i,j-1)\Big)C^{CD}(k,l)\\ &+\Big({}^{c}\eta^{CD}(k,l)+\bar{C}^{CD}(k-1,l)+\bar{C}^{CD}(k,l-1)\Big)C^{AB}(i,j)\ ,\end{aligned} \tag{61}$$

where ${}^{c}\eta^{AB}(i,j)={}^{c}\bar{\xi}^{AB}(i,j)-\mathbf{F}C^{AB}(i,j)$. If we consider $i+j+k+l$ as the order of $P_{AB,CD}(i;j,k,l)$ in the frequencies and $i+j$ as order of the vectors $C^{XY}(i,j)$, $\bar{C}^{XY}(i,j)$ we obtain $2n+1$ and $2n+2$ rules [30] for the second-order Cauchy C^{XY} vectors and their Lagrangian multiplier counterparts $\bar{C}^{XY}$: second-order Cauchy vectors C^{XY} up to order n are sufficient to calculate the $P_{AB,CD}(i,j;k,l)$ up to order $2n+1$ and from the vectors $\bar{C}^{XY}$ up to order n the $P_{AB,CD}(i,j;k,l)$ intermediates can be calculated up to order $2n+2$.

The coupled cluster expression for frequency-dependent second hyperpolarizability, Eq. (30), can now be expanded in a Taylor series in its frequency arguments around its static limit as:

$$\gamma_{ABCD}(\omega_A;\omega_B,\omega_C,\omega_D) = -\tfrac{1}{2}\hat{C}^{\pm\omega}\sum_{i,j,k,l=0}^{\infty}\omega_A^i\,\omega_B^j\,\omega_C^k\,\omega_D^l\,d_{ABCD}(i,j,k,l)\ , \tag{62}$$

with the coefficients $d_{ABCD}(i,j,k,l)$ defined as

$$\begin{aligned} d_{ABCD}(i,j,k,l) &= \hat{C}^{BCD}P_{AB,CD}(i,j;k,l) \\ &+\hat{P}^{ABCD}\Big\{\tfrac{1}{24}\,\mathbf{H}\,C^A(i)\,C^B(j)\,C^C(k)\,C^D(l) \\ &+\tfrac{1}{6}\,\bar{C}^A(i)\,\mathbf{C}\,C^B(j)\,C^C(k)\,C^D(l) \\ &+\delta_{j,0}\tfrac{1}{2}\bar{C}^A(i)\,\mathbf{B}^B\,C^C(k)\,C^D(l)\Big\}\ . \end{aligned} \tag{63}$$

Rewriting the redundant frequency ω_A as $-\omega_B-\omega_C-\omega_D$, we find:

$$\begin{aligned} \gamma_{ABCD}(\omega_A;\omega_B,\omega_C,\omega_D) &= -\tfrac{1}{2}\hat{C}^{\pm\omega}\sum_{pqr=0}^{\infty}\sum_{lmn=0}^{\infty}(-1)^{p+q+r}\frac{(p+q+r)!}{p!\,q!\,r!} \qquad (64)\\ &\times\omega_B^{p+l}\,\omega_C^{q+m}\,\omega_D^{r+n}d_{ABCD}(p+q+r,l,m,n)\ , \\ &= -\tfrac{1}{2}\hat{C}^{\pm\omega}\sum_{lmn=0}^{\infty}\omega_B^l\omega_C^m\omega_D^n\sum_{p=0}^{l}\sum_{q=0}^{m}\sum_{r=0}^{n}(-1)^{p+q+r} \qquad (65)\\ &\times\frac{(p+q+r)}{p!\,q!\,r!}d_{ABCD}(p+q+r,l-p,m-q,n-r)\ . \end{aligned}$$

From Eq. (65) we identify for the dispersion coefficients $D_{ABCD}(l,m,n)$ defined in Eq. (35) the expression

$$D_{ABCD}(l,m,n) = -\sum_{p=0}^{l}\sum_{q=0}^{m}\sum_{r=0}^{n}(-1)^{p+q+r}\frac{(p+q+r)!}{p!q!r!} \tag{66}$$

$$\times\frac{1}{2}\Big(d_{ABCD}(p+q+r, l-p, m-q, n-r)$$
$$+(-1)^{l+m+n} d_{ABCD}(p+q+r, l-p, m-q, n-r)^*\Big) \ .$$

Thus coefficients with an even total order $l + m + n$ are real and coefficients with an odd total order $l + m + n$ are pure imaginary. In the following we consider only dipole hyperpolarizabilities. In this case the four operators A, B, C and D are cartesian components of the dipole operator and the odd dispersion coefficients vanish.

3.2 Special expansions for $\gamma_{||}$, $\gamma_{\perp}$ and γ_K

For optical processes which can be parametrized in terms of a single laser frequency like third harmonic generation (THG), electric field induced second harmonic generation (ESHG), the dc-Kerr effect, and degenerate four wave mixing (DFWM), specialized versions of Eq. (35) can be derived, with dispersion coefficients that depend only on one order parameter. For example the third harmonic generation hyperpolarizability $\gamma^{THG}_{ABCD}(\omega) = \gamma_{ABCD}(-3\omega; \omega, \omega, \omega)$ can be expanded as:

$$\gamma^{THG}_{ABCD}(\omega) = \sum_{n=0}^{\infty} \omega^{2n} D^{THG}_{ABCD}(2n) \ , \tag{67}$$

with

$$D^{THG}_{ABCD}(n) = \sum_{m=0}^{n}\sum_{l=0}^{m} D_{ABCD}(n-m, m-l, l) \ . \tag{68}$$

The analogous dispersion coefficients for the ESHG hyperpolarizability $\gamma^{ESHG}_{ABCD}(\omega) = \gamma_{ABCD}(-2\omega; \omega, \omega, 0)$, the dc-Kerr effect $\gamma^{dc-Kerr}_{ABCD}(\omega) = \gamma_{ABCD}(-\omega; 0, 0, \omega)$ and DFWM $\gamma^{DFWM}_{ABCD}(\omega) = \gamma_{ABCD}(-\omega; \omega, \omega, -\omega)$ are found as:

$$D^{ESHG}_{ABCD}(n) = \sum_{m=0}^{n} D_{ABCD}(n-m, m, 0) \tag{69}$$

$$D^{dc-Kerr}_{ABCD}(n) = D_{ABCD}(0, 0, n) \tag{70}$$

$$D^{DFWM}_{ABCD}(n) = \sum_{m=0}^{n}\sum_{l=0}^{m} (-1)^l D_{ABCD}(n-m, m-l, l) \tag{71}$$

An alternative compact expansion with coefficients which are independent of the optical process can be derived for the isotropic parallel average of the second hyperpolarizability $\gamma_{||}$ defined as [13]

$$\gamma_{||} = \tfrac{1}{15}\sum_{\xi\eta}\left\{\gamma_{\xi\eta\eta\xi} + \gamma_{\xi\eta\xi\eta} + \gamma_{\xi\xi\eta\eta}\right\}, \qquad \xi,\eta = x,y,z, \tag{72}$$

which is the tensor component probed in the majority of hyperpolarizability measurements. It was first suggested by Shelton [18] based on a single resonance model that $\gamma_{||}$ can be expanded in low orders in terms of the effective frequency $\omega_L^2 = \omega_0^2 + \omega_1^2 + \omega_2^2 + \omega_3^2$ as

$$\gamma_{||}(\omega_0;\omega_1,\omega_2,\omega_3) = \gamma_0\left(1 + A\omega_L^2 + \ldots\right) . \tag{73}$$

As was proven later by Bishop [19], the coefficient A in the expansion (73) is the same for all optical processes. If the expansion (73) is extended to fourth-order [4, 19] by adding the term $B\omega_L^4$ the coefficient B is the same for the dc-Kerr effect and for electric field induced second-harmonic generation, but other fourth powers of the frequencies than ω_L^4 are in general needed to represent the frequency-dependence of $\gamma_{||}$ with process-independent dispersion coefficients [19]. Bishop and De Kee [20] proposed recently for the all-diagonal components $\gamma_{\alpha\alpha\alpha\alpha}$ the expansion

$$\gamma_{\alpha\alpha\alpha\alpha}(\omega_0;\omega_1,\omega_2,\omega_3) = \gamma_0 + AW_2 + BW_2^2 + B'W_4 + \ldots , \tag{74}$$

with $W_n = \sum_i \omega_i^n$ and B and B' process-independent. Employing the invariance of $\gamma_{||}(\omega_0;\omega_1,\omega_2,\omega_3)$ with respect to permutations of the frequency arguments, we have prooved [31] that the expansion (74) is valid also for the isotropic parallel average $\gamma_{||}$ and that it can be generalized to arbitrarly high orders in the frequency arguments, using only process-independent dispersion coefficients.However, the ansatz in Eq. (74) has the disadvantage that simplifications which occur if one or more of the external fields are static are not reflected in the effective frequencies W_n: Already for the dc-Kerr effect and also for the electric field induced second harmonic generation both fourth-order coefficients, B and B', enter the description of the dispersion. In Ref. [31] we proposed therefore for

$\gamma_{||}$ the alternative open-ended expansion:

$$\gamma_{||}(\omega_0;\omega_1,\omega_2,\omega_3) = \gamma_0 \left(1 + \sum_{n=1}^{\infty} \sum_{2k+3k+4m=2n} A_{k,l,m}\, \omega_{L2}^{2k}\, \omega_{L3}^{3l}\, \omega_{L4}^{4m} \right) , \tag{75}$$

where the coefficients $A_{k,l,m}$ are strictly process-independent and the effective frequencies ω_{L2}^2, ω_{L3}^3 and ω_{L4}^4 are defined as: [32]

$$\omega_{L2}^2 = \sum_i \omega_i^2 = \omega_L^2, \tag{76}$$

$$\omega_{L3}^3 = \sum_i \omega_i^3 = 3\Big(\omega_0\omega_1\omega_2 + \omega_0\omega_1\omega_3 + \omega_0\omega_2\omega_3 + \omega_1\omega_2\omega_3\Big) \tag{77}$$

$$\omega_{L4}^4 = \sum_i \left(\omega_i^4 - \frac{1}{2}\omega_i^2\right) = -4\omega_0\omega_1\omega_2\omega_3 \tag{78}$$

These effective frequencies are chosen such that ω_{L3}^3 contributes only for optical processes with at least three non-zero frequency arguments, while ω_{L4}^4 is only non-zero if all four frequency arguments are non-zero. During the implementation it was also found that this choice for the effective frequencies leads to the most compact expressions for the coefficients $A_{k,l,m}$ in terms of $D(n,m,l)$. Using a similar notation as in Eq. (73), $\gamma_{||}(\omega_0;\omega_1,\omega_2,\omega_3)$ can up to sixth order in the frequencies be expanded as

$$\begin{aligned}\gamma_{||}(\omega_0;\omega_1,\omega_2,\omega_3) &= \gamma_0\Big(1 + A\omega_{L2}^2 + B\omega_{L2}^4 + B'\omega_{L4}^4 \\ &\quad + C\omega_{L2}^6 + C'\omega_{L3}^6 + C''\omega_{L2}^2\omega_{L4}^4 \ldots\Big) .\end{aligned} \tag{79}$$

The hyperpolarizability dispersion coefficients A, B, B', C, C', C'' etc. can be calculated from the coefficients $D(l,m,n)$ of the Taylor expansion introduced in Eqs. (35) and (66). The required expressions are found by equating the coefficients for the individual frequency powers $\omega_1^l\omega_2^m\omega_3^n$ in the two expansions and were derived in Ref. [33]. In analogy to the parallel isotropic average of the hyperpolarizability, we introduce the parallel isotropic averages of the dispersion coefficients,

$$D_{||}(l,m,n) = \frac{1}{15}\sum_{\xi\eta}\Big\{D_{\xi\eta\eta\xi}(l,m,n)+D_{\xi\eta\xi\eta}(l,m,n)+D_{\xi\xi\eta\eta}(l,m,n)\Big\}, \tag{80}$$

with $\xi, \eta = x, y, z$. With the $D_{||}(l,m,n)$ coefficients f.x. A, B, and B' can be calculated as:

$$A = \frac{1}{2\gamma_0} D_{||}(2,0,0) \ , \tag{81}$$

$$B = \frac{1}{4\gamma_0} D_{||}(4,0,0) \ , \tag{82}$$

$$B' = \frac{1}{4\gamma_0}\Big(D_{||}(2,1,1) - 4D_{||}(4,0,0)\Big) \ . \tag{83}$$

Expressions for the sixth- through tenth-order coefficients are given in the Appendix. In ESHG experiments with the optical field polarized perpendicular to the static electric field, the measured second hyperpolarizability is: [13]

$$\gamma_\perp = \frac{1}{15}\sum_{\xi\eta}\Big\{2\gamma_{\eta\xi\xi\eta} - 3\gamma_{\xi\eta\xi\eta} + 2\gamma_{\xi\xi\eta\eta}\Big\}, \qquad \xi,\eta = x,y,z. \tag{84}$$

The component $\gamma_\perp$ is also related to the second hyperpolarizability measured in dc-Kerr experiments

$$\gamma_K = \frac{3}{2}\Big(\gamma_{||} - \gamma_\perp\Big) \tag{85}$$

To obtain for $\gamma_\perp$ and γ_K compact dispersion formulas similar as Eq. (79) for $\gamma_{||}$, these hyperpolarizability components must be written as sums of tensor components which are irreducibel with respect to the permutational symmetry of the operator indices and frequency arguments:

$$\gamma_\perp = \frac{1}{3}\gamma_{||} + \frac{2}{3}\gamma_{ms} \ , \tag{86}$$

$$\gamma_K = \gamma_{||} - \gamma_{ms} \ , \tag{87}$$

with

$$\gamma_{ms} = \frac{1}{6}\sum_{\xi\eta}\left\{\gamma_{\eta\xi\xi\eta} - 2\gamma_{\xi\eta\xi\eta} + \gamma_{\xi\xi\eta\eta}\right\}, \qquad \xi,\eta = x,y,z. \tag{88}$$

The index ms indicates that γ_{ms} transforms according to the mixed symmetry representation of the symmetric Group S_4 [33]. γ_{ms} is an irreducible tensor component which describes a deviation from Kleinman symmetry [34]. It vanishs in the static limit and for third harmonic generation ($\omega_1 = \omega_2 = \omega_3$). Up to sixth order in the frequency arguments it can be expanded as [33]:

$$\begin{aligned}\gamma_{ms}(\omega_0;\omega_1,\omega_2,\omega_3) &= \gamma_{||}(0)\Big(A_{ms}\omega_{M2}^2 + B_{ms}\omega_{M2}^2\omega_{L2}^2 + B'_{ms}\omega_{M4}^4 \\ &+C_{ms}\omega_{M2}^2\omega_{L2}^4 + C'_{ms}\omega_{M4}^4\omega_{L2}^2 + C''_{ms}\omega_{M2}^2\omega_{L4}^4 + \ldots\Big)\ ,\end{aligned} \tag{89}$$

with the mixed-symmetry effective frequencies ω_{M2}^2 and ω_{M4}^4 defined as:

$$\omega_{M2}^2 = (\omega_0 - \omega_1)(\omega_2 - \omega_3) + (\omega_0 - \omega_3)(\omega_2 - \omega_1) \tag{90}$$

$$\begin{aligned}\omega_{M4}^4 &= -3\Big[(\omega_0^2\omega_1 - \omega_1^2\omega_0)(\omega_2 - \omega_3) + (\omega_0 - \omega_1)(\omega_2^2\omega_3 - \omega_3^2\omega_2) \\ &+(\omega_0^2\omega_3 - \omega_3^2\omega_0)(\omega_2 - \omega_1) + (\omega_0 - \omega_3)(\omega_2^2\omega_1 - \omega_1^2\omega_2)\Big]\end{aligned} \tag{91}$$

We note that ω_{M4}^4 vanishs if two of the four the frequency arguments ω_0, ω_1, ω_2, ω_3 become zero. For the case that three frequency arguments are equal (third harmonic generation) both mixed-symmetry effective frequencies ω_{M2}^2 and ω_{M4}^4 vanish. The coefficients A_{ms}, B_{ms}, B'_{ms}, etc. can again be calculated from the coefficients $D(l,m,n)$ of the Taylor expansion in ω_1, ω_2, ω_3. Explicite expressions for the dispersion coefficients up to sixth order were derived in Ref. [33] and are listed in the Appendix. The approach outlined in Eqs. (35) – (89) has been implemented for the calculation of dispersion coefficients for cubic response functions for the coupled cluster model hierachy CCS, CC2 and CCSD into the coupled cluster response code described in Refs. [22–24, 29, 35–41].

4 Dispersion coefficients for the second hyperpolarizability of methane

The expressions (62) – (89) have been applied to calculate dispersion coefficients for the isotropic components of the second hyperpolarizability of methane using the three coupled cluster models CCS, CC2 and CCSD and five different basis sets taken from the series of correlation consistent basis sets developed by Dunning and coworkers [42]. The calculations were performed at the experimental equilibrium distance $r_{CH} = 2.052$ a.u. [43,44] For all three coupled cluster models we employed the d-aug-cc-pVDZ, t-aug-cc-pVDZ and d-aug-cc-pVTZ basis sets, which were selected to obtain basis set saturation to different levels of accuracy. The d-aug-cc-pVDZ and t-aug-cc-pVDZ basis sets do not contain f functions at the C atom and no d functions for H, but the t-aug-cc-pVDZ basis contains many diffuse functions. The d-aug-cc-pVTZ basis contains such polarization functions with higher angular momenta and the exponents of its most diffusest functions are intermediate between those in the d-aug-cc-pVDZ and t-aug-cc-pVDZ basis. This basis is therefore expected to provide a balanced compromise between diffusenes and polarization level. For the CCSD calculations we included in addition the t-aug-cc-pVTZ and the d-aug-cc-pVQZ basis sets to test saturation towards an improved description of the valence and diffuse region. The d-aug-cc-pVQZ basis comprises for methane 353 contracted functions and includes 3 g functions for the C atom and 3 f functions for H. We expect this basis to give results of benchmark quality near to the one-particle basis set limit. In a recent investigation of the second hyperpolarizabilities of Ne, Ar, N_2 and CH_4 [45] the d-aug-cc-pVQZ basis was employed to caculate the static hyperpolarizability of CH_4, which was then combined with dispersion curves for four different optical processes calculated pointwise at 6 frequencies using the t-aug-cc-pVTZ basis. Due to the large computational costs, a pointwise calculation of dispersion curves with the d-aug-cc-pVQZ basis was not feasible. The analytic calculation of the dispersion coefficients reduces the computational costs significantly if the dispersion expansion can be truncated in low order. The A and B coefficients for example can be calculated with approximately the same

order	THG	ESHG	DFWM	Kerr
0	2.3782(3)	2.3782(3)	2.3782(3)	2.3782(3)
2	2.1486(5)	1.0743(5)	7.1619(4)	3.5810(4)
4	1.4148(7)	3.5673(6)	1.6258(6)	3.9637(5)
6	8.4389(8)	1.0639(8)	3.4110(7)	3.8291(6)
8	4.8816(10)	3.0270(9)	7.0106(8)	3.4179(7)
10	2.8113(12)	8.4138(10)	1.4398(10)	2.8965(8)

Table 1: Coefficients for $\gamma_{||}(\omega)$ for third harmonic generation (THG), degenerate four wave mixing (DFWM), electric field induced second harmonic generation (ESHG), and Kerr effect in methane at the experimental geometry $r_{CH} = 2.052$ a.u. A CCSD wavefunction and the t-aug-cc-pVDZ basis were used. (Results given in atomic units, the number in parentheses indicate powers of ten.)

computational costs as for one frequency-dependent calculation for the electric field induced second harmonic generation (ESHG) process. This made it possible to calculate Taylor and Padé approximations to the d-aug-cc-pVQZ dispersion curves which are correct through fourth order in the optical frequencies.

4.1 Convergence with the order of the expansion

First, we examine for a fixed basis set and a fixed wavefunction model the convergence of truncated Taylor expansions and Padé approximants with the order of the dispersion coefficients. In Table 1 we compiled for the t-aug-cc-pVDZ basis set and a CCSD wavefunction the coefficients up to tenth order for the parallel isotropic average $\gamma_{||}$ for THG, ESHG, DFWM and dc-Kerr effect. From the tabulated coefficients we calculated the Taylor approximations of order 2 through 10 for two wavelength, 671.5 nm and 476.5 nm. The first wavelength has been chosen intermediate between the typical experimental wavelengths 632.8 and 694.3 nm, used in many dc-Kerr, THG and ESHG experiments. The second wavelength is the shortest which has been used in

ESHG measurements for methane and has been chosen as representative for the high-frequency limit of the usual experimental frequency range. The results are listed in Table 2 together with the results obtained with diagonal Padé approximants [46] which are correct through the same order in the frequency arguments and the series limits which were calculated using the implementation for the frequency-dependent response functions.

The convergence rate of the Taylor expansion and the series of diagonal Padé approximants is in the normal dispersion region determined by how close the frequency argument is to the first pole in the hyperpolarizability. For the dc-Kerr effect this occurs when ω equals the first dipole-allowed transition energy $\Delta\varepsilon_{01}$; for ESHG and THG the first pole occurs at $\omega = \Delta\varepsilon_{01}/2$ and $\omega = \Delta\varepsilon_{01}/3$, respectively. For degenerate four wave mixing the first pole occurs usually when ω equals half the transition energy for the first two-photon dipole-allowed transition.

The differences in the positions for the first poles are distinctly reflected by the different convergence rates for the four non-linear optical processes: For both frequencies $\gamma_{||}$ for the dc-Kerr effect converges very fast. Already the second-order Taylor approximations obtained from only γ_0 and A are within 1% (671.5 nm) and 1.5% (476.5 nm) of the series limit. The error is reduced by about a factor of two if instead the [0,1] Padé approximant is used, which is also calculated using only γ_0 and A. With the [1,1] Padé approximant, calculated from γ_0 and the A and B coefficients, the error is reduced to less than 0.1%. For degenerate four wave mixing already a fourth-order Taylor approximation is needed to get $\gamma_{||}^{DFWM}(\omega)$ at the two frequencies correct within 1%. Again the errors are drastically reduced if instead a [1,1] Padé approximant is used, which can be obtained from the same dispersion coeffcients. The convergence is again somewhat slower for the ESHG process: at the higher of the two frequencies (476.5 nm) already a sixth-order Taylor expansion would be needed to get the hyperpolarizability correct within 1%, alternatively this accuracy can be obtained using the [1,1] Padé approximant which requires only the dispersion coefficients up to fourth order, i. e. the A and B coefficients.

$\gamma_{||}$ at 671.5 nm (0.0678531 a.u.)

order/approximant	Kerr Taylor	Kerr Padé	DFWM Taylor	DFWM Padé	ESHG Taylor	ESHG Padé	THG Taylor	THG Padé
2 / [0,1]	2543.08	2555.36	2707.95	2761.02	2872.82	3002.69	3367.42	4071.93
4 / [1,1]	2551.48	2551.93	2742.41	2746.43	2948.43	2962.08	3667.33	3797.81
6 / [1,2]	2551.85	2551.87	2745.74	2746.08	2958.82	2960.37	3749.69	3778.96
8 / [2,2]	2551.87	2551.87	2746.05	2746.09	2960.18	2960.38	3771.62	3779.46
10 / [2,3]	2551.87	2551.87	2746.08	2746.09	2960.35	2960.38	3777.44	3779.55
∞	2551.87		2746.09		2960.38		3779.56	

$\gamma_{||}$ at 476.5 nm (0.0956209 a.u.)

order/approximant	Kerr Taylor	Kerr Padé	DFWM Taylor	DFWM Padé	ESHG Taylor	ESHG Padé	THG Taylor	THG Padé
2 / [0,1]	2705.63	2757.90	3033.05	3281.87	3360.47	4051.64	4342.73	13672.1
4 / [1,1]	2738.76	2742.50	3168.97	3204.57	3658.70	3788.73	5525.55	7315.26
6 / [1,2]	2741.69	2741.95	3195.04	3201.01	3740.03	3768.23	6170.61	6861.30
8 / [2,2]	2741.93	2741.95	3199.94	3201.07	3761.18	3768.36	6511.79	6884.80
10 / [2,3]	2741.95	2741.95	3200.86	3201.08	3766.56	3768.36	6691.45	6894.48
∞	2741.95		3201.08		3768.36		6896.81	

Table 2: Convergence of the Taylor series and the series of diagonal Padé approximants (CCSD response, t-aug-cc-pVDZ basis). The "inifinite" order results were calculated using the implementation for the frequency-dependent response function.

A similar convergence is found for the third harmonic generation process at the lower of the two frequencies, 671.5 nm. At the higher frequency, 476.5 nm, the Taylor approximations for the third harmonic generation hyperpolarizability converge only very slowly, even with a tenth-order Taylor approximation a one-percent accuracy is not obtained. This accuracy, however, is still achieved with a [1,2] Padé approximant calculated from the dispersion coefficients up to sixth order.

The much better performance of the diagonal Padé approximants compared to the Taylor expansion is not unexpected since similar convergence trends were found previously for the expansion of the first hyperpolarizabilities $\beta_{||}^{SHG}(\omega)$ and $\beta_{||}^{EOPE}(\omega)$ of ammonia. The good performance of Padé approximants for the approximation of dispersion curves for frequency-dependent properties is also well known from linear polarizabilities [47]. The present results are promising for further applications to other molecules. A truncation error of 1% in the electronic hyperpolarizability will be acceptable for most pratical applications. To achieve an overall accuracy of 1% in *ab initio* calculations would require very large basis sets, the inclusion of connected triples and a very accurate treatment of vibrational effects — a challenging task no yet met for polyatomic systems. Also, only for the most accurate ESHG and dc-Kerr measurements [4, 9, 11–14] the uncertainities are less than 1%; for THG, DFWM, ac-Kerr and CARS measurements the uncertainities are often larger than 10% [13, 48, 49].

4.2 Convergence with the basis set

In Table 3 we have listed the results of a basis set and correlation study for the hyperpolarizability dispersion coefficients. In a previous investigation of the basis set effects on the dispersion coefficients for the first hyperpolarizability β of ammonia [22] we found quite different trends for the static hyperpolarizability and for the dispersion coefficients. While the static hyperpolarizability was very sensitive to the inclusion of diffuse functions, the dispersion coefficients remained almost unchanged on augmentation of the basis set with additional diffuse functions, but the results obtained with the CC2 and CCSD models, which include dynamic electron correlation, showed large changes with an increase of the polarization level from double zeta to triple zeta basis sets. This basis set trend for the hyperpolarizability dispersion coefficients is corroborated by the CC2 and CCSD results for the second hyperpolarizability of methane, although the effects are less exaggerated than for $\beta_{||}$ of ammonia. For methane also the static hyperpolarizability is sensitive to the polarization level of the basis set.

The dispersion coefficients for the mixed-symmetry component γ_{ms} which describes the deviation from Kleinman symmetry are for methane more than an order of magnitude smaller than coefficients of the same order in the frequencies for $\gamma_{||}$. Their varations with basis sets and wavefunction models are, however, of comparable absolute size and give rise to very large relative changes for the mixed-symmetry dispersion coefficients.

From a basis set study at the CCSD level for the static hyperpolarizability we concluded in Ref. [45] that the d-aug-cc-pVQZ results for γ_0 is converged within 1 – 2% to the CCSD basis set limit. The small variations for the A, B and B' coefficients between the two triple zeta basis sets and the d-aug-cc-pVQZ basis, listed in Table 4, indicate that also for the first dispersion coefficients the remaining basis set error in d-aug-cc-pVQZ basis is only of the order of 1 – 2%. This corroborates that the results for the frequency-dependent hyperpolarizabilities obtained in Ref. [45] by a combination of the static d-aug-cc-pVQZ hyperpolarizability with dispersion curves calculated using the smaller t-aug-cc-pVTZ basis set are close to the CCSD basis set limit.

	CCS			CC2			CCSD		
	daD	*taD*	*daT*	*daD*	*taD*	*daT*	*daD*	*taD*	*daT*
γ_0[a]	2132.9	2174.7	2227.2	2526.1	2580.5	2555.5	2327.6	2378.2	2363.5
A	6.6855	6.6706	6.6614	7.7064	7.6795	7.5063	7.5573	7.5287	7.3776
B	33.521	33.349	32.985	43.532	43.232	41.048	41.976	41.667	39.754
B'	−2.4211	−2.4097	−2.4183	−4.8706	−4.7963	−4.3894	−4.3070	−4.2404	−3.8940
C	148.05	146.85	143.78	214.46	212.31	195.93	203.41	201.26	187.06
C'	25.768	25.472	24.542	37.662	37.141	33.708	35.624	35.112	32.090
C''	−57.356	−56.599	−54.989	−103.92	−101.98	−91.363	−93.170	−91.386	−82.401
D	606.59	600.10	582.33	975.58	963.29	865.45	910.20	898.24	814.60
D'	−570.96	−561.66	−542.10	−1080.1	−1057.7	−930.69	−958.13	−937.87	−832.83
D''	197.62	195.13	186.71	333.75	308.48	288.92	305.80	300.87	267.12
D'''	337.41	332.87	319.97	549.57	540.80	480.52	511.61	503.18	451.30
A_{ms}	−0.2320	−0.1763	0.0077	−0.2320	−0.3919	−0.1550	−0.4436	−0.3696	−0.1307
B_{ms}	0.1123	0.4387	1.3481	0.1123	−1.1510	0.1251	−1.3517	−0.8670	0.4147
B'_{ms}	−0.8197	−0.7818	−0.6555	−0.8197	−1.1717	−0.9662	−1.1536	−1.1061	−0.9097
C_{ms}	11.814	13.008	15.632	11.814	7.2571	10.919	6.8975	8.9250	12.687
C'_{ms}	−8.5246	−8.1038	−6.8990	−8.5243	−13.753	−11.271	−13.263	−12.672	−10.375
C''_{ms}	0.2520	0.1024	−0.3894	0.2520	−1.5512	−2.0347	−0.8649	−1.0837	−1.5779

Table 3: Comparison of the dispersion coefficients for $\gamma_{||}$ and γ_{ms} for different basis sets and coupled cluster models.

4.3 The effect of electron correlation on the dispersion coefficients

If we compare results obtained with the same basis sets with the three coupled cluster models CCS, CC2 and CCSD, we find similar trends as observed in Refs. [22,45]: The CCS model underestimates strongly the static hyperpolarizabilities and their dispersion. The results are usually of similar quality as those obtained with SCF. For methane, the CCS static hyperpolarizabilities are intermediate between the SCF and the CCSD values obtained in the same basis set. In Ref. [45] the CCS percentage dispersion contribution to the third harmonic generation (THG) hyperpolarizability of methane was found to be slightly smaller than for SCF, both underestimating significantly the dispersion obtained with the correlated coupled cluster models CC2 and CCSD. Accordingly the CCS dispersion coefficients listed in Table 3 are substantially smaller than the respective CCSD results obtained in the same basis sets.

The CC2 model performes very different for static hyperpolarizabilities and for their dispersion. For methane, CC2 overestimates γ_0 by a similar amount as it is underestimated by CCS, thus giving no improvement in accuracy relative to the uncorrelated methods CCS and SCF. In contrast to this, the CC2 dispersion coefficients listed in Table 3 are by a factor of 3 – 8 closer to the CCSD values than the respective CCS results. The dispersion coefficients should be sensitive to the lowest dipole-allowed excitation energy, which determines the position of the first pole in the dispersion curve. The substantial improvements in accuracy for the dispersion coefficients are thus consistent with the good performance of CC2 for excitation energies [35,37,50].

The CCSD model gives for static and frequency-dependent hyperpolarizabilities usually results close to the experimental values, provided that the effects of vibrational averaging and the pure vibrational contributions have been accounted for. Zero point vibrational corrections for the static and the electric field induced second harmonic generation (ESHG) hyperpolarizability of methane have recently been calculated by Bishop and Sauer using SCF and MCSCF wavefunctions [51].

	$\gamma_0^e(R_e)$ [a]	$\bar{\gamma_0}^e$ [b]	A [c]	B [c]	B' [c]	A_{ms}	B_{ms}	B'_{ms}
CCS d-aug-cc-pVTZ	2227	2512	6.661	32.99	−2.42	−0.01	1.35	−0.66
CC2 d-aug-cc-pVTZ	2555	2840	7.506	41.05	−4.39	−0.16	0.13	−0.97
CCSD d-aug-cc-pVTZ	2364	2649	7.378 (7.9)	39.75 (43)	−3.89	−0.13	0.41	−0.91
CCSD t-aug-cc-pVTZ	2365	2650	7.378 (7.9)	39.75 (43)	−3.90	−0.13	0.43	−0.91
CCSD d-aug-cc-pVQZ	2330	2615	7.338 (7.9)	39.30 (44)	−3.84	−0.06	0.46	−1.34
SCF [d]	1882	2186	7.049 (7.73)	36.68 (44.94)				
MCSCF [d]	2152	2438	7.057 (7.61)	36.85 (43.81)				
Experiment Ref. [14]		2607[e]	(7.38)[e]	(100.6)[e]		(-0.16 ± 0.014)[e]		
Experiment Ref. [53]		2590[f]	(7.46)[f]	(95.08)[f]				

Table 4: Comparison of various *ab initio* results and experimental estimates for the dispersion coefficients of the electronic hyperpolarizabilities $\gamma_{||}^e$ and γ_{ms}^e of methane. (All results in atomic units. Results for the dispersion coefficients refer to single point calculations at the equilibrium geometry. Where available, dispersion coefficients for the vibrational average are given in parentheses.)

Combining [52] the zero point vibrational corrections of Ref. [51] with the CCSD results obtained in the t-aug-cc-pVTZ and d-aug-cc-pVQZ basis sets we obtained the estimates for the ZPV corrected γ_0, A and B coefficients listed in Table 4. An experimental estimate for γ_0, A, and B has been derived by Shelton by fitting the results of ESHG measurements to the expresssion $\gamma(\omega_{L2}^2) = \gamma_0(1 + A\omega_{L2}^2 + B\omega_{L2}^4)$. This fit was later repeated by Bishop and Pipin [53] after subtraction of *ab initio* calculated results for the pure vibrational contribution to the hyperpolarizability from the orignal experimental values. These fits did not include higher powers of ω_{L2} than ω_{L2}^4. The results, in particular the B coefficient, obtained in this way are affected by higher-order terms. The contributions from the C and higher-order coefficients to the ESHG hyperpolarizability are substantial for wavelengths shorter than 671.5 nm ($\omega_{L2}^2 > 0.03$ a.u.), when a polynomial (Taylor) expansion is used, as follows from the discussion of the convergence of the dispersion expansion. The increasing importance of the sixth- and higher-order terms with increasing frequency can also be seen from the difference between the fourth-order Taylor approximation and the [1,1] Padé approximant for the CCSD hyperpolarizability in the d-aug-cc-pVQZ basis, which are both plotted in Figure 1. The dispersion coefficients obtained from the two fits to the experimental ESHG data are thus not strictly comparable with analytically calculated dispersion coefficients, as have been calculated in the present work, or with results obtained by least square fits to higher polynomials in ω_{L2} as done by Bishop and Sauer in Ref. [51] for the SCF and MCSCF results included in Table 4. Nevertheless, the *ab initio* calculated A coefficients agree with the values deduced from the experimental ESHG data within $\approx 10\%$. The B coefficients derived by Shelton and by Bishop and Pipin from their fits to the experimental ESHG data are about a factor of 2 – 2.5 larger than the *ab initio* calculated values, which is likely an effect of the inclusion of higher-order terms in the "experimental" estimates for B, since the variations of B with basis set, *ab initio* method and zero point vibrational correction are only small. This is also corroborated by the closeness (within 1 – 3%) of the CCSD dispersion curves to the results of the ESHG experiments of Refs. [4, 54] and to the dispersion curve fitted to this data (see Fig.

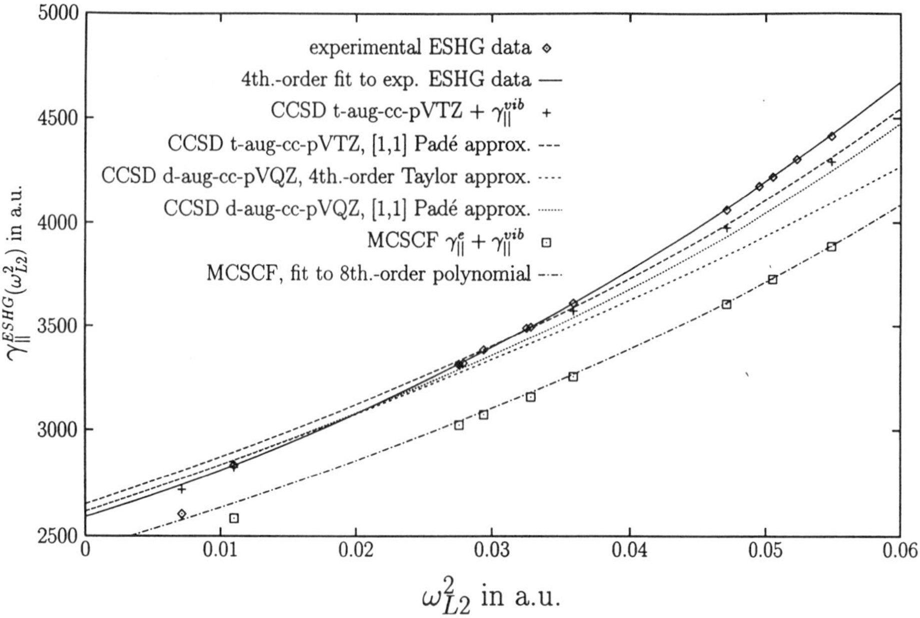

Figure 1: The ESHG hyperpolarizability of CH_4.

1). Compared with the CCSD results, the MCSCF values for γ_0, A and B obtained in Ref. [51] are close the uncorrelated SCF results in the same basis set. Both underestimate the static hyperpolarizability and the dispersion (cmp. Fig. 1).

A crude experimental estimate for the A_{ms} coefficient is obtained from Sheltons fit of ESHG measurements for the hyperpolarizability ratio $\gamma_{||}/\gamma_{\perp}$ to the expression $3(1+A^{*}\omega_{L2}^2)$. Using the dispersion expansions (79) and (89), we obtain for the ratio $\gamma_{||}/\gamma_{\perp}$ up to second order in the frequencies the expansion

$$\frac{\gamma_{||}}{\gamma_{\perp}} = 3(1 - 2A_{ms}\omega_{M2}^2 + \ldots) \ . \tag{92}$$

For ESHG the effective frequency ω_{M2}^2 is $-3\omega^2 = \omega_{L2}^2/2$, thus $A^{*,\mathrm{ESHG}} = A_{ms}$. Since the fit made by Shelton did not include higher than second powers of ω_{L2}^2, the experimental estimate for A_{ms} is similar as the B

coefficient affected by higher-order terms. Nevertheless, this estimate agrees in sign and in the order of magnitude with the CCSD results in the best basis sets. The agreement is as good as can be expected taking into account the smallness of A_{ms} and its sensitivity to basis variations and correlation treatment. To our knowledge this is the first *ab intio* calculation for the frequency-dependence of the hyperpolarizability ratio $\gamma_{||}/\gamma_{\perp}$ for a many-electron system. The only other *ab intio* calculation for $\gamma_{||}/\gamma_{\perp}$, we are aware of, are CI-Hylleraas calculations for helium, H_2 and D_2 by Bishop and Pipin [55] and Bishop et al. [56].

5 Conclusions

We have derived and implemented analytic dispersion coefficients for second hyperpolarizabilities for the coupled cluster models CCS, CC2 and CCSD. The dispersion coefficients are defined through a power series expansion of the cubic response function in its frequency arguments and are thus applicable to general frequency-dependent fourth-order properties and are independent of the non-linear optical process. For the irreducible tensor components of the second hyperpolarizability γ, the dispersion can be expressed in a compact form using permutation symmetry-adapted linear combinations of the optical frequencies. In an application to the second hyperpolarizability of methane we find good convergence of the Taylor expansion in the typical experimental frequency range for the dc-Kerr effect, DFWM and ESHG. For third harmonic generation a good convergence of the Taylor expansion is only found for small frequencies. With Padé approximants the convergence of the dispersion expansion is improved significantly for all four investigated optical processes and the application range of the dispersion expansion is considerably extended to higher frequencies. For many molecules, in particular such with the first transition frequencies lying as high as in methane, the first two dispersion coefficients will be sufficient to obtain the dispersion curves within a one percent accuracy in the usual experimental frequency range, provided that the dispersion curve is calculated using a [1,1] Padé approximant.

A comparison of the dispersion coefficients obtained with different basis sets and the coupled cluster models CCS, CC2 and CCSD shows that the dispersion coefficients are as sensitive to the choice of the basis set and the correlation treatment as are the static hyperpolarizabilities. Similar as in a previous study on the dispersion coefficients for the first hyperpolarizability of ammonia, we find that the CC2 model performs considerably better for dispersion coefficients than it does for static hyperpolarizabilities.

Our results indicate that dispersion coefficients obtained from fits of pointwise given frequency-dependent hyperpolarizabilities to low order polynomials can be strongly affected by the inclusion of high-order terms. A and B coefficients derived from a least square fit of experimental frequency-dependent hyperpolarizibility data to a quadratic function in ω_i^2 are therefore not strictly comparable to dispersion coefficients calculated by analytical differentiation or from fits to higher-order polynomials. *Ab initio* calculated dispersion curves should therefore be compared with the original frequency-dependent experimental data.

The present study demonstrates that the analytic calculation of hyperpolarizability dispersion coefficients provides an efficient alternative to the pointwise calculation of dispersion curves. The dispersion coefficients provide additional insight into non-linear optical properties and are transferable between the various optical processes, also to processes not investigated here as for example the ac-Kerr effect or coherent anti-Stokes Raman scattering (CARS), which depend on two independent laser frequencies and would be expensive to study with calculations explictly frequency-dependent calculations.

Acknowledgements

One of the authors (CH) thanks the European Commission for financial support through the Training and Mobility of Researchers (TMR) Programm (Grant ERBFMBICT 96.1066). This work has been supported by the Danish Natural Science Research Council (Grant No 9600856).

A Formulas for the sixth- through tenth-order dispersion coefficients for $\gamma_{||}$ and γ_{ms}

The coefficients for the sixth- through tenth-order terms of the dispersion expansion Eq. (75) for $\gamma_{||}$ are found as:

$$C = A_{3,0,0} = \frac{1}{8\gamma_{||}(0)} D_{||}(6,0,0) \tag{93}$$

$$C' = A_{0,2,0} = \frac{1}{9\gamma_{||}(0)} \Big(D_{||}(4,2,0) - 6D_{||}(6,0,0)\Big) \tag{94}$$

$$C'' = A_{1,0,1} = \frac{1}{8\gamma_{||}(0)} \Big(D_{||}(4,1,1) - 2D_{||}(4,2,0) + 3D_{||}(6,0,0)\Big) \tag{95}$$

$$D = A_{4,0,0} = \frac{1}{16\gamma_{||}(0)} D_{||}(8,0,0) \tag{96}$$

$$D' = A_{2,0,1} = \frac{1}{16\gamma_{||}(0)} \Big(D_{||}(6,1,1) - 2D_{||}(6,2,0) + D_{||}(7,1,0)\Big) \tag{97}$$

$$D'' = A_{0,0,2} = \frac{1}{16\gamma_{||}(0)} \Big(D_{||}(4,2,2) - D_{||}(4,3,1) - D_{||}(5,2,1) + 4D_{||}(7,1,0)\Big) \tag{98}$$

$$D''' = A_{1,2,0} = \frac{1}{18\gamma_{||}(0)} \Big(D_{||}(6,2,0) - 10D_{||}(8,0,0)\Big) \tag{99}$$

$$E = A_{5,0,0} = \frac{1}{32\gamma_{||}(0)} D_{||}(10,0,0) \tag{100}$$

$$E' = A_{3,0,1} = \frac{1}{32\gamma_{||}(0)} \Big(D_{||}(8,1,1) - 2D_{||}(10,2,0) + D_{||}(9,1,0)\Big) \tag{101}$$

$$E'' = A_{1,0,2} = \frac{1}{32\gamma_{||}(0)} \Big(D_{||}(6,2,2) - 2D_{||}(6,3,1) + 3D_{||}(8,1,1) + 10D_{||}(8,2,0) - 29D_{||}(9,1,0)\Big) \tag{102}$$

$$E'''' = A_{2,2,0} = \frac{1}{36\gamma_{||}(0)}\Big(D_{||}(8,2,0) - 15D_{||}(10,0,0)\Big) \tag{103}$$

$$E''''' = A_{0,2,1} = \frac{1}{36\gamma_{||}(0)}\Big(D_{||}(6,3,1) - 9D_{||}(8,1,1) - 14D_{||}(8,2,0) + 61D_{||}(9,1,0)\Big) \tag{104}$$

The dispersion coefficients for γ_{ms} are obtained as:

$$A_{ms} = \frac{1}{\gamma_{||}(0)} D_{ms}(2,0,0) \ , \tag{105}$$

$$B_{ms} = \frac{1}{2\gamma_{||}(0)} D_{ms}(4,0,0) \ , \tag{106}$$

$$B'_{ms} = -\frac{1}{9\gamma_{||}(0)}\Big(D_{ms}(3,1,0) + D_{ms}(4,0,0)\Big) \ , \tag{107}$$

$$C_{ms} = \frac{1}{4\gamma_{||}(0)} D_{ms}(6,0,0) \ , \tag{108}$$

$$C'_{ms} = -\frac{1}{18\gamma_{||}(0)} D_{ms}(5,1,0) \ , \tag{109}$$

$$C''_{ms} = \frac{1}{24\gamma_{||}(0)}\Big(3D_{||}(3,2,1) - 2D_{||}(5,1,0)\Big) \ , \tag{110}$$

with the intermediates $D_{ms}(l,m,n)$ defined as:

$$D_{ms}(l,m,n) = \frac{1}{6}\sum_{\xi\eta}\Big\{D_{\eta\xi\xi\eta}(l,m,n) - 2D_{\xi\eta\xi\eta}(l,m,n) + D_{\xi\xi\eta\eta}(l,m,n)\Big\}, \tag{111}$$

where the summation is over $\xi, \eta = x, y, z$.

References

[1] D. N. Zubarev, Usp. Fiz. Nauk **71** (1960); (English translation Soviet Phys. Uspekhi **3** (1960) 320.).
J. Linderberg and Y. Öhrn, Proc. Roy. Soc. (London) A. **285**, 445 (1965).
Y. Öhrn and J. Linderberg, Phys. Rev. A. **139**, 1063 (1965).

[2] J. Linderberg and Y. Öhrn. *Propagators in Quantum Chemistry.* Academic, London, 1973.

[3] D. P. Shelton, Phys. Rev. Lett. **62**, 2660 (1989).

[4] D. P. Shelton, Phys. Rev. A. **42**, 2578 (1990).

[5] D. P. Shelton and Z. Lu, Phys. Rev. A. **37**, 3813 (1988).

[6] M. Jaszuński and D. L. Yeager, Phys. Rev. A. **40**, 1651 (1989).

[7] D. M. Bishop, Phys. Rev. Lett. **65**, 1688 (1990).

[8] H. J. A. Jensen, P. Jørgensen, H. Hettema and J. Olsen, Chem. Phys. Lett. **187**, 387 (1991).

[9] D. P. Shelton and E. A. Donley, Chem. Phys. Lett. **195**, 591 (1992).

[10] C. Hättig and P. Jørgensen, Chem. Phys. Lett. **283**, 109 (1998).

[11] Y. Mizrahi and D. P. Shelton, Phys. Rev. Letters. **55**, 696 (1985).

[12] Y. Mizrahi and D. P. Shelton, Phys. Rev. A. **31**, 3145 (1985).

[13] D. P. Shelton and J. E. Rice, Chem. Rev. **94**, 3 (1994).

[14] D. P. Shelton and J. J. Palubinskas, J. Chem. Phys. **104**, 2482 (1996).

[15] J. E. Rice, J. Chem. Phys. **96**, 7580 (1992).

[16] Y. Luo, O. Vahtras, H. Ågren and P. Jørgensen, Chem. Phys. Lett. **205**, 555 (1993).

[17] M. Jaszuński, P. Jørgensen and H. J. Å. Jensen, Chem. Phys. Lett. **191**, 293 (1992).

[18] D. P. Shelton, J. Chem. Phys. **84**, 404 (1986).

[19] D. M. Bishop, J. Chem. Phys. **90**, 3192 (1989).

[20] D. M. Bishop and D. W. De Kee, J. Chem. Phys. **104**, 9876 (1996).

[21] E. K. Dalskov, H. J. Å. Jensen and J. Oddershede, Mol. Phys. **90**, 3 (1997).

[22] C. Hättig and P. Jørgensen, Theor. Chim. Acc. (1998). in press.

[23] C. Hättig, O. Christiansen, H. Koch and P. Jørgensen, Chem. Phys. Lett. **269**, 428 (1997).

[24] C. Hättig, O. Christiansen and P. Jørgensen, Chem. Phys. Lett. **282**, 139 (1998).

[25] O. Christiansen, P. Jørgensen and C. Hättig, Int. J. Quantum Chem. **68**, 1 (1998).

[26] J. Olsen and P. Jørgensen. *Time-Dependent Response Theory with Applications to Self-Consistent Field and Multiconfigurational Self-Consistent Field Wave Functions*, in *Modern Electronic Structure Theory*, edited by D. R. Yarkony, volume 2, chapter 13, pp. 857–990. World Scientific, Singapore, 1995.

[27] H. Sambe, Phys. Rev. A. **7**, 2203 (1972).

[28] P. W. Langhoff, S. T. Epstein and M. Karplus, Rev. Mod. Phys. **44**, 602 (1972).

[29] C. Hättig, O. Christiansen and P. Jørgensen, J. Chem. Phys. **107**, 10592 (1997).

[30] T. Helgaker and P. Jørgensen, Theor. Chim. Acta. **75**, 111 (1989). P. Jørgensen and T. Helgaker, J. Chem. Phys. **89**, 1560 (1988).

[31] C. Hättig, Mol. Phys. **94**, 455 (1998).

[32] In the notation used in Ref. [31] the effective frequencies ω_{L2}^2, ω_{L3}^3 and ω_{L4}^4 were denoted as Q_2, Q_3 and Q_4.

[33] C. Hättig, Chem. Phys. Lett. (1999). in press.

[34] D. A. Kleinman, Phys. Rev. **126**, 1977 (1962).

[35] O. Christiansen, H. Koch and P. Jørgensen, Chem. Phys. Lett. **243**, 409 (1995).

[36] O. Christiansen, H. Koch and P. Jørgensen, J. Chem. Phys. **103**, 7429 (1995).

[37] H. Koch, O. Christiansen, P. Jørgensen and J. Olsen, Chem. Phys. Lett. **244**, 75 (1995).

[38] O. Christiansen, H. Koch, A. Halkier, P. Jørgensen, T. Helgaker and A. M. Sanchez de Meras, J. Chem. Phys. **105**, 6921 (1996).

[39] A. Halkier, H. Koch, O. Christiansen, P. Jørgensen and T. Helgaker, J. Chem. Phys. **107**, 849 (1997).

[40] C. Hättig, O. Christiansen and P. Jørgensen, J. Chem. Phys. **108**, 8355 (1998).

[41] O. Christiansen, A. Halkier, H. Koch, P. Jørgensen and T. Helgaker, J. Chem. Phys. **108**, 2801 (1998).

[42] T. H. Dunning, J. Chem. Phys. **90**, 1007 (1989). R. A. Kendall, T. H. Dunning and R. J. Harrison, J. Chem. Phys. **96**, 6796 (1992). D. E. Woon and T. H. Dunning, J. Chem. Phys. **98**, 1358 (1993). D. E. Woon and T. H. Dunning, J. Chem. Phys. **100**, 2975 (1994). D. E. Woon and T. H. Dunning, J. Chem. Phys. 103, 4572 (1995). The basis sets were obtained from the Extensible Computational Chemistry Environment Basis Set Database, Version 1.0, as developed and distributed by the Molecular Science Computing Facility, Environmental and Molecular Sciences Laboratory which is part of the Pacific Northwest Laboratory, P.O. Box 999, Richland, Washington 99352, USA, and funded by the U.S. Department of Energy. PNL is a multiprogram laboratory operated by Batelle Memorial Institute for the U.S. Department of Energy under contract DE-AC06-76RLO 1830.

[43] K. P. Huber and G. Herzberg. *Molecular Spectra and Molecular Structure: IV. Constants of Diatomic Molecules.* Van Nostrand, New York, 1979.

[44] D. L. Gray and A. G. Robiette, Mol. Phys. **37**, 1901 (1979).

[45] C. Hättig and P. Jørgensen, J. Chem. Phys. **109**, 2762 (1998).

[46] G. A. Baker. *Essentials of Padé Approximants.* Academic Press, Inc., New York, 1975.

[47] P. W. Langhoff and M. Karplus, J. Opt. Soc. Am. **59**, 863 (1969). P. W. Langhoff and M. Karplus, J. Chem. Phys. **52**, 1435 (1970).

[48] H. J. Lehmeier, W. Leupacher and A. Penzkofer, Opt. Coummun. **56**, 67 (1985).

[49] X. F. Li, A. L'Huillier, M. Ferray, L. A. Lompré and G. Mainfray, Phys. Rev. A. **39**, 5731 (1989).

[50] O. Christiansen, H. Koch, P. Jørgensen and J. Olsen, Chem. Phys. Lett. **256**, 185 (1996).

[51] D. M. Bishop and S. P. A. Sauer, J. Chem. Phys. **108**, 8502 (1997).

[52] The pointwise given MCSCF results for zero point vibrational corrections for the ESHG hyperpolarizability were added to the CCSD results obtained from the dispersion coefficients and than fitted to a fourth-order polynomial in ω_{L2}^2.

[53] D. M. Bishop and J. Pipin, J. Chem. Phys. **103**, 4980 (1995).

[54] D. P. Shelton, Phys. Rev. A. **34**, 304 (1986).

[55] D. M. Bishop and J. Pipin, J. Chem. Phys. **91**, 3549 (1989).

[56] D. M. Bishop, J. Pipin and S. M. Cybulski, Phys. Rev. A **43**, 4845 (1991).

On the Extensivity Problem in Coupled-Cluster Property Evaluation

Hideo Sekino and Rodney J. Bartlett

Quantum Theory Project

P.O. Box 118435

University of Florida

Gainesville, Florida 32611–8435

There is a dilemma that occurs in the coupled-cluster treatment of second- and higher-order properties, because the usual perturbation (polarization propagator) expressions are inconsistent with the energy derivative. This is a consequence of the energy derivative formulas not satisfying the higher-order analogues of the Hellman-Feynman theorem. However, the equation-of-motion (EOM) coupled-cluster theory retains the propagator form for properties, but contrary to the derivative theory, the results for second- and higher-order properties are not extensive, this is because EOM-CC has certain "CI-like" elements. Models with elimination of unlinked terms are proposed to rectify this. Illustrative calculations on dipole polarizabilities and transition moments of noninteracting LiH molecules are performed. Complete elimination of unlinked terms through solving a connected left-hand equation for the auxiliary operator Λ leads to a model that is strictly size extensive yet adheres to the usual perturbation or propagator form. Numerically, this model gives values close to those obtained by the derivative theory.

ADVANCES IN QUANTUM CHEMISTRY, VOLUME 35

0065-3276/99 $30.00

Introduction

Coupled-cluster (CC) theory has been recognized [1] as a powerful and efficient method for the investigation of molecular electronic structure where correlation effects play an important role. Much theoretical and computational effort has been devoted to applying the method to calculate the ground-state energy. The great success of the method in elucidating the electronic ground state has encouraged its application to excited-state energies and properties [2–7]. The equation-of-motion coupled-cluster (EOM-CC) method, originally developed for excitation energies [2–6], also provides higher-order properties that can be computed straightforwardly by *formally* using sum-over-state *unsymmetric* polarization propagator (or perturbation theory) expressions, where the EOM-CC excited states are the intermediate states [7, 8], and the ground state is that obtained from CC theory and its auxiliary (Λ) state [9]. Similarly, transition probabilities are readily obtained as *generalized expectation values*. That means the poles in the dynamic polarizability and associated residues are exactly those obtained by the EOM-CC eigenvalue equation.

Though derived differently, the coupled cluster linear response (CCLR) [11–15] first presented by Monkhorst without any consideration of the Λ auxiliary function, provides identical excitation energies to EOM-CC. However, in the CCLR theory, the critical quantities are always energy derivatives instead of generalized expectation values. Since the ground state CC energy is size extensive, any such derivative property also has to be size extensive [8, 16]. To the contrary, CCLR does not retain the usual polarization propagator [8] form for second- and higher-order properties, and this is an undesirable feature. In fact, CCLR does not provide *wavefunctions* for excited states, which in EOM-CC, leads to its generalized expectation values.

This raises a dilemma in treating second- and higher-order properties in coupled-cluster theory. In the EOM-CC approach, which is basically a CI calculation for a non-Hermitian Hamiltonian $H = e^{-\hat{T}} H e^{\hat{T}}$ that incorporates ground-state CC information via the $\hat{T}$ operator, the usual propagator form for higher-order properties is obtained as long as $\bar{H}$ is treated as an entity. Also, at the full CI limit, these are the exact results. However, once the CC nature of the wavefunction is fully included, as in CCLR [11], this convenient propagator form is lost, only to be regained, again, at the full CI limit. The equivalence of the propagator form and the derivative of the energy requires the satisfaction of the *higher-order* analogs of the (generalized) Hellman-Feynman theorem. We mean by this that there is *no* dependence of the nth order energy on the nth order derivative wavefunctions. That is, $E^{(1)}$ depends only on ψ_o, $E^{(2)}$ only on $\psi^{(1)}$, $E^{(3)}$ only on $\psi^{(2)}$ or $\psi^{(1)}$ in a $2n + 1$ rule formula, and so on. This will not necessarily hold for approximate wavefunctions. Consequently, we can explore a series of approximations that lie between EOM-CC and CCLR that attempt to regain extensivity while maintaining the highly desirable propagator form for the property.

We study three different approximations for removing unlinked diagrams in EOM-CC and show that these models provide second-order properties and transition probabilities that are close to those provided by CCLR in isolated molecular systems, but in a more convenient computational structure.

Theory

The coupled-cluster equations may be derived by variation of the functional

$$\varepsilon = \langle 0|(1+\Lambda)e^{-T}He^{T}|0\rangle = \langle 0|(1+\Lambda)\bar{H}|0\rangle, \tag{1}$$

with respect to T and Λ, where Λ is a de-excitation operator and T is the usual cluster excitation operator [17]. The CC equations are $Q\bar{H}|0\rangle = 0$ and the Λ equations, $P(1+\Lambda)\bar{H}Q - PE\Lambda Q = 0$ where $P = |0\rangle\langle 0|$, $Q = (1 - |0\rangle\langle 0|) = |\mathbf{f}\rangle\langle\mathbf{f}|$, $|\mathbf{h}\rangle = |0, \mathbf{f}\rangle$ and $E_{cc} = \langle 0|\bar{H}|0\rangle$.

In EOM-CC, we solve the eigenvalue equation for the nonsymmetric Hamiltonian,

$$\bar{H} = \mathrm{e}^{-T}H\mathrm{e}^{T}. \tag{2}$$

This has all ground-state CC information in it via the coupled-cluster T amplitudes. Its eigenvalues and eigenvectors are given for $k > 0$ by

$$\begin{aligned} \bar{\mathrm{H}}\mathbf{r_k} &= \omega_\mathbf{k}\mathbf{r_k} \\ \mathbf{l_k}\bar{\mathrm{H}} &= \mathbf{l_k}\omega_\mathbf{k}, \end{aligned} \tag{3}$$

where $\bar{\mathbf{H}} = \langle\mathbf{h}|H|\mathbf{h}\rangle$, with $\langle l_o| = \langle 0|(1+\Lambda)$, and $|r_o\rangle = |0\rangle$, and $\langle\mathbf{l_k}| = \mathbf{l_k}\langle\mathbf{f}|$, $k \neq 0$. We choose $\langle l_p|r_q\rangle = \delta_{pq}$. The resolution of the identity is $1 = \sum_p |r_p\rangle\langle l_p|$, where $|\mathbf{f}\rangle$ consists of the single and double excitations in a CCSD calculation, eg.

Consider ordinary perturbation theory for the nonsymmetric Hamiltonian with a perturbation θ, then $\bar{H} = \bar{H}_o + \lambda\bar{\theta}$, and

$$\bar{\theta} = \mathrm{e}^{-T}\theta\mathrm{e}^{T}, \tag{4}$$

where

$$\begin{aligned} \bar{H}_o|0\rangle &= E_o|0\rangle = E_o|r_o\rangle \\ \langle 0|(1+\Lambda)\bar{H}_o &= \langle 0|(1+\Lambda)E_o = \langle l_o|E_o, \\ E_o &= E_{cc}. \end{aligned} \tag{5}$$

Then it follows from the inhomogeneous equations of RSPT that

$$\begin{aligned} (E_o - \bar{H}_o)|\psi^{(1)}\rangle &= \left(\bar{\theta} - E^{(1)}\right)|\psi_o\rangle \\ E^{(1)} &= \langle l_o|\bar{\theta}|r_o\rangle \end{aligned} \tag{6}$$

where $E^{(1)}$ is the generalized expectation value of CC theory, since $|\psi_o\rangle = |r_o\rangle = |0\rangle$. Note $E^{(1)} = \langle l_o|\frac{\partial \bar{H}}{\partial \lambda}|r_o\rangle$ since the Hellman-Feynman theorem is satisfied, as both Λ and T are stationary in the functional Eqn. (1). Introducing the resolvent $\mathcal{R}_o = \left(E_o - \bar{H}_o\right)^{-1}Q$, $|\psi^{(1)}\rangle = \mathcal{R}_o(\bar{\theta} - E^{(1)})|r_o\rangle$. We will initially choose $Q = 1 - |0\rangle\langle 0| = |\mathbf{f}\rangle\langle\mathbf{f}|$,, where $|\mathbf{f}\rangle$ represents all configurations involved in the CC equations orthogonal to $|0\rangle$. This makes

$$\mathcal{R}_o = |\mathbf{f}\rangle\langle\mathbf{f}|E_o - \bar{H}_o|\mathbf{f}\rangle^{-1}\langle\mathbf{f}|. \tag{7}$$

Continuing in higher order we have

$$\begin{aligned}
\left(E_o - \bar{H}_o\right)|\psi^{(2)}\rangle &= \left(\bar{\theta} - E^{(1)}\right)|\psi^{(1)}\rangle - E^{(2)}|\psi_o\rangle \\
E^{(2)} &= \langle l_o|\left(\bar{\theta} - E^{(1)}\right)\mathcal{R}_o\left(\bar{\theta} - \mathrm{E}^{(1)}\right)|r_o\rangle \\
\psi^{(2)} &= \mathcal{R}_o\left(\bar{\theta} - E^{(1)}\right)|\psi^{(1)}\rangle,
\end{aligned} \tag{8}$$

and

$$\begin{aligned}
E^{(3)} =&\langle l_o|\left(\bar{\theta} - E^{(1)}\right)\mathcal{R}_o\left(\bar{\theta} - E^{(1)}\right)\mathcal{R}_o\left(\bar{\theta} - E^{(1)}\right)|r_o\rangle \\
&- E^{(2)}\langle l_o|\mathcal{R}_o\left(\bar{\theta} - E^{(1)}\right)|r_o\rangle \\
\psi^{(3)} =&\mathcal{R}_o\left(\bar{\theta} - E^{(1)}\right)|\psi^{(2)}\rangle - E^{(2)}\mathcal{R}_o|\psi^{(1)}\rangle.
\end{aligned} \tag{9}$$

Note that unlike standard perturbation theory in intermediate normalization., $\langle l_o|\psi^{(n)}\rangle \neq 0$. In fourth order,

$$\begin{aligned}
\left(E_o - \bar{H}_o\right)\psi^{(4)} =&\left(\bar{\theta} - E^{(1)}\right)\psi^{(3)} - E^{(2)}\psi^{(2)} - E^{(3)}\psi^{(1)} - E^{(4)}\psi_o \\
E^{(4)} =&\langle l_o|\left(\bar{\theta} - E^{(1)}\right)\mathcal{R}_o\left(\bar{\theta} - E^{(1)}\right)\mathcal{R}_o\left(\bar{\theta} - E^{(1)}\right)\mathcal{R}_o\left(\bar{\theta} - E^{(1)}\right)|r_o\rangle \\
&- E^{(2)}\langle l_o|\left(\bar{\theta} - E^{(1)}\right)\mathcal{R}_o \cdot \mathcal{R}_o\left(\bar{\theta} - E^{(1)}\right)|r_o\rangle \\
&- E^{(2)}\langle l_o|\mathcal{R}_o\left(\bar{\theta} - E^{(1)}\right)\mathcal{R}_o\left(\bar{\theta} - E^{(1)}\right)|r_o\rangle \\
&- E^{(3)}\langle l_o|\mathcal{R}_o\left(\bar{\theta} - E^{(1)}\right)|r_o\rangle.
\end{aligned} \tag{10}$$

The above development considers $\bar{H}_o$ to be an entity, and develops the nonsymmetric form of RSPT accordingly [18].

Alternatively, we could insist that our perturbed wavefunctions be represented by the complete set of the EOM-CC right (or left) eigenvectors,

$$\begin{aligned}
|\psi^{(n)}\rangle &= \sum_{k=1} |r_k\rangle c_k^{(n)} = |\mathbf{r}\rangle\mathbf{c}^{(n)} \\
\langle\psi^{(n)}| &= \sum_{k=1} \langle l_k|d_k^{(n)} = \mathbf{d}^{(n)}\langle\mathbf{l}|
\end{aligned} \tag{11}$$

where, without restriction, we can exclude $|r_o\rangle$ and $\langle l_o|$ from their respective expansions since $(E_o - \bar{H}_o)|r_o\rangle = 0$ and $\langle l_o|(E_o - \bar{H}_o) = 0$. Since $\langle l_o|r_k\rangle = 0, k \neq 0$, $\langle l_o|\psi^{(n)}\rangle = 0$ and we have a kind of intermediate normalization. This gives the spectral expansion form of the reduced resolvent,

$$\begin{aligned} \mathcal{R}_o' &= |\mathbf{r}\rangle\langle \mathbf{l}|E_o - \bar{H}_o|\mathbf{r}\rangle^{-1}\langle \mathbf{l}| \\ &= \sum_{k=1} |r_k\rangle\langle l_k|\omega_k^{-1}. \end{aligned} \tag{12}$$

where $\mathcal{R}_o' = \mathcal{R}_o - |0\rangle\langle 0|l_o R_o$.

Using this form of the resolvent, we have

$$\begin{aligned} E^{(2)} &= \sum_{k=1}\langle l_o|\left(\bar{\theta} - E^{(1)}\right)|r_k\rangle\langle l_k|\left(\bar{\theta} - E^{(1)}\right)|r_o\rangle\omega_k^{-1} \\ E^{(3)} &= \sum_{k,l=1}\langle l_o|\left(\bar{\theta} - E^{(1)}\right)|r_k\rangle\langle l_k|\left(\bar{\theta} - E^{(1)}\right)|r_l\rangle\langle l_l|\left(\bar{\theta} - E^{(1)}\right)|r_o\rangle\omega_k^{-1}\omega_l^{-1} \\ E^{(4)} &= \sum_{k,l,m=1}\langle l_o|\left(\bar{\theta} - E^{(1)}\right)|r_k\rangle\langle l_k|\left(\bar{\theta} - E^{(1)}\right)|r_l\rangle\langle l_l|\left(\bar{\theta} - E^{(1)}\right)|r_m\rangle\times \\ &\quad \langle l_m|\left(\bar{\theta} - E^{(1)}\right)|r_o\rangle\omega_k^{-1}\omega_l^{-1}\omega_m^{-1} \\ &\quad - E^{(2)}\sum_{k=1}\langle l_o|\left(\bar{\theta} - E^{(1)}\right)|r_k\rangle\langle l_k|\left(\bar{\theta} - E^{(1)}\right)|r_o\rangle\omega_k^{-2}. \\ &\vdots \end{aligned} \tag{13}$$

which are non-symmetric generalizations of higher-order theory. From the propagator viewpoint [19, 20], we would choose a reference function and a set of intermediate states, hoping to make a consistent choice. In EOM-CC these are the left and right eigenfunctions of $\bar{H}_o$, which is natural, consistent choice.

Alternatively, rather than treating $\bar{H}_o$ as an entity, we can recognize explicitly the form of the CC wavefunctions. In this case we have

$$\begin{aligned} \psi^{(0)} &= e^T\varphi^{(0)} = e^T|0\rangle \\ \psi^{(1)} &= e^T\varphi^{(1)} = e^T T^{(1)}|0\rangle \\ \psi^{(2)} &= e^T\varphi^{(2)} = e^T(\frac{1}{2}T^{(1)}T^{(1)} + T^{(2)})|0\rangle. \end{aligned} \tag{14}$$

From the usual inhomogeneous equations of RSPT, projecting against $\langle 0|(1 + \Lambda)e^{-T} = \langle l_o|e^{-T}$ provides $E^{(1)} = \langle 0|(1 + \Lambda)\bar{\theta}|0\rangle = \langle l_o|\bar{\theta}|r_o\rangle$, just as above. Yet in the second-order inhomogeneous equation,

$$\langle l_o|(E_o - \bar{H}_o)(\frac{1}{2}T^{(1)}T^{(1)} + T^{(2)})|r_o\rangle = \langle l_o|(\bar{\theta} - E^{(1)})T^{(1)}|0\rangle - E^{(2)}\langle l_o|r_o\rangle, \tag{15}$$

which gives

$$E^{(2)} = \langle l_o|\left(\bar{\theta} - E^{(1)}\right)T^{(1)}|r_o\rangle - \frac{1}{2}\langle l_o|(E_o - \bar{H}_o)\left(T^{(1)}\right)^2|r_o\rangle$$
$$= \langle l_o|\left(\bar{\theta} - E^{(1)}\right)\mathcal{R}_o\left(\bar{\theta} - E^{(1)}\right)|r_o\rangle - \frac{1}{2}\langle l_o|(E_o - \bar{H}_o)T^{(1)}\mathcal{R}_o\left(\bar{\theta} - E^{(1)}\right)|r_o\rangle \quad (16)$$

The $T^{(2)}$ disappears from Eqn. (15) since it generates only single and double excitations, and $\langle l_o|$ is the left-hand eigenfunction of $E_o - \bar{H}_o$ in that space.

However, the second term in Eqns (16) causes the expression to differ from that in Eqn (8). This difference arises from the fact that when T is restricted to certain excitation levels like singles and doubles, the product $\left(T^{(1)}\right)^2$ can introduce triples and quadruples, and these have not been included in the excitation space for the CC problem, so they do not vanish upon projection by $\langle l_o|$. If we have only two electrons, then there are no non-vanishing higher excitations and we regain Eqn (8). Similarly, if we do the full CI EOM-CC problem, we also regain Eqn. (8) since the $T^{(1)^2}$ term cannot introduce any more excitations. The appearance of the second term in Eqn. (16) is an example of the generalized Hellman-Feynman theorem *not* being satisfied as defined above, as a component of the second-order wavefunction contributes to the second-order energy. From another viewpoint, the derivative formulas properly introduce the non-Hellman-Feynman corrections. Extensive comparisons of Eqn (16) and Eqn (8) have been made for dynamic polarizabilities [21] and spin-spin coupling constants [22].

We can readily derive the higher-order expressions by simply retaining the nonlinear terms that arise from perturbed wavefunctions in the inhomogeneous equations. Thus,

$$\psi^{(3)} = \mathrm{e}^T\left(\frac{1}{3!}\left(T^{(1)}\right)^3 + T^{(1)}T^{(2)} + T^{(3)}\right)|0\rangle$$
$$\psi^{(4)} = \mathrm{e}^T\left(\frac{1}{4!}\left(T^{(1)}\right)^4 + \frac{1}{2}\left(T^{(1)}\right)^2T^{(2)} + \frac{1}{2}\left(T^{(2)}\right)^2 + T^{(1)}T^{(3)} + T^{(4)}\right)|0\rangle$$
$$\vdots \quad (17)$$

and we have

$$E^{(3)} = \langle l_o|\left(\bar{\theta} - E^{(1)}\right)\mathcal{R}_o\left(\bar{\theta} - E^{(1)}\right)\mathcal{R}_o\left(\bar{\theta} - E^{(1)}\right)|r_o\rangle - E^{(2)}\langle l_o|\mathcal{R}_o(\bar{\theta} - E^{(1)})|r_o\rangle$$
$$- \frac{1}{3!}\langle l_o|(E_o - \bar{H}_o)\left(T^{(1)}\right)^3|r_o\rangle - \langle l_o|(E_o - \bar{H}_o)T^{(1)}T^{(2)}|r_o\rangle \quad (18)$$

and

$$
\begin{aligned}
E^{(4)} =&\langle l_o|\left(\bar{\theta} - E^{(1)}\right)\mathcal{R}_o\left(\bar{\theta} - E^{(1)}\right)R_o\left(\bar{\theta} - E^{(1)}\right)\mathcal{R}_o\left(\bar{\theta} - E^{(1)}\right)|r_o\rangle \\
&- E^{(2)}\langle l_o|(\bar{\theta} - E^{(1)})\mathcal{R}_o^2(\bar{\theta} - E^{(1)})|r_o\rangle - E^{(3)}\langle l_o|\mathcal{R}_o(\bar{\theta} - E^{(1)})|r_o\rangle \\
&- E^{(2)}\langle l_o|R_o(\bar{\theta} - E^{(1)})\mathcal{R}_o(\bar{\theta} - E^{(1)})|r_o\rangle \\
&- \frac{1}{4!}\langle l_o|(E_o - \bar{H}_o)\left(T^{(1)}\right)^4|r_o\rangle - \frac{1}{2}\langle l_o|(E_o - \bar{H}_o)\left(T^{(1)}\right)^2 T^{(2)}|r_o\rangle \\
&- \frac{1}{2}\langle l_o|(E_o - \bar{H}_o)\left(T^{(2)}\right)^2|r_o\rangle - \langle l_o|(E_o - \bar{H}_o)T^{(1)}T^{(3)}|r_o\rangle. \\
&\vdots
\end{aligned} \tag{19}
$$

As these expressions correspond to the CC energy derivative, they must give size-extensive results. However, the price we pay is that the energy of a given order requires wave function contributions of the same order. Furthermore, these non linear terms are difficult to evaluate. The quadartic in $T^{(1)}$ term in second-order, requires comparable difficulty to the quadratic terms in a CCSD calculation

Specifying $\bar{H}_o$ into its open and closed (E_o) part, Eqn. (16) can be written as

$$
\begin{aligned}
E^{(2)} = \langle 0|(1+\Lambda)\bar{H}_{open}\left(\frac{1}{2}T^{(1)^2}\right)|0\rangle - \\
\langle 0|(1+\Lambda)(E^{(1)} - \bar{\theta}^{(1)})T^{(1)}|0\rangle
\end{aligned} \tag{20}
$$

The first term in Eqn. (20) can then be rewritten to be

$$
\frac{1}{2}\langle 0|(1+\Lambda)[[\bar{H}_{open}, T^{(1)}], T^{(1)}]|0\rangle + \langle 0|(1+\Lambda)T^{(1)}\bar{H}_{open}T^{(1)}|0\rangle. \tag{21}
$$

Here we used $\bar{H}_{open}|0\rangle = 0$. With the same separation for $\bar{\theta}$,

$$
\begin{aligned}
\bar{H}_{open}T^{(1)}|0\rangle &= ((\bar{H}_{open}T^{(1)})_{closed} + (\bar{H}_{open}T^{(1)})_{open})|0\rangle \\
&= ((\bar{H}_{open}T^{(1)})_{closed} - \bar{\theta}_{open})|0\rangle \\
&= ((E^{(1)} - \langle\bar{\theta}\rangle - \bar{\theta}_{open})|0\rangle.
\end{aligned} \tag{22}
$$

The second equivalence in the above equation resulted from the fact that $T^{(1)}$ fulfils the CC equation of the first-order with respect to the perturbation $\theta^{(1)}$. The second term of Eqn. (21), is further decomposed as

$$
\begin{aligned}
&\langle 0|(1+\Lambda)T^{(1)}\bar{H}_{open}T^{(1)}|0\rangle \\
&\quad = (E^{(1)} - \langle 0|\bar{\theta}|0\rangle)\langle 0|\Lambda T^{(1)}|0\rangle \\
&\qquad - \langle 0|(1+\Lambda)T^{(1)}\bar{\theta}_{open}|0\rangle \\
&\quad = (E^{(1)} - \langle 0|\bar{\theta}|0\rangle)\langle 0|\Lambda T^{(1)}|0\rangle \\
&\qquad - \langle 0|(1+\Lambda)(\bar{\theta}_{open}, T^{(1)})_{disc}|0\rangle \\
&\quad = E^{(1)}\langle 0|\Lambda T^{(1)}|0\rangle - \langle 0|(1+\Lambda)(\bar{\theta}, T^{(1)})_{disc}|0\rangle.
\end{aligned} \tag{23}
$$

For the last equivalence, we used $(\bar{\theta}, T^{(1)})_{disc}|0\rangle = ((\bar{\theta}_{close} + \bar{\theta}_{open}), T^{(1)})_{disc}|0\rangle$. The second term in Eqn. (20) is

$$- E^{(1)}\langle 0|\Lambda T^{(1)}|0\rangle + \langle 0|(1+\Lambda)[\bar{\theta}, T^{(1)}]_{dis}|0\rangle + \langle 0|(1+\Lambda)[\bar{\theta}, T^{(1)}]|0\rangle. \tag{24}$$

The first and second terms cancel with Eqn. (23) and we finally arrive at

$$E^{(2)} = \langle 0|(1+\Lambda)([\bar{\theta}, T^{(1)}] + \frac{1}{2}[[\bar{H}_o, T^{(1)}], T^{(1)}])|0\rangle. \tag{25}$$

We here see the structure of the second-order energy. All the $\bar{H}_o$, $\bar{\theta}$, and $T^{(1)}$ are connected to each other, but Λ is not necessarily connected to each of those. As pointed out, the term involving $T^{(2)}$ in Eqn. (15) vanishes, as long as the Λ equation is solved in the same excitation manifold as the CC equation. However, the term involving $T^{(1)}T^{(1)}$ in Eqn. (15) creates a larger excitation manifold and cannot be automatically neglected. It contains unlinked and disconnected terms as well as connected terms. In the above derivation, we cancelled the unlinked and disconnected terms with unlinked terms in the second term in Eqn. (20) and arrived at the completely connected expression Eqn. (25). This expression is identical to that in CCLR theory. Since Eqn. (25) has a completely connected structure, it is not possible to have unlinked terms, therefore it is size extensive, and the CCSD second-order property thus calculated is equivalent to the CCSD second-order energy without orbital relaxation. This means the CCSD second-order properties are *exact* for two-electron problems. In the EOM-CC approach, we use the linear ansatz for representing the transition between the states of concern which corresponds to using only the second term in Eqn. (20). With no truncation in the excitation manifold, this is exactly equivalent to Eqn. (25), but in a truncated scheme, it is not precisely equivalent [8]. Rather, using the linear ansatz provides a CI-like scheme for the transition quantity. The non-vanishing contribution from the term involving $T^{(1)}T^{(1)}$ in Eqn. (20) is from the higher excitation manifold and the second term in Eqn (25) must be correspondingly small. As can be seen in Eqn (25), however, the second-order energy of the EOM-CC in its CI-like approximation contains unlinked contributions, so results are *not* exactly size-extensive when the excitation manifold is truncated. The first terms in Eqn. (24) are inherently unlinked since Λ itself has a disconnected structure [9]. We propose several models for evaluating the second-order properties. First, we use Eqn. (25), which provides the exact CC second-order derivative energy without orbital relaxation, as a reference.

Model (I): Evaluate properties by Eqn. (8). This is the EOM-CCSD scheme originally implemented.

Model (II): Evaluate properties by Eqn. (24) after eliminating the first term, then

$$E^{(2)} = \langle 0|(1+\Lambda)\bar{\theta}T^{(1)}|0\rangle \tag{26}$$

Model (II′): Evaluate properties by Eqn. (24) after eliminating the first and second term. Then

$$E^{(2)} = \langle 0|(1+\Lambda)|[\bar{\theta}, T^{(1)}]|0\rangle \tag{27}$$

Model (III): Evaluate properties by Eqn. (26) but with a new Λ' amplitude, which is a solution of the equation without disconnected terms.

$$E^{(2)} = \langle 0|\Lambda'|\bar{\theta}T^{(1)}|0\rangle, \tag{28}$$

where

$$\langle 0|(\Lambda'\bar{H}_o)_{connected}|\mathbf{f}\rangle = 0. \tag{29}$$

In this case, of course, Λ' is not consistent with the stationary condition of Eqn. (1). Model (II) eliminates an apparent unlinked term, so is not completely size-extensive. On the other hand, the models (II′) and (III) do not contain any unlinked terms and are therefore exactly extensive. However, unlike EOM-CCSD itself, none of these models is exact for two-electron systems.

Dynamic Properties

The excitation energy and dynamic properties are evaluated from the time-averaged derivatives of the corresponding time-dependent energy functionals [11, 23–25]. However, a more straightforward way to define dynamic properties is through an expectation value of the corresponding properties over a state *I*:

$$\begin{aligned} P(t) &= \langle I|\theta|I\rangle \\ &= Tr\langle \theta\rho(t)\rangle. \end{aligned} \tag{30}$$

If we choose $\langle I| = \langle l_I|$ and $|J\rangle = |r_I\rangle$ we will get the EOM form. Otherwise, we have a perfectly general form. Here θ is the operator corresponding to the property and the density obeys the equation of motion:

$$[H, \rho(t)] = -i\frac{\partial\rho(t)}{\partial t}. \tag{31}$$

Since our objective is to evaluate the dynamic properties corresponding to a certain process of particular order with respect to an externally applied field rather than the dynamics of the system itself, it is much more convenient to express the density by its Fourier components in the frequency domain. The density is expanded in the Fourier series as

$$\rho(t) = \rho_0 + \int d\omega \rho(\omega) e^{\pm i\omega t} + \frac{1}{2}\int\int d\omega_1 d\omega_2 \rho(\omega_1, \omega_2) e^{\pm i(\omega_1+\omega_2)t} + \frac{1}{6}\int\int\int d\omega_1 d\omega_2 d\omega_3 \rho(\omega_1, \omega_2, \omega_3) e^{\pm i(\omega_1+\omega_2+\omega_3)t} + \ldots. \tag{32}$$

Here, ρ_0 is time independent density matrix and can be defined for initial state I. The excitation of electrons caused by absorption of a single photon is regarded as a polarization of the electron density, which is measured by the linear polarizability $P(-\omega;\omega) = Tr\langle\rho(\omega)\theta\rangle$. The equation of motion for the total system, including the photon field

$$\mathbf{E}(t) = \sum_j E_j \cdot e^{\pm i\omega_j t} \tag{33}$$

of frequency ω_j, is

$$-i\frac{\partial\rho(t)}{\partial t} = [(H + \theta \cdot \mathrm{E}(t)), \rho(t)], \tag{34}$$

and the property θ here is a dipole operator. The process can be regarded as a linear response of the system described by the first-order density $\rho^{(1)}(t) = \int d\omega\rho(\omega)e^{\pm i\omega t}$, where the Fourier component is

$$\rho(\omega) = \frac{[\theta, \rho_0]}{\omega - [H_N, \rho_0]}. \tag{35}$$

The component can be represented over the initial $|\mathrm{I}\rangle$ and intermediate $|\mathrm{J}\rangle$ states of the system defined by H_N, which is a normal-ordered Hamiltonian with respect to the $|\mathrm{I}\rangle$ state:

$$\rho^{JI}(\omega) = \frac{\theta^{JI}}{\omega - \omega_{JI}} \tag{36}$$

and

$$\rho^{IJ}(\omega) = -\frac{\theta^{IJ}}{\omega + \omega_{JI}}. \tag{37}$$

this is to say, a sum-over-state (SOS) representation. Here the excitation energies ω_JI are evaluated from the poles of the first-order density $\rho(\omega)$, while the transition probability is determined by the strength of the density oscillation at the corresponding resonances. Since the linear component of the density can be expressed on the complete set of the basis using the state wavefunctions, the dynamic polarizability is thus expressed as

$$P(-\omega;\omega) = \sum_J \left(\frac{\theta^{IJ}\theta^{JI}}{\omega - \omega_{JI}} - \frac{\theta^{IJ}\theta^{JI}}{\omega + \omega_{JI}}\right), \tag{38}$$

and the transition probability between I and J is proportional to

$$\theta^{IJ} \cdot \theta^{JI}, \tag{39}$$

where the transition moment is defined by

$$\begin{aligned} \theta^{IJ} &= \langle I|\theta|J\rangle \\ &= \mathrm{Tr}(\rho^{\mathrm{IJ}}\theta). \end{aligned} \tag{40}$$

In the EOM-CC theoretical framework, the excitation energy ω_{IJ} between the states I and J is given as eigenvalues of the Hamiltonian $\bar{H}_N = e^{-T}He^T - \langle e^{-T}He^T\rangle$ and the (p,q) element of the transition density matrix ρ^{IJ} between I^{th} and J^{th} eigenstates is defined by

$$\rho_{pq}^{IJ} = \langle 0|l exp(-T)p^{\dagger}q exp(T)r|0\rangle. \tag{41}$$

Here, l_{I} and r_{J} are the Ith left- and the Jth right-hand eigenvectors of the non-Hermitian Hamiltonian $\bar{H}$. The operator is represented on the space spanned by the manifold created by the excitations out of a Hartree-Fock reference determinant, including the null excitation (the reference function). When we calculate the transition probability between a ground state $|\mathrm{g}\rangle$ and an excited state $|\mathrm{e}\rangle$, we need to evaluate ρ^{ge} and ρ^{eg}. The reference function is a right-hand eigenstate of the above Hamiltonian $\bar{H}_o$, $|r_g\rangle = |r_o\rangle = |0\rangle$, and Λ state is a left-hand eigenstate $\langle l_g| = \langle l_o|$, for the ground state. On the other hand, r_{e} and l_{e} are obtained from diagonalization of the $\bar{H}_o$ matrix and fulfill the biorthonormality relation, rather than providing intermediate normalization to the perturbed state. Therefore r_{e}, l_{e} as well as l_{g}, have disconnected structures. Note also that l_{e} has no component of the reference function. The right-hand transition density ρ^{eg} calculated by r_{g} and l_{e} is completely linked and therefore the right-hand transition moment θ^{eg} has a size-intensive nature, even in truncated calculations. On the other hand, r_{e} has a non-zero component r_0 on the reference function and creates unlinked contributions when contracted with l_{g}. The diagrammatic representation of the transition density matrix ρ^{ge} in the CCSD model is summarized in Fig. 1. The regular vertex lines at the bottom correspond to the T amplitudes, the regular vertex lines on top correspond to Λ amplitudes, and the bold vertex line indicates r^e. The terms $\rho_{\mathrm{ij}}(1)$, $\rho_{\mathrm{ij}}(2)$, $\rho_{\mathrm{ab}}(1)$, $\rho_{\mathrm{ab}}(2)$, $\rho_{\mathrm{ia}}(1)$, $\rho_{\mathrm{ai}}(1)$, $\rho_{\mathrm{ai}}(2)$, $\rho_{\mathrm{ai}}(3)$, $\rho_{\mathrm{ai}}(4)$, and $\rho_{\mathrm{ai}}(5)$ have the common

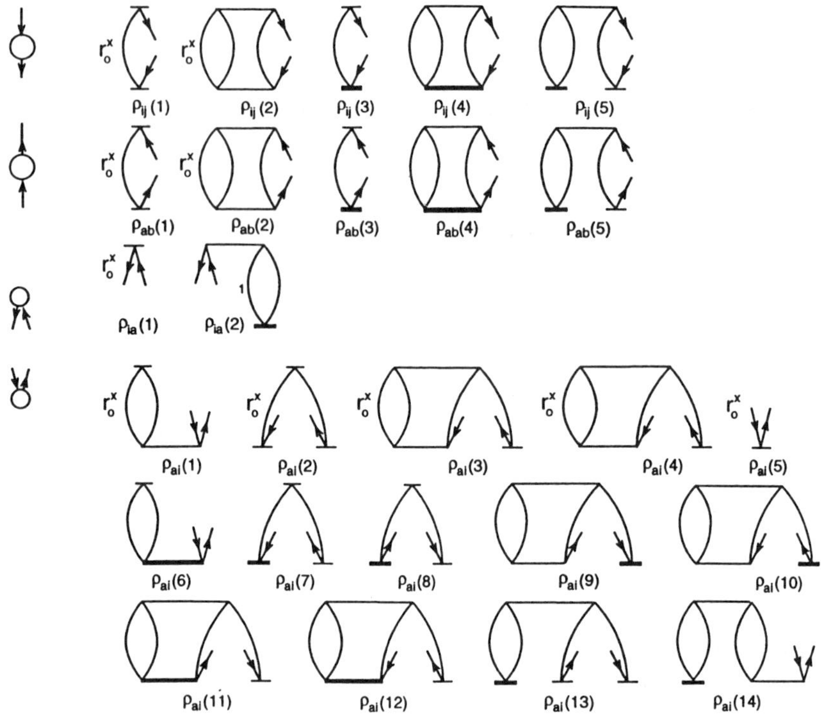
r_o^x
$\rho_{ij}(1)$ $\rho_{ij}(2)$ $\rho_{ij}(3)$ $\rho_{ij}(4)$ $\rho_{ij}(5)$
$\rho_{ab}(1)$ $\rho_{ab}(2)$ $\rho_{ab}(3)$ $\rho_{ab}(4)$ $\rho_{ab}(5)$
1
$\rho_{ia}(1)$ $\rho_{ia}(2)$
$\rho_{ai}(1)$ $\rho_{ai}(2)$ $\rho_{ai}(3)$ $\rho_{ai}(4)$ $\rho_{ai}(5)$
$\rho_{ai}(6)$ $\rho_{ai}(7)$ $\rho_{ai}(8)$ $\rho_{ai}(9)$ $\rho_{ai}(10)$
$\rho_{ai}(11)$ $\rho_{ai}(12)$ $\rho_{ai}(13)$ $\rho_{ai}(14)$

factor r_0, which arises from the above-mentioned normalization of r_e amplitudes and indicates the portion of the reference function in the right-hand excited-state wavefunction. Now we consider a state A^*B, represented by r_J, that consists of noninteractive subsystems: a subsystem A^* being locally excited within A and a subsystem B in its ground state. Since there is no component in the excitation manifold involving B, r_e in the combined system is the same as that of the subsystem A^*. Therefore, the factor r_0 is size intensive. Since the CC ground state is size extensive, the part involving the CC amplitude T directly connected to the perturbation operator (open part in the diagrams) is additive, and the T amplitude in the combined system is $T^A + T^B$. As a result, the terms factored by r_0 are apparently size extensive, not intensive. Elimination of these apparent unlinked terms corresponds to the model (II) in the second-order property evaluation. Since the left-hand amplitude Λ has a disconnected structure (see Fig. 2), the terms $\rho_{ij}(5)$, $\rho_{ab}(5)$, $\rho_{ia}(2)$, $\rho_{ai}(13)$, and $\rho_{ai}(14)$ include unlinked terms implicitly and violate size intensivity of the transition density, too. Simple elimination of the terms leads to a size-intensive theory, which corresponds to the model (II′) in the second-order property evaluation, but it also eliminates quite an amount of relaxation effects accounted for through Λ. Fig. 2 shows the disconnected contributions in the Λ equation. Using the Λ' amplitude instead of Λ while retaining the disconnected terms mentioned above, we can completely eliminate unlinked contributions. This corresponds to the model (III) in the second-order property evaluation. Note also that the unlinked contribution disappears for the property belonging to spatial or spin symmetry other than the totally symmetric ground state. Therefore, the size problem does not arise in the transition probabilities towards excited states of symmetry different from the ground state. This is the case for the dipole-allowed transitions in molecules that do not have permanent dipoles in their ground states. A spin-related perturbation, such as the Fermi contact perturbation, is an another example.

Multi-photon processes involve higher Fourier components of the electron density. For example, the density fluctuation caused by two photons with frequencies ω_1 and ω_2 can be described by $\rho^{(2)}(t) = \int\int d\omega_1 d\omega_2 \rho(\omega_1, \omega_2) e^{\pm i(\omega_1+\omega_2)t}$, which obeys a second-order equation of motion,

$$-i\frac{\partial \rho^{(2)}(t)}{\partial t} = [H, \rho^{(2)}(t)] + [\theta \cdot \mathbf{E}(\mathrm{t}), \rho^{(1)}(t)], \tag{42}$$

therefore,

$$\rho(\omega_1, \omega_2) = \mathrm{P}_{1,2}\frac{[\theta(\omega_2), \rho(\omega_1)]}{\omega_\sigma - [H_N, \rho_0]}. \tag{43}$$

where $\omega_\sigma = \omega_1 + \omega_2$ and $\theta(\omega_i)$ is property operator corresponding to the dynamic perturbation of frequency ω_i. The SOS representation over the initial $|\mathrm{I}\rangle$ and

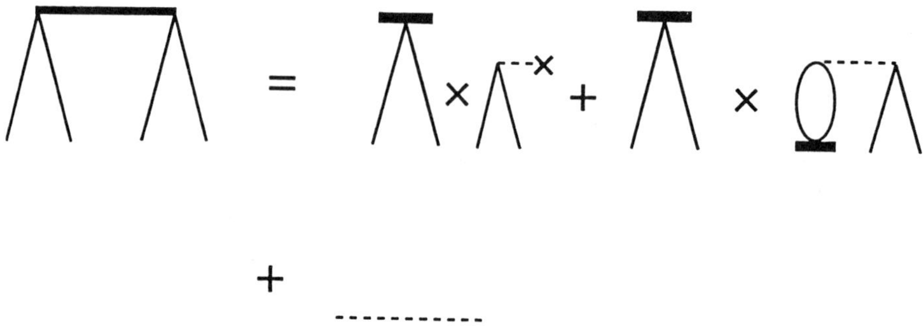

Figure 2

intermediate |J⟩ and |K⟩ states is

$$\rho^{JI}(\omega_1,\omega_2) = \mathrm{P}_{1,2}\sum_K \frac{\theta^{JK}(\omega_2)}{\omega_\sigma - \omega_{JI}} \cdot \frac{\theta^{KI}(\omega_1)}{\omega_1 - \omega_{KI}}, \tag{44}$$

$$\rho^{IJ}(\omega_1,\omega_2) = \mathrm{P}_{1,2}\sum_K \frac{\theta^{IK}(\omega_1)}{\omega_1 + \omega_{KI}} \cdot \frac{\theta^{KJ}(\omega_2)}{\omega_\sigma + \omega_{JI}}, \tag{45}$$

and

$$\begin{aligned}\rho^{JK}(\omega_1,\omega_2) = -\mathrm{P}_{1,2}[&\frac{\theta^{JI}(\omega_2)}{\omega_\sigma - \omega_{JK}} \cdot \frac{\theta^{IK}(\omega_1)}{\omega_1 + \omega_{KI}} \\ &+ \frac{\theta^{JI}(\omega_2)}{\omega_2 - \omega_{JI}} \cdot \frac{\theta^{IK}(\omega_2)}{\omega_\sigma - \omega_{JK}}].\end{aligned} \tag{46}$$

The third expression arises since the density matrix is second-order and the initial state |I⟩ can evolve after the second scattering caused in both bra and ket states caused by the applied field. Further, three-photon absorptions with frequencies ω_1, ω_2, and ω_3 can be described by $\rho^{(3)}(t) = \int\int\int d\omega_1 d\omega_2 d\omega_3 \rho(\omega_1,\omega_2,\omega_3) e^{\pm i(\omega_1+\omega_2+\omega_3)t}$, which obeys a third-order equation of motion,

$$-i\frac{\partial \rho^{(3)}(t)}{\partial t} = [H, \rho^{(3)}(t)] + [\bar{\theta}, \rho^{(2)}(t)] \tag{47}$$

$$\rho(\omega_1,\omega_2,\omega_3) = \mathrm{P}_{1,2,3}\frac{[\theta(\omega_3), \rho(\omega_2,\omega_1)]}{\omega_\sigma - [H_N, \rho_0]}, \tag{48}$$

and the first– and second- hyperpolarizabilities are

$$P(-\omega_\sigma;\omega_1,\omega_2) = Tr\langle \theta\rho(\omega_1,\omega_2)\rangle \tag{49}$$

and

$$P(-\omega_\sigma;\omega_1,\omega_2,\omega_3) = Tr\langle \theta\rho(\omega_1,\omega_2,\omega_3)\rangle, \tag{50}$$

respectively, where $\mathrm{P}_{1,2}$ and $\mathrm{P}_{1,2,3}$ denote the average of all terms generated by simultaneously permuting the frequencies ω_1, ω_2, and ω_3 and the Cartesian components of $\theta(\omega_1)$, $\theta(\omega_2)$ and $\theta(\omega_3)$.

Calculation

The EOM-CCSD (singles and doubles) calculations for the supermolecule of noninteracting LiH are performed using different models. The basis set is taken from the Dunning's double zeta basis set [26], and the experimental bond distance (3.015 au) is employed [27]. Molecules are aligned at a distance of 5000 au. Therefore the symmetry of the super molecules is the same as the individual LiH molecules. For the NMR spin-spin coupling constant calculation of C_2H_6, the [5s2p1d/2s] basis set [28] is used. For the polarizabilities of C_4H_6, we employ [3s3p1d/2s] 6–31G plus polarization and diffuse functions.

Results and Discussion

In Table 1, the static and dynamic polarizabilities of different models described above are summarized. The number in the first column is the CCSD second-order energy without orbital relaxation and can be obtained by using Eqn. (16) or Eqn. (25). The second-order energy thus calculated is size extensive and scales linearly with the number of the noninteracting molecules. The difference from the CCSD second-order energy with orbital relaxation is almost negligible. This is because CCSD can account for a large part of the orbital relaxation effects through the exponential ansatz [10]. The polarizabilities calculated by model I and model II are not size extensive and do not scale linearly with respect to the number of individual molecules. As has been discussed [8,16,27], this may cause problems when those methods are applied for extended or very large molecular systems. The model I implemented originally for the evaluation of second-order energies has unlinked terms because it is of CI type. Yet this model gives the usual perturbation theory form (satisfies the higher analogs of the Hellmann-Feynman theorem). Elimination of these apparent unlinked terms in model II drastically improves the problem. However, model II contains unlinked terms implicitly as explained above and is *not* precisely extensive. Complete elimination of the terms that introduce unlinked terms leads to an exactly size-extensive model (model II′), but this also takes out some important relaxation effects. In model III, these terms that may introduce an unlinked contribution are retained, but the unlinked terms are completely eliminated by using Λ' amplitudes, which is a solution of a completely connected equation. This model provides completely size-extensive second-order properties, but at the cost of introducing Λ' instead of the regular Λ operator. Although some of the EOMCC models are not exactly size-extensive, the calculated polarizabilities for an individual molecule are very close to the CCSD second-order energy with and without orbital relaxation (Table 3).

Table 2 shows transition moments calculated by the different EOM-CCSD models. As has been discussed above, the right-hand transition moment θ^{eg} is size intensive but the left-hand transition moment θ^{ge} in model I and model II is not size intensive. Model II is much improved as far as size intensivity is concerned because of the elimination of the apparent unlinked terms. The apparent unlinked terms are a product of the size-intensive quantity r_0 and size-extensive quantities and therefore are size extensive. The difference between the values of model I and model II, as summarized in the fifth column, reveals strict size extensivity. Complete elimination of unlinked diagrams by using Λ' amplitudes brings strict size intensivity for the transition moment and therefore the transition probabilities calculated by model III are strictly size intensive.

In Table 3, a comparison of several properties calculated by model I and model III are presented. For the calculation of individual molecular properties, both models provide very similar values. A comparison of the CCSD second-

Table 1 EOM-CCSD Dipole Polarizabilities[a] for $(LiH)_n$

Number of LiH	α_{zz} (ω=0)				α_{zz} (ω=0.1)		
	$E^{(2)a}$	I[b]	II[c]	III[d]	I[b]	II[c]	III[d]
1	24.25 (24.23)	24.31[e]	24.37[e]	24.36	63.43[e]	63.65[e]	63.60
2		48.47	48.73		126.38	127.27	
3		72.49	73.08		188.84	190.85	
4		96.37	97.42		250.82	254.39	
5		120.10	121.75		312.32	317.90	

[a]CCSD Energy Second Derivatives. The number in the parenthesis is CCSD with orbital relaxation.

[b]EOM-CCSD

[c]EOM-CCSD without apparent unlinked terms.

[d]EOM-CCSD without unlinked terms. $\alpha_{xx} = 33.24$, $\alpha_{xx}(\omega) = 56.84$. The numbers with elimination of the unlinked terms explained in the text are $\alpha_{zz} = 25.71$ and $\alpha_{xx} = 33.87$.

[e]$\alpha_{xx} = 33.25$ and $\alpha_{xx}(\omega) = 56.86$.

Table 2 EOM-CCSD Transition Moments for $(LiH)_n$

Number of LiH	$\langle\psi_e\|z\|\psi_g\rangle$	$\langle\psi_g\|z\|\psi_e\rangle$			
		I[a]	II[b]	I-II	III[c]
1	.97025	.90915	.91604	.00688	.91758
2	.97025	.90196	.91572 (.00032)	.01377	
3	.97025	.89476	.91541 (.00031)	.02065	
4	.97025	.88757	.91510 (.00031)	.02753	
5	.97025	.88037	.91479 (.00031)	.03442	

[a]EOM-CCSD.

[b]EOM-CCSD without apparent unlinked terms. The numbers in the parenthesis are increments from N-1.

[c]EOM-CCSD without unlinked terms.

Table 3 Second Order Properties Calculated by EOM-CCSD Methods

		I[a]	III[b]	CCSD	Exp.[c]		
$^1J_{c\text{-}c}$ of C_2H_6 [d] (Hz)		35.216	35.216	35.2	34.6		
Polarizabilities of C_4H_6 [e] (a.u.)							
		I[a]	III[b]	CCSD	CCSD(T)	MBPT(2)	HF
Static	xx	73.14	73.14	71.15	72.60	74.54	84.70
	yy	43.34	43.34	42.48	42.96	43.52	42.81
	zz	32.82	32.82	32.02	32.59	33.53	32.21
	xy	5.36	5.36				
ω=0.0656	xx	76.14	76.13				
	yy	44.12	44.12				
	zz	33.64	33.64				
	xy	5.72	5.72				

[a]EOM-CCSD.

[b]EOM-CCSD without unlinked terms.

[c]Ref. [41].

[d]The basis set [5s2p1d/2s] is employed. $R_{C\text{-}C} = 1.531$ Å, $R_{C\text{-}H}$ and $\theta_{H\text{-}C\text{-}H} = 107.8°$.

[e]6-31G + PD ($\zeta_p = \zeta_d = 0.05$) basis set.

order energies and experimental values indicates that both models are very good approximations for isolated molecules.

Size Extensivity of Properties and Connectivity of the Wavefunction

Both EOM-CC through Eqn. (8) or CCLR using Eqn. (25) can provide a size-extensive, second-order total energy. Eqn. (25) does not contain any unlinked terms and Eqn. (8) has no unlinked terms if no truncation is made. However, this does not mean that *all* the excitation energies obtained from the EOM-CC or CCLR are size-intensive. Koch et al. indicated [13] that *all* the excitation energies evaluated in CCLR were equivalent to the excitation energies of a supermolecule that consists of the non-interacting molecules (subsystems). It should be noted, however, that not *all* the excitation energies of the supermolecule are equivalent to those of the individual molecules. These extra excitations correspond to particle number nonconserving excitations in the corresponding subsystems and charge-transfer excitations in the combined system. Meissner and Bartlett proved that pure extensitivity of CCLR would require a generalization of CCLR to include the correct treatment of these nonparticle-number-conserving terms [29].

Let us take an example of an excitation consisting of a removal of an electron from subsystem A followed by an addition to subsystem B. In subsystem A, the CCSD amplitude can have three holes and a particle, while in subsystem B, the amplitude can have a hole and three particles. Therefore, in the supermolecule A-B, we need to have three-hole–three-particle amplitudes T_3 to recover the connectivity. This has been pointed out by Mukherjee et al. [30]. Indeed, the excitation energies corresponding to such charge-transfer states in EOM-CCSD or CCSDLR do not sustain its size-intensive characteristic. Superficially it sounds contradictory to the size extensivity of methods for second-order properties, but the statement that EOM-CC or CCLR provides a size-extensive second-order energy does not mean that all the information obtained from the methods are size-extensive or size-intensive. The reason that the second-order property obtained by summing over all the states from the methods reveals size-extensivity is that though all the intermediate states are *not* size-intensive, these have no intensity (no overlap with the ground state), and consequently provide no contribution to the second-order property. However, it should be noted that in the intermediate region, those charge-transfer-like states have intensity and contribute to the property evaluation. In those theories, although the properties expressed by the wavefunctions are connected, the effective Hamiltonian that provides the excitation energies has an unlinked structure for some excitations and therefore the corresponding wavefunction is disconnected. The extended CC (ECCM) theory proposed by Arponen and Bishop [17,31] provides completely connected wavefunctions, represented by left- and right-hand eigenvectors, and therefore the second-order properties calculated by the wavefunctions are size-extensive. However, in the method, the left- and right-hand vectors couple with each other and the general solution of ECCM are, in practice, very difficult to obtain. It should also be noted that the wavefunctions in the ECCM strictly reveal connectivity for the matrix elements between the ground state and the excited states. The connectivity is not automatically guaranteed in the matrix elements among excited states. The nonlinear response properties are defined as a more than second-order response of the energy functional and can be expressed in the SOS representation using the matrix elements among excited states. Therefore, the nonlinear response properties calculated by the ECCM are *not* size extensive because the ECCM wavefunctions are obtained from a stationary condition of an expectation value of properties over the ground state. To retain the *size extensivity* for the nonlinear response properties, another stationary condition may be imposed [32]. While several CC-based methods have been proposed [33–35] for calculating properties, the Fock-space coupled cluster (FSCC) approaches [36–38] (and the recently introduced STEOM-CC [39]) provide connected wavefunctions and maintain size extensivity. The strategy in FSCC is to evaluate the properties using the eigenvalues and the eigenfunctions of the effective Hamiltonian, which are created in a hierarchical manner. The effective Hamiltonian that provides a single excitation spectrum is constructed in the (1,1) sector. The effective Hamiltonians of lower sectors (1,0) and (0,1) that have been determined contribute as a constant and guarantee the connectivity of the (1,1) effective Hamiltonian. However, the effective Hamiltonian couples with the (1,1) T_2 amplitudes. The nonlinear ansatz thus created introduces the problem of *intruder states* for some cases. A method with a linear excitation ansatz, such as EOM-CC for the effective Hamiltonian $\bar{H}$, does not introduce the intruder state problem, but the wavefunctions for the effective Hamiltonian are *not* connected and cannot retain their size-extensive characteristics, as has been discussed above. An alternative way to avoid the problem may be a modified (dressed) EOM-CC-effective Hamiltonian scheme [40]. The modified EOM-CC effective Hamiltonian $\bar{H}$ for single excitations dressed by the contributions arising from the corresponding FSCC solutions in the intermediate space is completely linked and thus able to provide size extensivity for all the properties produced by the single excitations, while it sustains a linear structure in the operator space for the excitation [28,40]. It should be especially addressed that *not* all the properties produced by single excitations are size-extensive in CCSDLR nor EOM-CCSD. The reason that both methods can provide a size-extensive second-order property is because the excitations that violate extensivity do not have *any* intensity and therefore do not contribute in the summations. However, this is not a desirable feature of the theory. In finite systems, such as molecules, charge-transfer excitations may not be well described in those methods. Specifically, the theories may reveal their inability to correctly describe the charge-transfer excitations in a chemical process such as rotation, where the moieties of the molecule go through a noninteracting path. On the other hand, in extended systems, a size inconsistency of any kind may cause a serious problem for describing phenomena such as density waves or Wannier excitons, where the unit cell does not maintain the electron number.

Conclusion

A comparison of the theory for EOM-CC properties, which empahsize eigenstates and generalized expectation values, and the derivative approach of CCLR has been presented. The usual form of perturbation theory for properties, employ only lower-order wavefunctions in their determination. CCLR involves consideration of wavefunctions of the same order as the energy of interest, but this ensures extensivity of computed properties.

Different models in EOM-CCSD methods are proposed. Model I, originally implemented for transition probabilities and second-order properties, is *exact* for two-electron systems. Of course, the method is equivalent to the CCSDLR and the CCSD second-order energy with and without orbital relaxation for such cases. However, for cases with more than two electrons, the method is different from CCSDLR. The model contains unlinked terms and is therefore *not* size extensive for second-order properties nor size intensive for transition probabilities. In model II, where apparent unlinked terms are eliminated, the problem is diminished, but it is still not exactly size extensive nor size intensive. By complete elimination of unlinked terms using the connected Λ' amplitudes, we have a strictly size-extensive and size-intensive model. However, it should be noted that neither model II nor model III is *exact* for two-electron cases. Calculations of different properties for isolated molecules indicate that all the EOM-CC models are, in practice, accurate and convenient approximations to the CCSD second-order energy derivative calculation and that they reproduce experimental values quite well. Because of its simple form, which retains both size-extensive and size-intensive character, model III is recommended for applications to larger systems. No CCLR or EOMCC type method is completely linked in the sense that each lacks the capability of describing response properties where particle number is not conserved during excitation. It remains to formulate a completely consistent *generalized* coupled-cluster method that will unify perturbation theory and derivative theory for *all* properties for molecules and especially solids.

Acknowledgment

We dedicate this paper to Yngve Ohrn on the occasion of his 65th birthday. One of us (RJB) first learned about propagators from Yngve, and besides being his student, I thank him for bringing me back to join the QTP faculty. This work was supported by the US Office of Scientific Research (Grant No. AFOSR-F49620–95–1–0130). We thank Dr. Ajith Perera for all his help.

References

1. R. J. Bartlett, J. Phys. Chem. **93**, 1697 (1989); M. Urban, I. Cernusak, V. Kello, and J. Noga, in *Methods in Computational Chemistry* 2, edited by S. Wilson (Plenum, New York, 1987); J. Paldus, *Methods in Computational Molecular Physics* NATO, ASI, 1991.
2. K. Emrich, Nucl. Phys. **A351**, 392 (1981)
3. H. Sekino and R.J. Bartlett, Int. J. Quant. Chem. Symp., **18**, 255 (1984)
4. J. Geertsen, M. Rittby and R.J. Bartlett, Chem. Phys. Lett., **164**, 57 (1989)
5. D.C. Comeau and R.J. Bartlett, Chem. Phys. Lett., **207**, 414 (1993)
6. J.F. Stanton and R.J. Bartlett, J. Chem. Phys., **98**, 7029 (1993).
7. J. F. Stanton and R.J. Bartlett, J. Chem. Phys., 98, 9335 (1993).
8. a) H. Sekino and R. J. Bartlett, Chem. Phys. Lett. **225**, 476 (1994) ; b) S. A. Perera, H. Sekino and R. J. Bartlett, J. Chem. Phys., **101**, 2186 (1994)
9. E. A. Salter, G. W. Trucks and R.J. Bartlett, J. Chem. Phys., **920**, 1752 (1989). R. J. Bartlett in *Geometrical Derivatives of Energy Surfaces and Molecular Properties.*, edited by P. Jorgensen and J. Simons, (Reidel) 35–61 (1986). L. Adamowicz, W. D, Laidig and R. J. Bartlett, Int. J. Quant. Chem., **18**, 245, 1984.
10. E.A. Salter, H. Sekino, and R.J. Bartlett, J. Chem. Phys., **87**, 502 (1987); H. Sekino and R.J. Bartlett, Int. J. Quant. Chem. Symp., **21**, 487 (1987)
11. H. J. Monkhorst, Int. J. Quant. Chem. Symp., **11**, 421 (1977); E. Dalgaard and H.J. Monkhorst, Phys. Rev. **A28**, 1217 (1983).
12. D. Mukherjee and P. K. Mukherjee, J. Chem. Phys. **37**, 325 (1979); S. Ghosh, D. Mukherjee and S. N. Bhattacharyya, Mol. Phys. **43**, 173 (1981).
13. H. Koch and P. Jorgensen, J. Chem. Phys. **93**, 3333 (1990); H. Koch, H.J. Aa. Jensen, P. Jorgensen and T. Helgaker, J. Chem. Phys. **93**, 3345 (1990).
14. M. Takahashi and J. Paldus, J. Chem. Phys. **85**, 1486 (1986).
15. R. J. Rico and M. Head-Gordon, Chem. Phys. Lett. **213**, (1993).
16. R. Kobayashi, H. Koch and P. Jørgensen, Chem. Phys. Lett. **219**, 30 (1994).
17. P. Szalay, M. Nooijen and R. J. Bartlett, J. Chem. Phys. **103**, 281 (1995).
18. P. O. Lowdin, J. Math. Phys., **24**, 70 (1983).
19. J. Linderberg and N. Y. Ohrn, *Propagators in Quantum Chemistry*, Academic Press, 1973,
20. J. Oddershede, Adv. Chem. Phys, **67**, 201 (1987).
21. P. Rozyzcko, S. A. Perera, M. Nooijen and R. J. Bartlett, J. Chem. Phys. **107**, 6736 (1997).
22. S. A. Perera, M. Nooijen and R. J. Bartlett, J. Chem. Phys. **104**, 3290 (1996).
23. H. Sambe, Phys. Rev., **A7**, 2203 (1973).
24. K. Sasagane, F. Aiga and R. Itoh, J. Chem. Phys., **99**, 3738 (1993); F. Aiga, K. Sasagane and R. Itoh, ibid., **99**, 3779 (1993) ; F. Aiga, K. Sasagane and

R. Itoh, Int. J. Quant. Chem. **51**, 87 (1994).
25. J. Rice and N.C. Handy, J. Chem. Phys., **94**, 4959 (1991); Int. J. Quantum Chem., **43**, 91 (1992).
26. T. H. Dunning Jr, J. Chem. Phys., **53**, 2823 (1970).
27. H.Koch, R. Kobayashi, A. Sanchez de Merás and P. Jørgensen, J. Chem. Phys., **100**, 4393 (1994); The basis set and the geometry used in Ref 16 and 17 are different from the present work. We used commonly used basis set and equilibrium geometry for calibrating more realistic estimate of the effects.
28. I. Carmichel, J. Phys. Chem., **97**, 1789 (1993)
29. L. Meissner and R. J. Bartlett, J. Chem. Phys., **102**, 7490 (1995).
30. D. Mukherjee and S. Pal, Adv. Quantum Chem., **20**, 291 (1989); D. Mukhopadhyay, S. Mukhopadhyay, R. Chaudhuri and D. Mukherjee, Theor. Cim. Acta, **80**, 441 (1991)
31. J. S. Arponen, R. F. Bishop and E. Pajanne, Phys. Rev. **A36**, 2519, 2539 (1987)
32. N. Vaval, K. B. Ghose and S. Pal, J. Chem. Phys. **101**, 4914, 1194.
33. S. Pal, Phys. Rev., **A33**, 2240; **A34**, 2682 (1986)
34. B. Kundu and D. Mukherjee, Chem. Phys. Lett., **179**, 468 (1991), **A39**, 39 (1989), **A49**, 1623 (1994); Theoret. Chim. Acta, **66**, 151 (1984) ; Chem. Phys. Lett., **211**, 15 (1993)
35. D. Mukherjee, R. K. Moitra and A. Mukhopadhyay, Mol. Phys., **30**, 1861 (1975) ; M. D. Prasad, S. Pal and D. Mukherjee, Pramana, **15**, 531 (1980)
36. A. Haque and D. Mukherjee, J. Chem. Phys., **80**, 5058 (1984); D. Mukherjee, Chem. Phys. Lett., **125**, 207 (1986)
37. S. Pal, M. Rittby, R. J. Bartlett, D. Sinha and D. Mukherjee, J. Chem. Phys. **88**, 4357 (1988)
38. L. Z. Stolarczyk and H. J. Monkhorst, Phys. Rev. **A32**, 725 (1985); L. Z. Stolarczyk and H. J. Monkhorst, Phys. Rev. **A32**, 743 (1985)
39. M. Nooijen and R. J. Bartlett, J. Chem. Phys. **106**, 6441, 6449 (1997), M. Nooijen amd R. J. Bartlett, J. Chem. Phys. **107**, 6812 (1997).
40. L. Meissner and M. Nooijen, J. Chem. Phys. **102**, 9604 (1995).
41. R. M. Lynden-Bell and N. Sheppard, Proc. R. Soc. London Ser., **A269**, 385 (1962)

The Bethe Sum Rule and Basis Set Selection in the Calculation of Generalized Oscillator Strengths

R. Cabrera-Trujillo,[a,b] John R. Sabin,[a,c] J. Oddershede,[a,c] and Stephan P.A. Sauer[d]

a. Chemistry Department, Odense University, 5230 Odense M, Denmark
b. Department of Physics, Universidad Autonoma Metropolitana – Iztapalapa, 09340 Mexico D.F., Mexico
c. Quantum Theory Project, University of Florida, Gainesville, Florida 32611–8435, USA
d. Chemistry Laboratory IV, University of Copenhagen, 2100 Copenhagen Ø, Denmark

Abstract

Fulfillment of the Bethe sum rule may be construed as a measure of basis set quality for atomic and molecular properties involving the generalized oscillator strength distribution. It is first shown that, in the case of a complete basis, the Bethe sum rule is fulfilled exactly in the random phase approximation. For an incomplete (computational) basis, some guidelines are developed for constructing higher angular momentum contributions to bases that will optimize the sum of generalized oscillator strengths and thus make the basis well suited for the calculation of stopping cross sections.

ADVANCES IN QUANTUM CHEMISTRY, VOLUME 35

0065-3276/99 $30.00

1. Foreword

It is a pleasure for us who are friends, colleagues, and collaborators, to offer this contribution to a volume published in honor of Yngve Öhrn's 65th birthday. For most of his career, Yngve has been interested in response properties of various systems to various probes, and we offer this contribution in that spirit. The Generalized Oscillator Strength, the subject of this paper, is the materials property that describes the response of a medium to swift particle, and thus, perhaps, an appropriate subject for this volume. Mostly, we are happy to take this opportunity to thank Yngve for his help, inspiration, and friendship over the years.

2. Introduction

Inelastic collisions of swift, charged particles with matter are completely described by the distribution of generalized oscillator strengths (GOS's) characterizing the collision. These quantities, characteristic of excitation in the N-electron target (or, in fact, of a dressed projectile as well [1]) from some initial state $|0\rangle$ to a final state $|n\rangle$ and concomitant momentum transfer, can be written

$$F_{0n}(\mathbf{q}) = \frac{2\,E_{0n}}{q^2} \left| \langle 0| \sum_{j=1}^{N} e^{-i\mathbf{q}\cdot\mathbf{r}_j} |n\rangle \right|^2 \tag{1}$$

Here $E_{0n} = E_n - E_0$ is the energy exchange and $\mathbf{q}$ is the momentum transfer during the collision. The sum is over all target electrons with coordinates $\mathbf{r}_j$. [Hartree atomic units are used here and through out this paper.] A plot of the generalized oscillator strength as a function of the energy and momentum transferred generates the Bethe surface for the target system, and it is this surface that "...embodies all information concerning the inelastic scattering of charged particles by an atom or molecule within the first Born approximation" [2].

Our particular interest lies in the calculation of the linear energy deposition, or stopping power, of swift ions in materials, $S_0(\mathbf{v})$. In the first Born approximation, and for a fully stripped projectile, this quantity can be written [2–4]

$$S_0(\mathbf{v}) = \frac{4\pi Z_1^2}{v^2} \sum_{n>0} \int_{q_{min}(\mathbf{v})}^{q_{max}(\mathbf{v})} F_{0n}(\mathbf{q}) \frac{dq}{q} \tag{2}$$

Here Z_1 is the charge of the projectile with velocity $\mathbf{v}$. In order to calculate stopping powers for atomic and molecular targets with reliability, however, one must choose a one-electron basis set appropriate for calculation of the generalized oscillator strength distribution (GOSD). The development of reasonable criteria for the choice of a reliable basis for such calculations is the concern of this paper.

3. Basis Set Considerations

3.1. Background

The most common way to obtain a basis is *via* energy optimization. However, it is well known that bases optimized for a particular property such as energy are not always good for calculation of other, perhaps only tangentially related, properties. It is thus useful to have another figure of merit for a basis in connection with a particular application. We outline here a method that may be useful in the context of the calculation of GOS's.

The Bethe approximation [4] to eq. 2 , comprising the optical approximation ($e^{-i\mathbf{q}\cdot\mathbf{r}} \approx 1 - i\mathbf{q}\cdot\mathbf{r}$) and the approximation that the projectile speed (v) is much greater than that of the scattering electrons (v_e) allows one to carry out the integration over q. In this case, the stopping cross section is written

$$S_0(v) = \frac{4\pi Z_1^2 Z_2}{v^2}\left[\ln\frac{2v^2}{I_0} + \frac{C(v)}{Z_2}\right] \tag{3}$$

where

$$I_0 = \sum_{n>0} f_{0n} \ln E_{0n} \tag{4}$$

is the mean excitation energy, the $\{f_{0n}\}$ are the optical (dipole) oscillator strengths, and the C/Z's are the shell corrections which compensate for inner shell electrons not conforming to Bethe assumption concerning low projectile speed. The central quantities in the Bethe formulation are the dipole oscillator strengths, so a figure of basis set merit associated with calculation of dipole transition moments would be appropriate. One quantity used as such by many is the Thomas-Reiche-Kuhn (TRK) [5, 6] sum rule

$$\sum_{n\neq 0} f_{0n} = N \tag{5}$$

which states that *for a complete basis*, the exact dipole oscillator strengths for the system sum to the number of electrons, or in other words, that the energy transfer to the system summed over all internal modes of excitation should be the same as the energy transferred to N free electrons [2]. (The summation symbol implies summation over discrete states and integration over the continuum.) To calculate the dipole oscillator strength distribution (DOSD) necessary to evaluate the TRK sum rule, formal approximations (in this case a truncated perturbation expansion) are introduced. In addition, the basis sets used in the calculation are not complete. Thus two sources of inaccuracy are introduced into the problem. However, it has been shown that the TRK sum rule is fulfilled for complete bases in the random phase approximation (RPA) [7–10], removing the first of these. Thus the TRK sum rule in the RPA has frequently been used as a figure of merit for bases in calculations involving quantities related to dipole properties [11]. It should be noted, however, that the fulfillment of the TRK sum rule is a necessary, but not sufficient, condition for basis set completeness.

3.2. The Bethe Sum Rule

Calculations of stopping cross sections *via* eq. 2 require the GOSD and involve the same two approximations mentioned above for calculations of stopping cross sections. The TRK sum rule, however, applies only when zero momentum is transferred during a projectile/target collision. It is therefore useful only in the optical limit, that is, when one is concerned with the calculation of stopping powers using the Bethe or high velocity approximation (eq. 3), and thus is not sufficient for problems involving finite momentum transfer. Its generalization to the case of non-zero momentum transfer, the Bethe sum rule (BSR), [4]

$$\sum_{n \neq 0} F_{0n}(\mathbf{q}) = N \tag{6}$$

is valid for all kinematically allowed values of momentum transfer, and thus could provide a reasonable figure of merit for basis set choice when dealing with generalized oscillator strength distributions. We examine first the GOS in the RPA and show that the Bethe sum rule is also satisfied in the RPA for a complete basis.

In the random phase approximation, the transition amplitude from state $|0\rangle$ to $|l\rangle$ for any one electron operator $\mathcal{O}$ may be written as

$$\langle 0|\mathcal{O}|l\rangle = \sum_{h,p} \left[\langle p|\mathcal{O}|h\rangle^* X^l_{h,p} + \langle h|\mathcal{O}|p\rangle^* Y^l_{h,p} \right] \tag{7}$$

where $|h\rangle$ and $|p\rangle$ are the complete sets of occupied and unoccupied orbitals, respectively, which are solutions to the ordinary Hartree-Fock equations

$$[h_0 + \mathcal{J} - \mathcal{K}]|i\rangle = \varepsilon_i |i\rangle \tag{8}$$

Here h_0 is the kinetic energy and nuclear attraction operator while $\mathcal{J}$ and $\mathcal{K}$ are the coulomb and exchange operators, respectively. The coefficients X and Y are solutions of the RPA equations, which for the lth singlet transition with excitation energy ω_l can be written as

$$\begin{aligned} \omega_l X^l_{h,p} &= (\varepsilon_p - \varepsilon_h) X^l_{h,p} + \sum_{h',p'} \left[\langle ph'||hp'\rangle X^l_{h',p'} + \langle pp'||hh'\rangle Y^l_{h',p'} \right] \\ -\omega_l Y^l_{h,p} &= (\varepsilon_p - \varepsilon_h) Y^l_{h,p} + \sum_{h',p'} \left[\langle hp'||ph'\rangle Y^l_{h',p'} + \langle hh'||pp'\rangle X^l_{h',p'} \right] \end{aligned} \tag{9}$$

Combining eqs. 7 and 9, and noting that

$$\sum_p |p\rangle\langle p| = 1 - \sum_h |h\rangle\langle h| \tag{10}$$

one can write the matrix element as

$$\begin{aligned} \omega_l\langle 0|\mathcal{O}|l\rangle = \sum_{h,p}\{[(\varepsilon_p - \varepsilon_h)\langle p|\mathcal{O}|h\rangle^* - \langle p|[\mathcal{O},\mathcal{K}]|h\rangle^*]X^l_{h,p} \\ -[(\varepsilon_p - \varepsilon_h)\langle h|\mathcal{O}|p\rangle^* + \langle h|[\mathcal{O},\mathcal{K}]|p\rangle^*]Y^l_{h,p}\} \end{aligned} \tag{11}$$

The terms involving the commutator arise as one-electron operators do not commute with the exchange operator. Using the properties of $\mathcal{J}$ and $\mathcal{K}$, from eqs. 7 and 8 it can be shown that

$$(\varepsilon_p - \varepsilon_h)\langle h|\mathcal{O}|p\rangle = \langle h|[\mathcal{O}, h_o]|p\rangle - \langle h|[\mathcal{O},\mathcal{K}]|p\rangle \tag{12}$$

If relations similar to eq. 12 are substituted into eq. 11, the dependence on the exchange operator can be eliminated, yielding

$$\omega_l\langle 0|\mathcal{O}|l\rangle = \sum_{h,p}\left[\langle p|[h_0,\mathcal{O}]|h\rangle^* X^l_{h,p} + \langle h|[h_0,\mathcal{O}]|p\rangle^* Y^l_{h,p}\right] \tag{13}$$

It is useful to adopt a double vector notation, so that $|\mathbf{X}^l, \mathbf{Y}^l\rangle$ represents the sets of coefficients $\left\{X^l_{h,p}, Y^l_{h,p}\right\}$. In this notation

$$\omega_l\langle 0|\mathcal{O}|l\rangle = \langle \mathcal{C}^*, \tilde{\mathcal{C}^*}|\mathbf{X}^l, \mathbf{Y}^l\rangle \tag{14}$$

where

$$\mathcal{C} = [h_0, \mathcal{O}] \tag{15}$$

Making use of the fact that the energy solutions of eq. 9 come in pairs, one can write a sum rule for the energy weighted transition matrix elements as

$$2\sum_{l>0}\omega_l \mid \langle 0|\mathcal{O}|l\rangle \mid^2 = \sum_{\pm l}\langle \mathcal{C}^*, \tilde{\mathcal{C}^*}|\mathbf{X}^l, \mathbf{Y}^l\rangle\langle \mathbf{X}^l, -\mathbf{Y}^l|\mathbf{O}, -\tilde{\mathbf{O}}\rangle \tag{16}$$

where $\mathbf{O}$ represents the set $\{\langle p|\mathcal{O}|h\rangle\}$ Thus, for any one electron operator, the energy weighted sum of transition matrix elements can be written

$$2\sum_{l>0}\omega_l \mid \langle 0|\mathcal{O}|l\rangle \mid^2 = \langle \mathcal{C}^*, \tilde{\mathcal{C}^*}|\mathbf{O}, -\tilde{\mathbf{O}}\rangle \tag{17}$$

where we have used the normalization condition for the RPA eigenvectors

$$\sum_{\pm l} |\mathbf{X}^l, \mathbf{Y}^l\rangle\langle\mathbf{X}^l, -\mathbf{Y}^l| = 1 \quad . \tag{18}$$

Equation 17 can be viewed as the general form of a sum rule for an arbitrary one-electron operator $\mathcal{O}$ expressed in terms of the square of the transition moment of the operator and its excitation energies.

We treat two special cases.

First consider the dipole operator ($\mathcal{O} = \mathbf{r}$). The matrix elements on rhs of eq. 17 are thus just the dipole transition moments, and the commutator becomes $\mathcal{C} = -i\mathbf{p}$. As the exact solution (complete basis set limit) to the RPA is under consideration, we may use eq. 10 to obtain

$$\sum_{l>0} f_{0l} = i \sum_{h} \langle h|[\mathbf{p}, \mathbf{r}]|h\rangle \tag{19}$$

As $[\mathbf{p}, \mathbf{r}] = -3i$, the TRK sum rule (eq. 5) is immediately recovered.

Now let us consider the plane wave operator $\mathcal{O} = e^{-i\mathbf{q}\cdot\mathbf{r}}$, *i.e.* that which appears in the Bethe sum rule. The matrix elements on the rhs of eq. 17 are those over the plane wave operator. Using a procedure similar to that used in the previous example, and noting that

$$\left[\mathbf{p}, e^{-i\mathbf{q}\cdot\mathbf{r}}\right] = -\mathbf{q}e^{-i\mathbf{q}\cdot\mathbf{r}} \tag{20}$$

the commutator (eq. 15) becomes

$$\mathcal{C} = e^{-i\mathbf{q}\cdot\mathbf{r}}\left(\frac{q^2}{2} - \mathbf{q}\cdot\mathbf{p}\right) \tag{21}$$

Again using the completeness of the particle-hole states (eq. 10), we find that the Bethe sum rule (eq. 6) is fulfilled.

Thus the Bethe sum rule is fulfilled exactly in the RPA at all values of the momentum transferred, provided that a complete basis set is used. Therefore, as in the case of the TRK sum rule when optical transition properties ($\mathbf{q} = 0$) are considered, we expect that the BSR sum rule will be useful in evaluating basis set completeness when generalized oscillator strength distributions are calculated, for example for use in calculating stopping cross sections. It should be noted [12] that the completeness of the computational basis set is dependent on $\mathbf{q}$, and thus care needs be taken to evaluate the BSR at various values of $\mathbf{q}$.

3.3. Criteria for Basis Set Choice

It seems clear that it will be very difficult to satisfy the BSR for an atom or molecule over all allowable momentum transfers using a finite basis of manageable size. One would like, then, to be sure that the BSR is reasonably well satisfied over the range of $\mathbf{q}$ necessary for calculation of the property of interest. In the case of the stopping cross section, inspection of the Bethe surface [2] shows that the largest contribution comes from the low E_{0n}, small $\mathbf{q}$ region of the surface. Thus it is anticipated that it is not necessary to have a completely accurate GOS distribution out to large values of $\mathbf{q}$, and that an acceptable basis would cause the BSR to be satisfied up to some few atomic units in $\mathbf{q}$. In the following, we attempt to develop some guidelines for construction of a calculational basis set that will meet this criterion. We wish to develop relations that can be used to build a set of exponents for higher angular momentum basis functions from a set with lower angular momentum. For example, for an atom, one might choose a standard s-type basis which satisfies the normal energy criteria from the literature, and develop a higher angular momentum basis from it.

In order to determine the relations among orbital exponents in a basis which will follow these guidelines, we look at the matrix elements contributing to $F_{0n}(\mathbf{q})$. To that end we consider eq. 17 for the plane wave operator (eq. 21) which involves evaluation of terms of the sort $\left|\langle p|e^{-i\mathbf{q}\cdot\mathbf{r}}|h\rangle\right|^2$. We wish to determine how these matrix elements behave as a function of orbital exponent and momentum transfer, and we then propose a scheme for choice of orbital exponents that will keep the BSR satisfied to as high momentum transfer as possible.

If we consider a calculation where the orbitals are expanded in a basis

$$|\chi_j\rangle = \sum_k c_{jk}|\varphi_k\rangle \tag{22}$$

of Cartesian Gaussian functions of the form

$$\varphi_k = N_k x^{l_k} y^{m_k} z^{n_k} e^{-\alpha_k r^2} \tag{23}$$

then the integrals we need to evaluate (eq. 1) can all be written as sums of primitive integrals over basis functions

$$\langle p|e^{-i\mathbf{q}\cdot\mathbf{r}}|h\rangle = \sum_{j,k} c^*_{jp} c_{kh} \langle j|e^{-i\mathbf{q}\cdot\mathbf{r}}|k\rangle \tag{24}$$

To understand the criteria for basis set choice, then, we need consider only the behavior of the primitive integrals. The primitive integrals over the basis functions can be expressed in terms of Hermite polynomials

$$\langle j|e^{-i\mathbf{q}\cdot\mathbf{r}}|k\rangle = N'\, e^{-\frac{q^2}{4a}} H_u\left(\frac{q_x}{2\sqrt{a}}\right) H_v\left(\frac{q_y}{2\sqrt{a}}\right) H_w\left(\frac{q_z}{2\sqrt{a}}\right) \tag{25}$$

where $q^2 = q_x^2 + q_y^2 + q_z^2$, $u = l_j + l_k$, $v = m_j + m_k$, $w = n_j + n_k$, with $a = \alpha_j + \alpha_k$, and where

$$N' = N_j N_k \left(\frac{\pi}{a^2}\right)^{\frac{3}{2}} \left(\frac{i}{2\sqrt{a}}\right)^{u+v+w} \tag{26}$$

is a normalization constant. For simplicity, we restrict to momentum transfer along the $z-$axis, and take the square magnitude of the matrix element

$$\left| \langle j|e^{-iq_z z}|k\rangle \right|^2 = N'^2 \left[\frac{u!\,v!}{(\frac{u}{2})!(\frac{v}{2})!}\right]^2 G_w\left(\frac{q_z}{2\sqrt{a}}\right) \tag{27}$$

where

$$G_w\left(\frac{q_z}{2\sqrt{a}}\right) = e^{-q_z^2/2a}\, H_w^2\left(\frac{q_z}{2\sqrt{a}}\right) \tag{28}$$

with u, v even. Note that for the case of excitation out of a reference orbital, increasing w corresponds to increasing the angular momentum quantum number of the virtual orbital. Thus, for the BSR to be fulfilled for each value of q_z, eq. 27 must yield a constant when summed over all transitions. Thus, the compliance of a particular calculation will be determined by the properties of the function $G_w(Q)$.

Let us consider the structure and behavior of the function

$$G_w(Q) = e^{-2Q^2}\, H_w^2(Q) \tag{29}$$

In Fig. 1 we display the function $G_w(Q)$ for degree $w = 0 - 6$, normalized to unity. By differentiating eq. 29 with respect to Q and employing the recursion

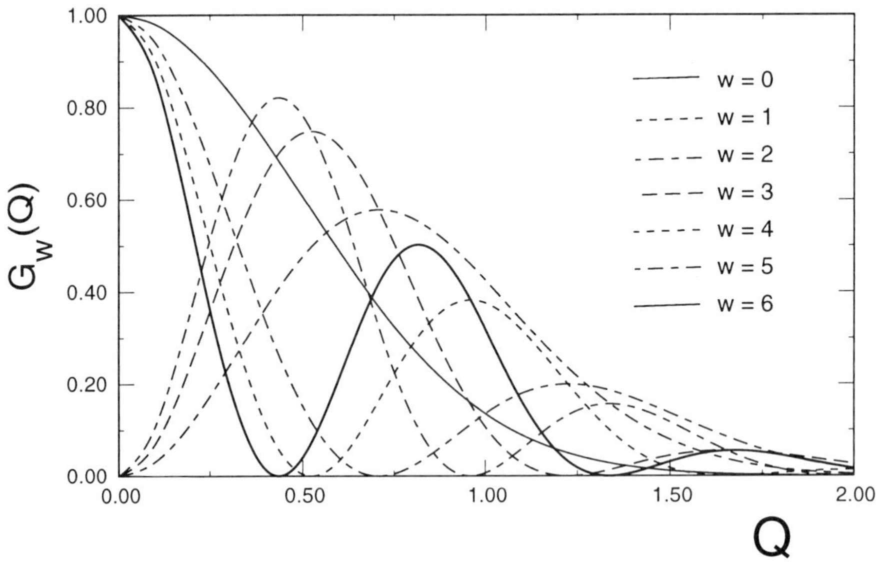

Figure 1: The function $G_w(Q)$ (eq. 29) as a function of Q for $w = 0 - 6$

relations for the Hermite polynomials, one can determine the positions of the extrema of $G_w(Q)$. A few of the lower maxima in the $G_w(Q)$ *vs.* Q are tabulated in Table 1. From Fig. 1 we note that the for lowest extrema in with

Table 1: Positions of the first few (n) maxima (Q^{max}) of $G_w(Q)$

w	n = 1	2	3	4
0	0			
1	0.7071			
2	0	1.2247		
3	0.5247	1.6507		
4	0	0.9586	2.0202	
5	0.4361	1.3359	2.3506	
6	0	0.8163	1.6734	2.6520
7	0.3812	1.1572	1.8917	2.9306

$Q \neq 0$ in $G_w(Q)$, the maxima in $G_w(Q)$ for w odd lie approximately at the same value of Q as do the minima $G_{w+1}(Q)$. In addition, the first $Q \neq 0$ (n=1) maximum in a w even curve lies at approximately twice the value of Q as does the first minimum. These two relations lead to a connection between the positions of the maxima of two curves of order w and $w+1$. In general, we can relate the n maxima by

$$\begin{aligned} \frac{1}{2}\, Q^{max}\Big)^{(n)}_{w} &\approx Q^{max}\Big)^{(n)}_{w+1} \qquad w \; even \\ 2\, Q^{max}\Big)^{(n)}_{w} &\approx Q^{max}\Big)^{(n)}_{w+1} \qquad w \; odd \end{aligned} \tag{30}$$

where $Q^{max}\Big)^{(n)}_{w}$ represents the value of the argument at the nth maximum in $G_w(Q)$. This behavior is dependent on the properties of the function $G_w(Q)$ only.

Let us now consider a transformation of the sort $Q = t_w q$ and ask for the relation of the behavior of $G_w(q)$ to that of $G_w(Q)$. Note that such a transformation may be w dependent, so that the transformation will be different for each value of w. Such a transformation will result in dilation of $G_w(q)$ along the abscissa when compared to $G_w(Q)$, as shown in Fig. 2. Here we have labeled the coordinate Q with w to emphasize that the transformation is different for each order of G.

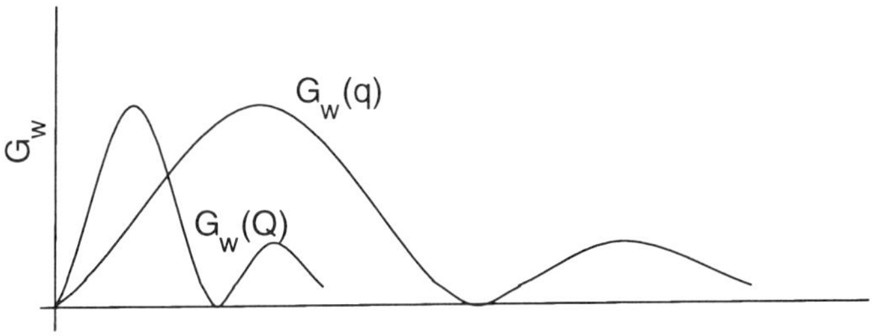

Figure 2: Sketch of the function G_w (w odd) of transformed and untransformed coordinates

As (*c.f.* eqs. 1, 28) the sums of $G_w(q_z)$'s at constant q_z are what will lead to the Bethe sum, the relationship between the positioning of the maxima and minima in $G_w(q_z)$ *via* choice of orbital exponent, can be used to optimize the BSR. As the maxima in $G_w(Q)$ for w odd lie approximately over the minima in $G_{w+1}(Q)$. For such pairings, sketched in Fig. 3 after the transformation, the sum $G_w(q_z) + G_{w+1}(q_z)$ should be maximum and as constant as possible for a given value of q_z. We thus choose the transformation t_w, that is the numerical

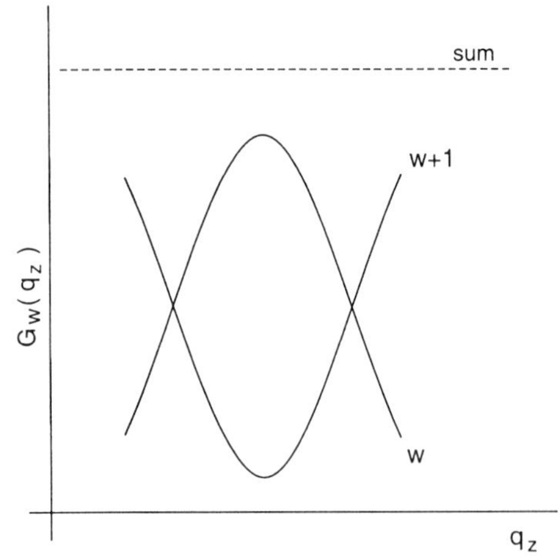

Figure 3: Sketch of the desired relationship between sum of$G_w(q_z)$ and $G_{w+1}(q_z)$ for odd w in order to satisfy thc BSR.

values of α_j and α_k, to shift the curves as in Fig. 2 so that the above mentioned geometrical relation among the $G_w(q_z)$ is preserved. That is, we wish to arrange the α_j and α_k such that the maximum in $G_w(q_z)$ occur at the same value of q_z as does the minimum in $G_{w+1}(q_z)$ (w odd).

The transformation to use is

$$Q_w = \frac{1}{2\sqrt{\alpha_j + \alpha_k}}\, q_z \qquad w = j + k \tag{31}$$

and as the $G_w(q_z)$'s have the same geometrical structure as do the $G_w(Q)$'s, the maxima and minima can be approximately related in the same way, namely

$$\begin{aligned} \frac{1}{2}\; q_z^{max}\Big)_w^{(n)} &\approx\; q_z^{max}\Big)_{w+1}^{(n)} \qquad w\ even \\ 2\; q_z^{max}\Big)_w^{(n)} &\approx\; q_z\Big)_{w+1}^{(n)} \qquad w\ odd \end{aligned} \tag{32}$$

This, together with eq. 31 leads to the relations

$$\begin{aligned} \frac{1}{2}\cdot 2\sqrt{\alpha_j + \alpha_k}\;\; Q^{max}\Big)_w^n &\approx 2\sqrt{\alpha_j + \alpha_{k+1}}\;\; Q^{max}\Big)_{w+1}^n \qquad w\ even \\ 2\cdot 2\sqrt{\alpha_j + \alpha_k}\;\; Q^{max}\Big)_w^n &\approx 2\sqrt{\alpha_j + \alpha_{k+1}}\;\; Q^{max}\Big)_{w+1}^n \qquad w\ odd \end{aligned} \tag{33}$$

As an example, consider an atomic case where the occupied orbital is an s-orbital. Then, due to the symmetry of Gaussians, at $q_z = 0$, $G_w(q_z)$ will be unity for even values of w and will vanish for odd values. For $q_z \neq 0$, all values of w can contribute. For the case of $w = 0$ there is no maximum, so we choose the point where the $w = 0$ and $w = 1$ curves cross. At this point $\frac{q_z}{2\sqrt{\alpha_s+\alpha_k}}$ is the same for the two curves (with $k = s$ for $w = 0$ and $k = p$ for $w = 1$) for the same fixed value of q_z, leading to the condition for the orbital exponents: $\alpha_s = \alpha_p$. Equation 33 in conjunction with the values of the Hermite polynomial roots tabulated in Table 1 lead to a set of relations connecting the orbital exponents of basis orbitals of higher azimuthal quantum number to those in an $l = 0$ set. For example, to determine the d-orbital exponents from those for p–orbitals, eq. 33 gives $\left.\frac{q_z}{2\sqrt{\alpha_s+\alpha_p}}\right)_p^{(1)} = \frac{1}{2}\left.\frac{q_z}{2\sqrt{\alpha_s+\alpha_d}}\right)_d^{(1)}$. From Table 1, the maxima of the p and d curves come at 0.707 and 1.225, respectively, leading to $0.707\sqrt{\alpha_s+\alpha_p} = \frac{1}{2}1.225\sqrt{\alpha_s+\alpha_d}$, or $\alpha_d = 1.67\,\alpha_s$. Following the same procedure for higher angular momentum orbitals, one obtains $\alpha_p = \alpha_s$, $\alpha_d = 1.67\alpha_s$, $\alpha_f = 2.63\alpha_s$, $\alpha_g = 3.35\alpha_s$, $\alpha_h = 4.26\alpha_s$ ... Although these rules are based on relations which apply best to only the lowest roots of $G_w(q_z)$, we choose to maintain these relations for the higher Hermite roots ($n > 1$) as well. Thus we have a set of criteria or guidelines for choice of basis function exponents, such that the exponents for higher angular momentum functions are written in terms of those of the s-orbital set, and should provide a Bethe sum that is reasonably free of wiggles, at least for not too large q_z. It should be noted that as we go to higher Hermite roots, the maxima begin to spread out. Including more roots and choosing exponents giving maxima in G_w closer together can eventually result in linear dependencies.

Finally we observe from Fig. 1 the magnitude of $G_w(q_z)$ decreases for increasing q_z at fixed w. Thus, the only way to fulfill the Bethe sum rule at arbitrarily large values of q_z will be to include basis functions of arbitrarily high angular momentum. This confirms a previously reached conclusion [12].

4. A Numerical Example

As an example, we choose the simple case of He, for which we have previously calculated the GOSD and Bethe sum rule [12]. We consider the [13s,11p,4d,3f] basis of that study (here designated basis A), which was derived from an energy minimized basis of Jaszunski and Roos [13]. The s-orbital exponents from that basis are tabulated in Table 2. We first ran a Hartree-Fock calculation [14], and then choose the core s-orbitals with largest coefficients in the 1s state to form the basis for constructing the higher angular momentum sets. In this case we chose the five orbitals having coefficients larger than 0.1, including orbitals 6s to 10s of Table 2. The relationships among exponents deduced above were then applied, resulting in a [13s,5p,5d,5f] basis (basis B).

Table 2: s-Orbital Basis for Helium [13]

orbitals	exponent
1s	4840.900
2s	723.1100
3s	164.3000
4s	46.63600
5s	15.27800
6s	5.526900
7s	2.132900
8s	0.849670
9s	0.343640
10s	0.138710
11s	0.055484
12s	0.022193
13s	0.008877

Finally, a set of 5 g-orbitals, again, generated according to the relations presented above, was added generating a [13s,5p,5d,5f,5g] basis (basis C). The Bethe sum rule calculated [15] using these bases is presented in Fig. 4. As was noted

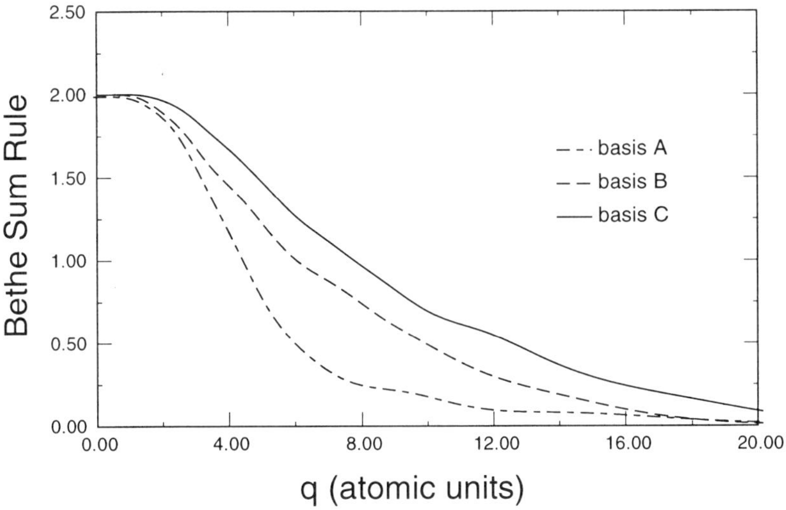

Figure 4: Bethe sum rule for He calculated with a standard basis (basis A) and with two consistent bases (bases B and C) generated with the guidelines suggested here.

earlier, the Hartree-Fock basis (A) constructed on the basis of energy, even though it continues to quite reasonable values of the angular momentum, falls off at rather small q from its theoretical value. In addition, by $q \approx 6$ it has fallen to approximately 25% of the correct value. Changing to the consistent basis constructed using the guidelines above (basis B), we find that the sum rule continues somewhat further out before departing from two, but falls off much more slowly, being approximately 50% of the proper value at $q \approx 6$. However, as we remarked above, it is clear that the basis must contain ever higher angular momentum basis functions. Including the next higher angular momentum state in the basis gives the basis C, and the Bethe sum rule calculated using this basis comes closer yet to fulfilling the Bethe sum rule. As can be seen, the improvement results more from increasing the sum rule at larger values of q than by making it exact at smaller q. It is thus clear that higher angular momentum basis functions are necessary to fulfill the Bethe sum rule at larger values of q.

In addition, it is possible that differences in the BSR may arise due to choice of the s-basis from which the higher angular momentum basis functions are constructed. However, on comparison of the Bethe sum rule for basis C, with the sum rule generated by the same method described above, from another energy optimized, frequently used basis (that of van Duijneveldt [16]), we find only small differences in the BSR, and only at large q values.

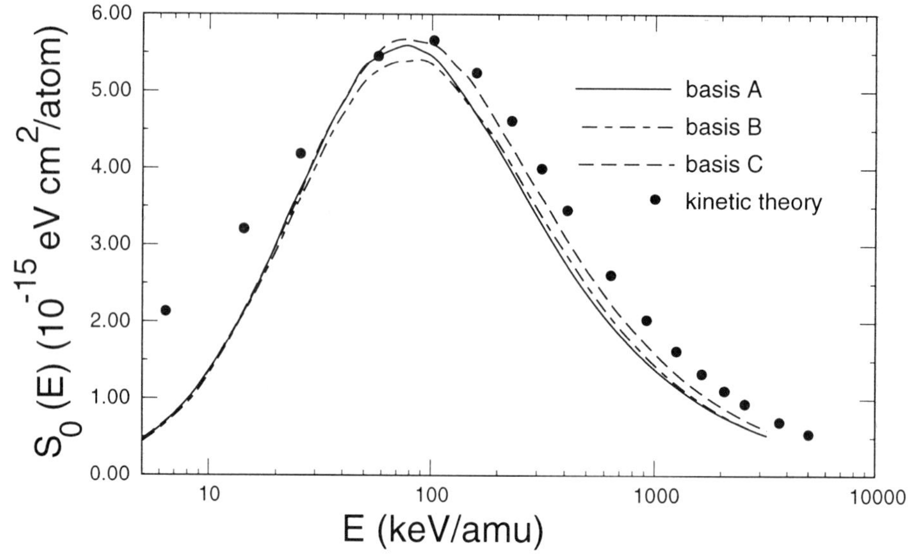

Figure 5: Stopping power for protons on He calculated with the standard basis (basis A), with two consistent bases (B and C), and in the Bethe approximation using the kinetic theory [17, 18].

Finally, the question arises as to whether or not fulfillment of the Bethe sum rule is important for calculation of stopping cross sections. In Fig. 5, we present the stopping cross section as a function of projectile energy for protons on helium, calculated in bases A, B, and C. For purposes of comparison, we also graph the Bethe theory result obtained using the kinetic theory of stopping [17, 18]. All three bases agree well at low projectile energies. At higher energies, there is divergence (*ca.* 10% at $E_p \approx 250\ keV$) with an increase of S_0 when these guidelines are used to construct the higher angular momentum basis functions, and further increase with increasing angular momentum in the basis. If one considers that the stopping cross section is an integral over the Bethe surface (eq. 2), then one would expect that the farther out that the surface is properly represented, the better, and larger, will be the resulting stopping cross section. This is consistent with the results in Fig. 5.

5. Summary

In this contribution, we have shown that the Bethe sum rule, like the Thomas-Reiche-Kuhn sum rule, is satisfied exactly in the random phase approximation for a complete basis. Thus, in calculations that are related to the generalized oscillator strengths of a system, the Bethe sum rule may be used as an indicator of completeness of the basis set, much as the Thomas-Reiche-Kuhn sum rule is for optical properties. In addition, considering that no computational basis sets are complete, and that there are no criteria extant for constructing them for use in generating generalized oscillator strengths, we have proposed a set of guidelines for construction of higher angular momentum basis functions to be included in the basis. These are generated by a set of relations determined from a set of s-orbitals only.

Acknowledgments

This work was supported in part by grants from the U.S. ARO (#DAA-H04-95-1-0326 to JRS), from CONACyT (#1405–E9207 to RCT) and from the Danish Natural Science Research Council (#9600856 and #9313314 to JO).

References

[1] R. Cabrera-Trujillo, S. Cruz, J. Oddershede, and J. Sabin, Phys. Rev. A **55**, 2864 (1997).

[2] M. Inokuti, Rev. Mod. Phys. **43**, 297 (1971).

[3] M. Inokuti, Y. Itikawa, and J. E. Turner, Rev. Mod. Phys. **50**, 23 (1978).

[4] H. Bethe, Ann. Phys. (Leipzig) **5**, 325 (1930).

[5] W. Kuhn, Z. Physik **33**, 408 (1925).

[6] F. Reiche and F. Thomas, Z. Physik **34**, 510 (1925).

[7] A. D. McLachlan and M. A. Ball, Rev. Mod. Phys. **36**, 844 (1964).

[8] R. A. Harris, J. Chem. Phys. **50**, 3947 (1969).

[9] P. Jørgensen and J. Linderberg, Int. J. Quantum Chem. **4**, 587 (1970).

[10] P. Jørgensen and J. Oddershede, J. Chem. Phys. **78**, 1898 (1983).

[11] J. Geertsen, J. Oddershede, and J. R. Sabin, Phys. Rev. A **34**, 1104 (1986).

[12] E. H. Mortensen, J. Oddershede, and J. R. Sabin, Nucl. Instrum. and Meth. B **69**, 24 (1992).

[13] M. Jaszunski and B. O. Roos, Mol. Phys. **52**, 1209 (1984).

[14] DALTON - an electronic structure program, T. Helgaker, H. J. Aa. Jensen, P Jørgensen, H. Koch, J. Olsen, H. Ågren, K. L. Bak, V. Bakken, O. Christiansen, P. Dahle, E. K. Dalskov, T. Enevoldsen, A. Halkjer, H. Heiberg, H. Hettema, D. Jonsson, S. Kirpekar, R. Kobayashi, A. S. de Meras, K. V. Mikkelsen, P. Norman, M. J. Packer, K. Ruud, T. Saue, P. R. Taylor, and O. Vahtras.

[15] For a description of the method of calculation of the BSR, see J.R. Sabin and J. Oddershede, Nucl. Inst. and Meth. **B115**, 79 (1996).

[16] F. B. van Duijneveldt, Gaussian basis sets for the atoms H - Ne for use in molecular calculations, Technical Report Report RJ 945, IBM Research Laboratory San Jose, 1971.

[17] P. Sigmund, Phys. Rev. A **26**, 2497 (1982).

[18] J. Oddershede and J. R. Sabin, Phys. Rev. A **26**, 3209 (1982).

The Molecular Magnetic Shielding Field: Response Graph Illustrations of the Benzene Field[†]

Morten Østergaard Jensen and Aage E. Hansen*
Chemical Laboratory IV, H.C. Ørsted Institute
Universitetsparken 5, DK-2100 Copenhagen Ø, Denmark

Dedicated to Professor Yngve Öhrn
on the occasion of his 65th birthday

[†]Presented in part at International Symposium on Atomic, Molecular, and Condensed Matter Theory, and Computational Methods, 37th Sanibel Symposium, (St. Augustine, Florida, USA, 1-7 March, 1997).

*Author to whom correspondence should be addressed. e-mail: aaeh@rpac.ki.ku.dk

0065-3276/99 $30.00

I. Introduction

When molecularly bound electrons are subjected to a static magnetic induction field, $\mathbf{B}$, their response is to generate an additional field $\mathbf{B}^{in}$ [1], which can be thought of as the result of an electronic current induced by the external magnetic perturbation [2–5]. This response field, the socalled magnetic shielding field, varies across the molecule in a structurally characteristic way, and in NMR spectroscopy magnetic nuclei serve as highly localized chromophores [6], sampling $\mathbf{B}^{in}$at the position of these nuclei. However, as the molecular geometry and the ground state electronic structure are reflected in the overall spatial variation of $\mathbf{B}^{in}$, this field can in itself be considered a molecular response property amenable most directly, perhaps, to theoretical and computational studies. The multipolar treatment of the shielding along the ring axis of the benzene molecule reported in ref. [7] with a view towards the ring current model [8, 9], and the related use of computed shieldings at molecular centers, proposed recently under the name "Nucleus-Independent Chemical Shifts" [10] as indicators of aromatic/antiaromatic character, are examples of theoretical studies of the induced field at manifestly nucleus-free locations for the purpose of electronic or structural information. The multipolar field expansions studied, e.g., in refs. [11–13] are closely related to this perspective, since the presence or absence of a nucleus is incidental to the theoretical developments of these expansions.

In the present contribution, we focus on the dual perspective of $\mathbf{B}^{in}$suggested above, namely as a molecular response field detached from the question of nuclei sampling the field, and as a quantity capable of providing molecular and electronic structure information through computational, rather than experimental, studies, and we advocate the use of response graphs [14] for pictorial illustration of the variation of the field. As a concrete example, we also resort to the benzene molecule, the archetypical aromatic system, and show pictorial representations of the field at various locations, selected to relate the present graphical approach to conventional pictures of the induced field used to rationalize the ring current model. For the computational versus experimental perspective we find it interesting to quote Robert S. Mulliken, who made the following remark in his acceptance speech for the Nobel Prize in 1965 [15]: "In conclusion, I would like to emphasize strongly my belief that the era of computing chemists, when hundreds if not thousands of chemists will go to the computing machine instead of the laboratory for increasingly many facets of chemical information, is already at hand."

More specifically, in Section II we discuss the induced field along the lines of the general formalism of molecular electromagnetic response properties [1, 2], emphasizing its character as a continuous vector field with a corresponding shielding tensor relating induced and applied fields at arbitrary locations. In Section II, we also discuss symmetry aspects of the field as reflected in the shielding tensor. Section III contains a discussion of the characteristics and display of the field, including the shielding vector and its secular part, and intensity measures for the field and its various contributions. The formulation presented in Section II is based on a central (common) magnetic gauge origin, while the numerical results used for the illustration of the field are obtained in the LORG (localized orbital/local origin) ab-initio approach [16–18] where local gauge origins associated with localized molecular orbitals replace the central gauge origin. The working equations for the LORG method are presented in Section IV, again emphasizing the continuous field perspective, and including a modification [19] that simulates results of a common origin gauge approach, for the purpose of the analysis of the results obtained within the LORG approach in terms of magnetic contributions defined relative to a central origin. The numerical and pictorial representations of the magnetic shielding field in the benzene molecule are presented in Section V, including an analysis into sigma and pi electron contributions. Section VI contains the discussion, and a few concluding remarks are added in Section VII.

II. The Induced Magnetic Field

In the Hamiltonian conventionally used for derivations of molecular magnetic properties, the applied fields are represented by electromagnetic vector and scalar potentials [1, 20] and if desired, canonical transformations are invoked to change the magnetic gauge origin and/or to introduce electric and magnetic fields explicitly into the Hamiltonian, see e.g. refs. [1, 20, 21]. Here we take as our point of departure the multipolar Hamiltonian derived in ref. [22] without recourse to vector and scalar potentials.

In a molecule-fixed coordinate system, the multipolar Hamiltonian of ref. [22] takes the form

$$\hat{H} = \frac{1}{2m}\hat{\mathbf{p}}^2 - \frac{q}{2m}\mathbf{B}\cdot[\mathbf{r}\times\hat{\mathbf{p}}] + \frac{q^2}{8m}|\mathbf{B}\times\mathbf{r}|^2 + V(\mathbf{r}) \tag{1}$$

for a bound particle of charge q and mass m in the presence of a uniform static magnetic induction field $\mathbf{B}$. $V(\mathbf{r})$ is the electrostatic binding potential, and we neglect electronic spin angular momentum and nuclear-electronic coupling terms resulting from transformation to a molecule-fixed coordinate system [23]. In eq.(1) the magnetic gauge origin coincides with the coordinate origin, and in order to introduce an arbitrary magnetic gauge origin, this multipolar Hamiltonian can be subjected to the canonical transformation [21],

$$\begin{aligned}\hat{K} &= e^{iX/\hbar}\hat{H}e^{-iX/\hbar} \\ &= \hat{H} + \frac{i}{\hbar}[X, \hat{H}] - \frac{1}{2\hbar^2}[X, [X, \hat{H}]] + \ldots.\end{aligned} \tag{2}$$

with the generator

$$X = -\frac{q}{2}[\mathbf{B} \times \mathbf{R}_O] \cdot \mathbf{r} \tag{3}$$

where $\mathbf{R}_O$ locates the gauge origin. The commutators become

$$\frac{i}{\hbar}[X, \hat{H}] = \frac{q}{2m}\mathbf{B} \cdot [\mathbf{R}_O \times \hat{\mathbf{p}}] - \frac{q^2}{4m}[\mathbf{B} \times \mathbf{R}_O] \cdot (\mathbf{B} \times \mathbf{r}) \tag{4}$$

$$-\frac{1}{2\hbar^2}[X, [X, \hat{H}]] = \frac{q^2}{8m}|\mathbf{B} \times \mathbf{R}_O|^2 \tag{5}$$

applying the commutator relation $[\mathbf{r}, \mathbf{B} \cdot (\mathbf{r} \times \hat{\mathbf{p}})] = i\hbar[\mathbf{B} \times \mathbf{r}]$ along with fundamental commutator relations. The higher order commutators vanish, and the transformed Hamiltonian becomes

$$\hat{K} = \frac{1}{2m}\hat{\mathbf{p}}^2 - \frac{q}{2m}\mathbf{B} \cdot [(\mathbf{r} - \mathbf{R}_O) \times \hat{\mathbf{p}}] + \frac{q^2}{8m}|\mathbf{B} \times (\mathbf{r} - \mathbf{R}_O)|^2 + V(\mathbf{r}) \tag{6}$$

providing the following single particle velocity operator

$$\hat{\mathbf{v}} \equiv \frac{d}{dt}\mathbf{r} = \frac{i}{\hbar}[\hat{K}, \mathbf{r}] = \frac{1}{m}\left\{\hat{\mathbf{p}} - \frac{q}{2}\mathbf{B} \times [\mathbf{r} - \mathbf{R}_O]\right\} \tag{7}$$

For a single electron, the velocity operator in eq.(7) corresponds to the current operator $\hat{\mathbf{j}}(\mathbf{R}_O, \mathbf{B}) = -e\hat{\mathbf{v}}$, and hence to the following operator representing the magnetic field induced at a field point $\mathbf{R}$

$$\begin{aligned}\hat{\mathbf{B}}^{\text{in}}(\mathbf{R}, \mathbf{R}_O, \mathbf{B}) &= -\frac{e\mu_0}{4\pi}\frac{(\mathbf{r} - \mathbf{R}) \times \hat{\mathbf{v}}}{|\mathbf{r} - \mathbf{R}|^3} \\ &= -\frac{e\mu_0}{4\pi m_e}\frac{(\mathbf{r} - \mathbf{R}) \times \hat{\mathbf{p}}}{|\mathbf{r} - \mathbf{R}|^3} - \frac{e^2\mu_0}{8\pi m_e}\frac{(\mathbf{r} - \mathbf{R}) \times [\mathbf{B} \times (\mathbf{r} - \mathbf{R}_O)]}{|\mathbf{r} - \mathbf{R}|^3}\end{aligned} \tag{8}$$

For a many electron system, eq.(8) can be cast into the following dyadic form

$$\hat{\mathbf{B}}^{\text{in}}(\mathbf{R},\mathbf{R}_O,\mathbf{B}) = -\frac{e\mu_0}{4\pi m_e}\sum_s \hat{\mathbf{m}}_s(\mathbf{R}) - \frac{e^2\mu_0}{8\pi m_e}\sum_s \{[(\mathbf{r}_s-\mathbf{R}_O)\cdot\hat{\mathbf{e}}_s(\mathbf{R})]\mathbf{1} - (\mathbf{r}_s-\mathbf{R}_O):\hat{\mathbf{e}}_s(\mathbf{R})\}\cdot\mathbf{B} \tag{9}$$

$\mathbf{1}$ is a unit tensor, and for the s'th electron we have introduced the operators [1].

$$\hat{\mathbf{e}}_s(\mathbf{R}) = \frac{(\mathbf{r}_s-\mathbf{R})}{|\mathbf{r}_s-\mathbf{R}|^3} \tag{10}$$

$$\hat{\mathbf{m}}_s(\mathbf{R}) = \hat{\mathbf{e}}_s(\mathbf{R})\times\hat{\mathbf{p}}_s = \frac{\hat{\mathbf{l}}_s}{|\mathbf{r}_s-\mathbf{R}|^3}, \quad \hat{\mathbf{l}}_s = (\mathbf{r}_s-\mathbf{R})\times\hat{\mathbf{p}}_s \tag{11}$$

Focussing on terms linear in the applied field $\mathbf{B}$, the induced magnetic field at the field point $\mathbf{R}$ obtains as the expectation value of $\hat{\mathbf{B}}^{\text{in}}(\mathbf{R},\mathbf{R}_O,\mathbf{B})$ with respect to the first order wave function corresponding to eq.(6), yielding

$$\begin{aligned}\langle\, \hat{\mathbf{B}}^{\text{in}}(\mathbf{R},\mathbf{R}_O,\mathbf{B})\,\rangle = \\ &- \frac{e^2\mu_0}{8\pi m_e}\langle 0|\sum_s \{[(\mathbf{r}_s-\mathbf{R}_O)\cdot\hat{\mathbf{e}}_s(\mathbf{R})]\mathbf{1} - (\mathbf{r}_s-\mathbf{R}_O):\hat{\mathbf{e}}_s(\mathbf{R})\}\,|0\rangle\cdot\mathbf{B} \\ &+ \frac{e^2\mu_0}{4\pi m_e^2\hbar}\sum_{q\neq 0}\omega_{q0}^{-1}\,Re\left\{\langle 0|\sum_s\hat{\mathbf{m}}_s(\mathbf{R})|q\rangle : \langle q|\sum_s\hat{\mathbf{l}}_s^O|0\rangle\right\}\cdot\mathbf{B} \qquad (12) \\ &\equiv -\left\{\boldsymbol{\sigma}^d(\mathbf{R},\mathbf{R}_O)+\boldsymbol{\sigma}^p(\mathbf{R},\mathbf{R}_O)\right\}\cdot\mathbf{B} \equiv -\boldsymbol{\sigma}(\mathbf{R},\mathbf{R}_O)\cdot\mathbf{B} \qquad (13)\end{aligned}$$

where

$$\hat{\mathbf{l}}_s^O = (\mathbf{r}_s-\mathbf{R}_O)\times\hat{\mathbf{p}}_s \tag{14}$$

reflects the first order perturbation term in eq.(6), and where dependence on field point, $\mathbf{R}$, and gauge origin, $\mathbf{R}_O$, is emphasized in the notation.

In accord with NMR conventions [8, 24], eq.(13) introduces the dimensionless shielding tensor $\boldsymbol{\sigma}(\mathbf{R},\mathbf{R}_O)$, relating the induced and applied fields at the point $\mathbf{R}$, although the term "shielding" probably should be considered a misnomer in this context since we, manifestly, make no assumptions about presence or absence of a magnetic nucleus being shielded. Eq.(13) also introduces a separation of the shielding tensor into diamagnetic and paramagnetic contributions, denoted by superscripts d and p, respectively. Explicit reference to the gauge origin $\mathbf{R}_O$ is included in the notation for the three tensors, since the total tensor may be gauge dependent in approximate calculations [24], and the separation into diamagnetic and paramagnetic contributions is gauge dependent, regardless of computational approach and accuracy. In eqs.(12,13), diamagnetic and paramagnetic hence simply designate a ground state expectation term and a sum-over-states term, respectively.

The shielding tensor, and its diamagnetic and paramagnetic components, are not necessarily symmetric in the Cartesian indices [25–29], and the shielding tensor can in general be decomposed into a symmetric and an antisymmetric component, i.e.

$$\boldsymbol{\sigma} = \boldsymbol{\sigma}^s + \boldsymbol{\sigma}^a \tag{15}$$

$$= \sigma^{\text{iso}}\mathbf{1} + \Delta\boldsymbol{\sigma}^s + \boldsymbol{\sigma}^a \tag{16}$$

yielding a maximum of 6 distinct elements of the Cartesian form $\sigma^s_{ij} = \frac{1}{2}(\sigma_{ij} + \sigma_{ij})$ for the symmetric tensor, and a maximum of 3 distinct elements of the form $\sigma^a_{ij} = \frac{1}{2}(\sigma_{ij} - \sigma_{ij})$ for the antisymmetric tensor. In eq.(16) we have separated the symmetric tensor into an isotropic part, $\sigma^{\text{iso}} = 1/3Tr(\boldsymbol{\sigma}^s)$, and a traceless anisotropy tensor $\Delta\boldsymbol{\sigma}^s$ with a maximum of 5 distinct elements. The elements of the antisymmetric tensor can be collected into an axial antisymmetry vector $\boldsymbol{\xi}$ with the Cartesian components [25]

$$\xi_k \equiv \frac{1}{2}\sum_{ij} \epsilon_{ijk}\sigma^a_{ij} \tag{17}$$

where ϵ_{ijk} is an element of the Levi-Civita alternating tensor. If desired, the decompositions represented by eqs.(15-17) can also be applied to the diamagnetic and paramagnetic tensors individually.

The expressions for the shielding tensor and its components in eqs.(12,13) and (15-17) suggest symmetry analyses from several perspectives. Choosing the coordinate origin at the molecular center-of-mass and coordinate axes reflecting the overall molecular point group symmetry, the paramagnetic contributions can be decomposed into electronic spectroscopic symmetry species, as illustrated in ref. [28] within the LORG approach. Focussing on the site symmetry at the location of a given nucleus, selection rules for the Cartesian elements of the symmetric and antisymmetric components of the shielding tensor were derived in ref. [25], see also refs. [28, 29]. In ref. [25], the coordinate system and gauge origin are located at a nucleus of interest, i.e. $\mathbf{R} = \mathbf{R}_O = \mathbf{0}$ in the notation of eqs.(12,13), in which case the diamagnetic tensor is symmetric, regardless of the site symmetry. For general locations of a field point $\mathbf{R}$

relative to the molecular coordinate system, the results listed for the paramagnetic contributions in ref. [25] apply for the Cartesian elements of the total tensor expressed relative to central coordinate system, assuming the direction of the coordinate axes appropriate for the site symmetry group at the point $\mathbf{R}$ coincide with the direction of the axes of the central system. For the antisymmetric part of the shielding tensor in particular, the axial vector character of eq.(17) implies that all three components of the antisymmetry vector are non-vanishing at field points of C_1 and C_i site symmetry. In addition, the results obtained in ref. [25] show that all components of the antisymmetry vector vanish for field points located at the intersection of two or more symmetry planes, while field points at C_n and C_{nh} site symmetry allow a non-vanishing component of the vector along the C_n axis, and field points at C_s site symmetry allow a non-vanishing component perpendicular to the symmetry plane. For the diamagnetic and paramagnetic tensors, selection rules and index symmetry will depend on the location of the gauge origin $\mathbf{R}_O$ in addition to site symmetry relative to $\mathbf{R}$.

III.Field Characteristics and Display

The dependence of the induced field $\mathbf{B}^{\text{in}}$, eq.(12), on the applied field $\mathbf{B}$, of course simply implies that we have a new induced field for each magnitude and direction of $\mathbf{B}$. For the purpose of interpretation and display, the linear dependence on the magnitude of $\mathbf{B}$ is circumvented by considering the dimensionless shielding vector which is defined as the negative of the induced field per unit applied field [14, 28, 30], i.e.

$$\mathbf{T}_B(\mathbf{R};\mathbf{u}_B) \equiv -\frac{\langle\, \hat{\mathbf{B}}^{\text{in}}(\mathbf{R},\mathbf{R}_O,\mathbf{B})\,\rangle}{B} = \boldsymbol{\sigma}(\mathbf{R})\cdot\mathbf{u}_B \tag{18}$$

$$= \boldsymbol{\sigma}^s(\mathbf{R})\cdot\mathbf{u}_B - \boldsymbol{\xi}(\mathbf{R})\times\mathbf{u}_B \tag{19}$$

$$= \sigma^{\text{iso}}(\mathbf{R})\mathbf{u}_B + \Delta\boldsymbol{\sigma}^s(\mathbf{R})\cdot\mathbf{u}_B - \boldsymbol{\xi}(\mathbf{R})\times\mathbf{u}_B \tag{20}$$

where $\mathbf{u}_B$ is a unit vector along the direction of the applied field, B is the magnitude of the field, and we suppress the gauge reference in shielding vector and tensors at this point. Note that the shielding vector, eq.(18), follows the sign of the shielding tensor in accord with our original treatment of this quantity [28, 30], implying that positive shielding makes the direction of the shielding vector oppose that of the applied field. Eqs.(19,20) show that while the symmetric part of the tensor in general scales and rotates the shielding vector relative to the applied field, the component of the shielding vector generated by the antisymmetric part of the tensor is strictly orthogonal to the applied field as well as to the antisymmetry vector. From eq.(18) we can define a dimensionless intensity of the shielding vector, i.e. a measure of the intensity of the induced field per unit intensity of the applied field, as

$$I_B(\mathbf{R};\mathbf{u}_B) = T_B(\mathbf{R};\mathbf{u}_B)^\dagger \cdot T_B(\mathbf{R};\mathbf{u}_B) = \mathbf{u}_B^\dagger \cdot \left\{\boldsymbol{\sigma}(\mathbf{R})^\dagger \boldsymbol{\sigma}(\mathbf{R})\right\} \cdot \mathbf{u}_B \tag{21}$$

where the dagger denotes transposition. The decompositions of the shielding tensor into diamagnetic and paramagnetic parts, or into the components in eqs.(15-17), allow corresponding (additive) break-downs of the shielding vector, eqs.(18-20). Eq.(21) similarly allows the evaluation of intensities corresponding to the various decompositions; however by the quadratic nature of the intensity expressions, cross terms will in general vitiate additivity.

In an NMR context, i.e. when the field point coincides with the position of a magnetic nucleus, the implications of the separation of the response fields in eqs.(18-20) into parts generated by the various components of the shielding tensor, are appreciated by first noting that the resonance frequency for the nucleus can be written as [28]

$$\omega_N(\mathbf{u}_B) = B\gamma_N[1 - T_B(\mathbf{R}_N;\mathbf{u}_B)] \tag{22}$$

to first order in the components of the shielding tensor. Here γ_N is the gyromagnetic ratio of the nucleus under consideration, $\mathbf{R}_N$ is the position vector of the nucleus, and

$$T_B(\mathbf{R}_N;\mathbf{u}_B) \equiv \mathbf{u}_B^\dagger \cdot \boldsymbol{\sigma}(\mathbf{R}_N) \cdot \mathbf{u}_B = \mathbf{u}_B^\dagger \cdot \boldsymbol{\sigma}^s(\mathbf{R}_N) \cdot \mathbf{u}_B \tag{23}$$

$$= \{\sigma_{xx}\cos^2\phi + \sigma_{yy}\sin^2\phi + \sigma_{xy}^s \sin 2\phi\}\sin^2\theta + \{\sigma_{xz}^s\cos\phi + \sigma_{yz}^s\sin\phi\}\sin 2\theta + \sigma_{zz}\cos^2\theta \tag{24}$$

is the projection of the shielding vector $\mathbf{T}_B(\mathbf{R}_N,\mathbf{u}_B)$, eq.(18), onto the direction of the applied field. θ and ϕ are spherical polar angles of the unit vector $\mathbf{u}_B$, relative to the axes directions of the Cartesian system. Eq.(22) shows how the nuclear magnetic resonance frequency for this nucleus varies as the applied field samples all directions in space relative to the fixed molecular framework. As apparent in eqs.(23,24), only the part of the field that is generated by the symmetric component of the shielding tensor contributes to the position of the resonance line, as consequence of the symmetric sampling invoked by the projection in eq.(23), and the field corresponding to the projected shielding vector $T_B(\mathbf{R}_N;\mathbf{u}_B)$ in eqs.(23,24) is referred to as the secular part of the shielding [27].

The part of the field that is generated by an antisymmetric component of the shielding tensor therefore makes no observable contribution to the position of the resonance. However, an antisymmetric shielding component contributes along with the traceless part of the symmetric tensor, $\Delta\boldsymbol{\sigma}^s$, to the mechanisms causing relaxation, as discussed in ref. [29]. In fact, for rapidly tumbling molecules, the T_1 and T_2 relaxation times are governed by the isotropic parts of the intensity quantities $I_B^{\Delta\sigma}(\mathbf{R};\mathbf{u}_B)$ and $I_B^a(\mathbf{R};\mathbf{u}_B)$, obtained from eq.(21) using the tensors $\Delta\boldsymbol{\sigma}^s$ and $\boldsymbol{\sigma}^a$ respectively, and the scalar quantities

$$||\Delta\boldsymbol{\sigma}^s(\mathbf{R})|| = \left\{Tr[\Delta\boldsymbol{\sigma}^s(\mathbf{R})^\dagger\Delta\boldsymbol{\sigma}^s(\mathbf{R})]\right\}^{\frac{1}{2}} \equiv \sqrt{2}S \tag{25}$$

$$||\boldsymbol{\sigma}^a(\mathbf{R})|| = \left\{Tr[\boldsymbol{\sigma}^a(\mathbf{R})^\dagger\boldsymbol{\sigma}^a(\mathbf{R})]\right\}^{\frac{1}{2}} = \sqrt{2}|\boldsymbol{\xi}(\mathbf{R})| \equiv A \tag{26}$$

hence serve as measures of the fields induced by the traceless anisotropy and the antisymmetry, respectively; see ref. [29] for the S and A notation used in the theory of relaxation.

In the display of the field in Section V, we apply the above separation into parts generated by the various components of the shielding tensor. The secular part of the shielding vector, eqs.(23,24), and the intensity quantities $I_B^{\Delta\sigma}(\mathbf{R};\mathbf{u}_B)$ and $I_B^a(\mathbf{R};\mathbf{u}_B)$, obtained from eq.(21) as discussed above, lend themselves immediately to the response graph technique described in ref. [14], as illustrated in Figures (1-3,5), while the antisymmetry vector is illustrated in Figure 4.

IV. Computational Approach

The LORG (localized orbital/local origin) method [16–18] is an ab-initio approach at the Coupled Hartree-Fock level, i.e. correct to first order in the fluctuation potential [31]; its extension to second order is referred to as SOLO [32, 33]. The LORG method is a member of a family of methods where the central gauge origin in the expressions for the field and tensors in eqs.(12,13) is replaced by local gauge origins, with the result that convergence towards complete basis set limits is accelerated significantly relative to the notoriously slow basis set convergence characteristic of common origin methods [34, 35], and gauge dependence is removed or reduced significantly; see refs. [36–38] for representative accounts and later developments. In our experience with shielding calculations for the benzene molecule [17, 33], second order effects are small, and we therefore remain at the first order level.

In previous presentations [16–19, 28], the LORG equations are formulated in a nucleus centered coordinate system. Explicit reference to a field point $\mathbf{R}$, can be introduced following eq.(13), and the resulting LORG equations for the i, j'th element of the shielding tensor become

$$\begin{aligned}\sigma_{ij}^{(1)}(\mathbf{R},\mathbf{R}_\alpha) = \frac{1}{c^2}\sum_\alpha\Big\{&\langle\alpha|(\mathbf{r}-\mathbf{R}_\alpha)\cdot\hat{\mathbf{e}}(\mathbf{R})\delta_{ij} - [\mathbf{u}_i\cdot(\mathbf{r}-\mathbf{R}_\alpha)][\mathbf{u}_j\cdot\hat{\mathbf{e}}(\mathbf{R})]|\alpha\rangle \\ &+ 2i\sum_\beta[\mathbf{u}_i\cdot\langle\alpha|\hat{\mathbf{m}}(\mathbf{R})|\beta\rangle][\mathbf{u}_j\cdot(\mathbf{R}_\alpha\times\langle\beta|\mathbf{r}|\alpha\rangle)]\Big\} \\ -\frac{2}{c^2}\sum_\alpha\Big\{&\sum_m\sum_{\beta n} t[\mathbf{u}_i\cdot\hat{\mathbf{m}}(\mathbf{R})]^{(1)}_{\alpha m}\Omega^{(1)}_{\alpha m,\beta n}t[\mathbf{u}_j\cdot\hat{\mathbf{l}}^{(\alpha)}]^{(1)}_{n\beta}\Big\} \end{aligned} \quad (27)$$

$$\begin{aligned}&\equiv \sum_\alpha \sigma^I_{ij}(\mathbf{R},\mathbf{R}_\alpha;\alpha) + \sum_\alpha \sigma^{II(1)}_{ij}(\mathbf{R},\mathbf{R}_\alpha;\alpha) \\ &\equiv \sigma^I_{ij}(\mathbf{R};\mathrm{LORG}) + \sigma^{II(1)}_{ij}(\mathbf{R};\mathrm{LORG}) = \sigma_{ij}(\mathbf{R};\mathrm{LORG})\end{aligned} \quad (28)$$

Here $\mathbf{u}_i$ and $\mathbf{u}_j$ are Cartesian unit vectors, $|\alpha\rangle$ and $|\beta\rangle$ are localized orbitals that are doubly occupied in the HF ground state, $|m\rangle$ and $|n\rangle$ are virtual orbitals. $\mathbf{R}_\alpha$ is the position vector of the local gauge origin assigned to orbital $|\alpha\rangle$ and $\hat{\mathbf{l}}^{(\alpha)} = (\mathbf{r}-\mathbf{R}_\alpha)\times\hat{\mathbf{p}}$ is the angular momentum relative to $\mathbf{R}_\alpha$. Superscript 1 denotes terms to first order in the fluctuation potential, and $[\mathbf{\Omega}^{(1)}]^{-1} = [\mathbf{A}^{(1)} - \mathbf{B}^{(1)}]$ is the principal propagator at the zero energy limit [18, 31], while $t[\mathbf{u}_i\cdot\hat{\mathbf{m}}(\mathbf{R})]^{(1)}_{\alpha m}$ and $t[\mathbf{u}_j\cdot\hat{\mathbf{l}}^{(\alpha)}]^{(1)}_{n\beta}$ stand for transition moments corresponding to the operators $\hat{\mathbf{m}}(\mathbf{R})$ and $\hat{\mathbf{l}}^{(\alpha)}$ respectively. In eq.(28), the resulting tensor is expressed as sums of single orbital contributions, allowing useful decompositions into structural terms, such as bond or group contributions. The terms labelled I contain contributions from occupied orbitals only, while the terms labelled II contain contributions from virtual orbitals as well.

Classical notions of magnetism traditionally assume a gauge origin at a molecular symmetry center, normally center-of-mass, for the division into diamagnetism and paramagnetism [39, 40], while the corresponding terms for nuclear shielding traditionally assume a gauge origin at the nucleus under consideration [8, 24, 25]. The local gauge structure will not in general allow interpretation of the LORG expressions, eqs.(27,28), in classical magnetic terms. However, as discussed in ref. [19], numerical results that simulate shielding results obtained in common origin calculations can in fact be obtained within the LORG technology, and, if desired, the connection to classical magnetic notions can then be established by an appropriate choice of this common origin.

The required manipulation follows by observing that the numerical problem in a common origin calculation lies in the evaluation of the paramagnetic term in eqs.(12,13). The corresponding diamagnetic shielding contribution can be calculated quite accurately from $\sigma^{I}_{ij}(\mathbf{R};\mathrm{LORG})$, eq.(28), in basis sets of the quality typically used for LORG calculations, using $\mathbf{R}_\alpha = \mathbf{R}_O$ for all α, i.e.

$$\sigma^{d}_{ij}(\mathbf{R};\mathbf{R}_O) = \sum_{\alpha} \sigma^{I}_{ij}(\mathbf{R},\mathbf{R}_\alpha = \mathbf{R}_O;\alpha) \tag{29}$$

A common origin paramagnetic contribution can subsequently be obtained as the difference between the LORG total shielding tensor and the diamagnetic contribution from eq.(29), i.e.

$$\sigma^{p}_{ij}(\mathbf{R};\mathbf{R}_O) \equiv \sigma^{(1)}_{ij}(\mathbf{R};\mathrm{LORG}) - \sigma^{d}_{ij}(\mathbf{R};\mathbf{R}_O) = \sum_{\alpha} \sigma^{II}_{ij}(\mathbf{R},\mathbf{R}_\alpha = \mathbf{R}_O;\alpha) \tag{30}$$

As indicated, the possibility of orbital decomposition is retained in the diamagnetic and paramagnetic terms in eqs.(29,30). The value for the paramagnetic shielding contribution extracted via eq.(30) does not, of course, correspond to a well-defined basis set, but is at the same level of numerical quality as the LORG calculation used for the total shielding.

V.The Magnetic Shielding Field in Benzene

The LORG calculations in the present study of the shielding field in the benzene molecule were performed with Version11.X of Program RPAC [41], and the response surface graphs shown in the figures are produced with the GRAPHICS modules in RPAC.11X; Program Gaussian 94 [42] generated the SCF results. We used the experimental geometry and a [3s3p1d/2s1p] atomic basis set [16], yielding a RHF ground state energy of -230.74549968 a.u. All shielding results reported here refer to nucleus-free field points, i.e. locations of relatively low electron density. In this case the socalled FULL LORG option, where all local origins are identified with centroids of the localized orbitals, i.e.

$$\mathbf{R}_\alpha \equiv \langle\alpha|\mathbf{r}|\alpha\rangle \qquad \forall \quad \alpha \tag{31}$$

is the appropriate gauge choice [16]. For the localization of the molecular orbitals, the core and sigma canonical orbitals were block localized using straightforward Foster-Boys localization [16, 43], while the pi orbitals were left delocalized, the resulting orbital scheme respecting the symmetry of the molecule.

The decompositions into diamagnetic and paramagnetic contributions reported below are obtained using eqs.(29,30), identifying the common gauge origin $\mathbf{R}_O$ with the center of the ring, alias the center-of-mass $\mathbf{R}_{\mathrm{COM}}$, and the additional decompositions into core+sigma and pi contributions are welldefined since the localized orbitals respect symmetry with respect to the plane of the molecule.

Numerical results for the shielding field of the benzene molecule are collected in Table 1 for the center of the molecule (labelled COM), and for points along a quarter circle of radius 2.47 Å from the z-axis to the x-axis, see Figure 3 for specification of axes. The radius of the circle corresponds to the distance from the ring center to a proton but, as defined, the points lie in the entirely nucleus-free xz-plane. Except for COM, the entries in the table are labelled by the angle between the z-axes and the direction to the field point. The table includes the isotropic part of the shielding, σ^{iso}, and the principal

Table 1 Isotropic and Principal Value Shieldings, and Intensity Measures for the Benzene Molecule. See text.

		σ^{iso}	σ_{11}	σ_{22}	σ_{33}	$\|\|\Delta\boldsymbol{\sigma}^s\|\|$	$\|\|\boldsymbol{\sigma}^a\|\|$
0	c+s	3.1	2.6	2.6	4.2	1.3	0.0
	pi	1.8	-1.8	-1.8	8.9	8.7	0.0
	total	4.9	0.8	0.8	13.1	10.0	0.0
$\pi/6$	c+s	3.3	1.9	2.9	5.2	2.4	1.4
	pi	1.0	-3.5	-2.3	8.7	9.5	2.8
	total	4.3	-0.6	-0.4	13.9	11.8	4.2
$\pi/3$	c+s	1.4	-1.3	0.3	5.2	4.8	6.5
	pi	0.7	-5.3	-0.5	8.0	9.4	0.3
	total	2.1	-5.0	-1.8	13.2	13.7	6.8
$\pi/2$	c+s	-3.9	-10.7	-1.7	0.6	8.4	0.0
	pi	3.9	-2.6	6.1	8.4	8.2	0.0
	total	0.0	-13.3	4.4	9.0	16.7	0.0
COM	c+s	-6.3	-18.5	-0.2	-0.2	14.9	0.0
	pi	21.1	13.0	13.0	37.2	19.8	0.0
	total	14.8	12.8	12.8	18.7	4.8	0.0

axes shieldings for the total shielding tensor and for its core+sigma (c+s) and pi contributions. The principal axes shieldings result from diagonalization of the symmetric part of the respective tensors, and we follow the convention $\sigma_{33} \geq \sigma_{22} \geq \sigma_{11}$, for indexing the principal axes [44]. The directions of the principal axes are apparent in Figure 1, as discussed below. The table also includes the field strength measures defined in eqs.(25,26), for the traceless anisotropy, $\Delta\boldsymbol{\sigma}^s$, and the antisymmetric part, $\boldsymbol{\sigma}^a$ of the tensor. Since the field points labelled COM, $\pi/2$, and 0 lie at the intersection of two or more symmetry planes, the antisymmetry vector, eq.(17), and hence the intensity measure in eq.(21) vanish identically, see Section II, as apparent in the table. The C_s site symmetry symmetry of the other field points allows non-vanishing components perpendicular to the C_s plane, as also discussed in Section II, and hence non-vanishing intensities according to eq.(21); see also Figure 4 and the discussion of this figure below.

Looking at Table 1 in more detail, the c+s and pi contributions to the isotropic shielding at the various field points trivially sum to the corresponding total values, as do the principal axes shieldings at COM, $\pi/2$ and 0, since the principal axes directions for these points coincide with the Cartesian axes for symmetry reasons, although the principal axes at COM clearly are reordered as consequence of the convention for indexing the axes. For the other field points, only the direction of the principal axis perpendicular to the sampling plane is fixed by symmetry, and the corresponding principal values ($\sigma_{yy} = \sigma_{33}$ in all cases) again sum correctly. The directions of the principal axes in the xz-plane for these points, and the corresponding principal values, result from the tensor diagonalizations, and are not fixed by symmetry. The fact that the corresponding components approximately sum to the total values, for all three points after reordering, reflects that the principal axes directions for the c+s and pi components lie within a few degrees of each other, see below. As discussed in Section III, the separate contributions to the intensities can not in general be expected to sum to the corresponding total values regardless of site symmetry. Note in particular, that the total intensity of the traceless anisotropy at COM is significantly smaller than the c+s and pi contributions because of large negatively signed cross terms.

Figure 1 shows cuts in the xz-plane of the response graphs [14, 28, 30] for the secular shielding $T_B(\mathbf{R};\mathbf{u}_B)$, eq.(24), for the core+sigma and pi results listed in Table 1. Only the three carbon atoms lying in front of the plot plane are shown in Figures 1 and 2. For each surface, the distance along a given direction from the center of this surface to a point on the surface provides the sign and magnitude of $T_B(\mathbf{R};\mathbf{u}_B)$ at the location of the center of the surface, induced by an external field applied along the direction considered, and negative values of $T_B(\mathbf{R};\mathbf{u}_B)$ are indicated by stippled lines. Recall that

Figure 1 2-Dimensional Response Graphs of Shielding Vector.
Left panel; pi shielding. Right panel; sigma shielding 1 Å ~ 10 ppm.
COM shielding is fourfold reduced. See text.

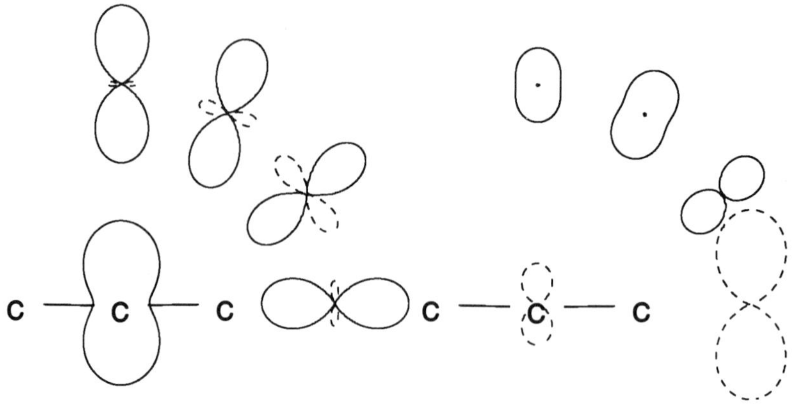

the sign of the shielding field, by definition, follows the sign of the shielding tensor, see comments following eqs.(18,19). The scale of the surfaces is such that 1 Å corresponds to a shielding of 10 ppm, except for the surfaces at COM where 1 Å corresponds to 40 ppm (representing a fourfold reduction of the shielding at COM relative to the other field points). The value of $T_B(\mathbf{R}; \mathbf{u}_B)$ at an extremum of a given surface is equal to a principal value for the shielding represented by this surface, cfr. Table 1, while the direction of the corresponding principal axis is given by the direction from the center of the surface to this particular extremum. As alluded to above, the directions of the c+s and pi principal axes are strictly parallel for the three high symmetry locations, while they lie within a few degrees of each other for the other two points.

Figure 2 is a central result of the present investigation, showing cuts in the xz-plane of the surfaces for $T_B(\mathbf{R}; \mathbf{u}_B)$, for the pi contributions (upper panels) and core+sigma contributions (lower panels), decomposed into diamagnetic parts (left hand panels) and paramagnetic parts (right hand panels). The pi contributions are shown on the same scale as in Figure 1, including the fourfold reduction of the shielding at COM relative to the other field points, while the scale for the core+sigma contributions is such that 1 Å corresponds to a shielding of 80 ppm, except that a fourfold reduction again is applied for of the shielding at COM. Hence all core+sigma shieldings are reduced by a factor of eight relative to the pi shieldings in this figure. The corresponding numer-

Figure 2 2-Dimensional Response Graphs of Shielding Vector.
Upper left panel: diamagnetic pi shielding. 1 Å ~ 10 ppm.
Upper right panel: paramagnetic pi shielding. 1 Å ~ 10 ppm.
Lower left panel: diamagnetic core+sigma shielding. 1 Å ~ 80 ppm.
Lower right panel: paramagnetic core+sigma shielding. 1 Å ~ 80 ppm.
COM shielding is fourfold reduced in all panels. See text.

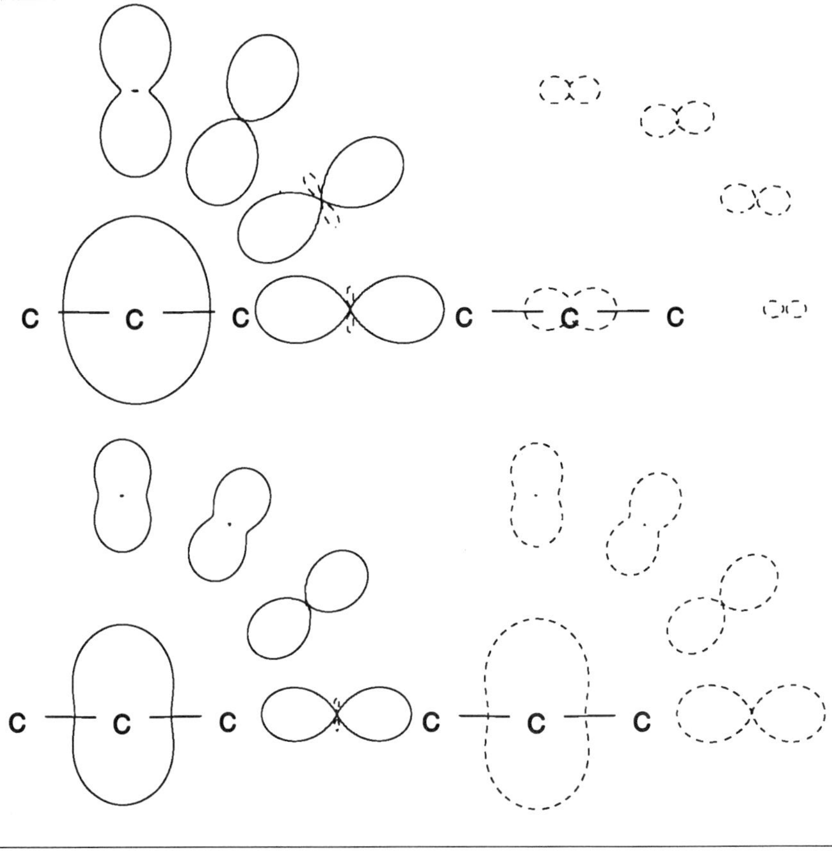

ical results are not reported for fear of tabulatory overcrowding. However as an example, we note that the perpendicular diamagnetic and paramagnetic shieldings at COM for the core+sigma contributions in the lower panels of Figure 2 are 326 ppm and -345 ppm, respectively, adding up to the total value of -19 ppm for this perpendicular shielding shown in Table 1 and Figure 1. The overall pattern of the contributions displayed in Figure 2 is that for the pi electrons, the paramagnetic shieldings are significantly smaller than the diamagnetic shieldings, making the diamagnetic parts dominate the total pi shieldings displayed in Figure 1. On the other hand, for the core+sigma electrons, the paramagnetic and diamagnetic shieldings are large, but oppositely signed, leading to the significant cancellations evident by comparing the scales in Figures 1 and 2 for the core+sigma surfaces.

The fact that the antisymmetry vector, eq.(17), vanishes for symmetry reasons for the field points in the molecular plane, and for the point on the z-axis, but retains a non-vanishing component, ξ_y, perpendicular to the xz-plane for the other points, is discussed above. The latter feature is illustrated in Figure 3, which shows the total antisymmetry vector for the $\pi/3$ field point projected onto the molecular plane. The vector accordingly extends from the x-axis, and is found to be negatively signed. The scale such that 1 Å corresponds to 2.5 ppm. The other vectors in the figure correspond to points generated by repeated reflections in the symmetry planes containing the z-axis and a C-H bond, and the rotatory behaviour of the vector reflects its axial vector (pseudovector) character.

Figure 3 Antisymmetry vector, $\boldsymbol{\xi}$, at $\pi/3$ field points. 1 Å $\sim$ 2.5 ppm. See text.

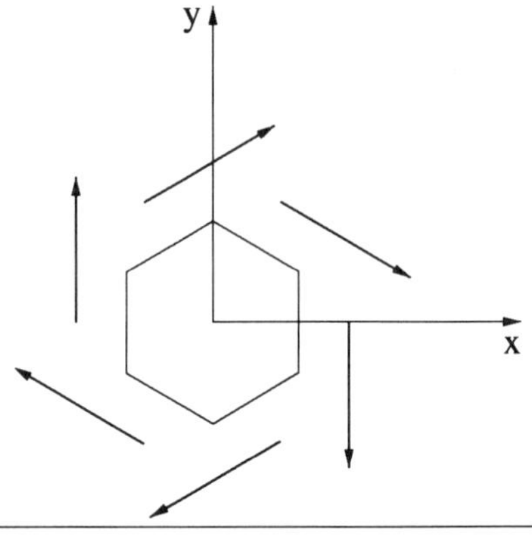

Figure 4 finally shows 3-dimensional response graphs for the intensities $I_B^a(\mathbf{R};\mathbf{u}_B)$ and $I_B^{\Delta\sigma}(\mathbf{R};\mathbf{u}_B)$ of the shielding vectors corresponding to the antisymmetric tensor $\boldsymbol{\sigma}^a$ and the traceless anisotropy tensor $\Delta\boldsymbol{\sigma}^s$, respectively, at the $\pi/3$ field point. To bring out the 3-dimensional perspective these figures are rotated relative to the 2-dimensional response graphs, showing clearly the doughnut-like shape of the intensity for the antisymmetric component. In fact

Figure 4 3-Dimensional Response Graphs of Intensity.
Left panel; antisymmetry intensity, I_B^a, right panel; anisotropy intensity, $I_B^{\Delta\sigma}$.
1 Å ~ 80 ppm^2. See text.

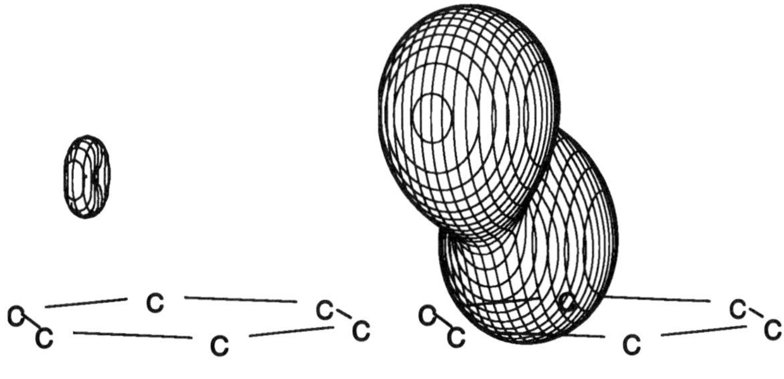

$I_B^a(\mathbf{R}; \mathbf{u}_B)$ is identically zero for this field point when the applied field is perpendicular to the C_s symmetry plane containing the point. This result holds for all field points in the C_s plane, since the antisymmetry vector is perpendicular to the plane, and since the field generated by the antisymmetric part of the shielding is strictly orthogonal both to the antisymmetry vector and to the applied field. The surface for $I_B^{\Delta\sigma}(\mathbf{R}; \mathbf{u}_B)$ shows the characteristic elongated response seen also in Figures 1 and 2.

VI.Discussion

The ring current model has played a dominating role in the discussion of magnetization and proton shieldings in aromatic systems, as reviewed in ref. [9], see ref. [19] for references to later developments. In Pauling's original formulation [40], the central point is the absence of paramagnetic contributions from the pi electrons for magnetic fields applied perpendicularly to the plane of the aromatic ring, due to the almost cylindrical symmetry of the pi orbitals. This leaves a diamagnetic pi electron current responsible for the characteristic magnetic anisotropy of aromatic systems, since Pauling further assumed "normal" (i.e. almost isotropic) contributions from sigma and core electrons. The model is often illustrated by diagrams like Figure 5, showing that the induced field generated by the pi electrons opposes/enhances the perpendicularly applied external field inside/outside the aromatic ring, see e.g. ref. [8].

Figure 5 The diamagnetic ring current effect in benzene.

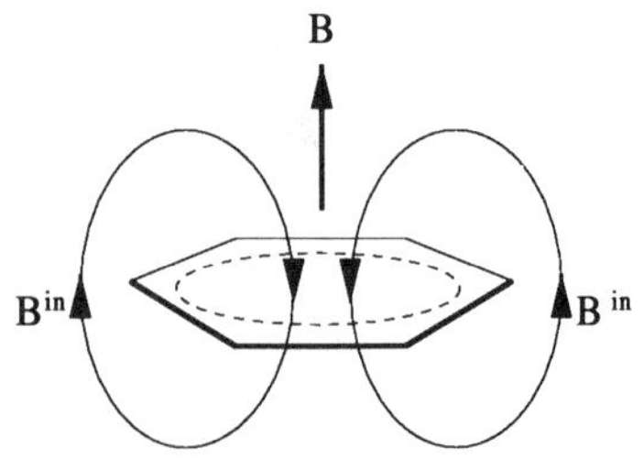

The present perspective for the contributions to the shielding field is brought out by Table 1 and in particular by Figures 1 and 2, and with the common gauge origin at COM in eqs.(29,30), our diamagnetic and paramagnetic contributions relate directly to Pauling's terminology. As noted in Section V, the upper panels of Figure 2 demonstrate first of all that, for all field points, the paramagnetic response for the pi electrons is almost vanishing when the external field is applied perpendicularly to the molecular plane, and they are also considerably smaller than the diamagnetic response, for all directions of the applied field, making the total pi shielding field in Figure 1 strongly dominated by the diamagnetic contributions, in complete accord with Pauling's assumptions. See e.g. ref. [19] for the selection rule arguments rationalizing this result. For the core + sigma contributions shown in the lower panels in Figure 2, the diamagnetic and paramagnetic contributions are oppositely signed, but numerically almost identical. These almost cancelling magnetic contributions therefore echo Pauling's assumption of "normal" contributions from core and sigma electrons. Among the more general aspects of the response features shown in Figures 1 and 2, it is seen that the response at COM is notably larger than the corresponding response at the other field points, probably in large parts due to the larger electron density at COM. In addition it is apparent that, except for the paramagnetic pi electron shieldings which are dominated by the almost vanishing perpendicular components, all other response graphs are strongly dominated by the radial direction.

To relate the pi electron ring current diagram in Figure 5 to the response surfaces of Figure 2, we note that the diamagnetic pi electron response for a field applied perpendicularly to the molecular plane at the $\pi/3$ field point is almost zero, implying according to eq.(19,23) that the shielding vector is (almost) orthogonal to the applied field at this point. This location therefore corresponds to a top point, i.e. a point of horizontal tangent, at a field line in Figure 5. Pursuing this line of thought, the negative perpendicular diamagnetic component at the $\pi/2$ in-plane field point corresponds to a positive induced field component, recalling the sign convention in eq.(18), in accord with Figure 5. For the other field points, the response to a perpendicular field are all positive, corresponding to induced field components opposing the applied field, also in accord with Figure 5. Since Figure 5 only relates to a field applied perpendicularly to the ring, the radial dominance of the diamagnetic pi shielding noted above, of course is not apparent in this diagram.

In ref. [19], we applied the same computational and response graph approach to the shielding of protons inside and outside one of the aromatic annulene systems studied in ref. [45]. As expected many of the above features, in particular the characteristics of the pi electron contributions, are found also for the proton shieldings. Larger differences appear in the core + sigma contributions, because of the increased electron density at the protons compared to the present sampling of the field at locations of low electron density. With regard to the ring current effect, we demonstrate in ref. [19], that the diamagnetic pi electron contributions are almost insensitive to ring closure, and we argue that rationalization of the unique magnetic anisotropy of aromatic systems in terms of diamagnetic *ring* currents seems a misrepresentation of the actual mechanism. In place of "ring current effect", and we therefore propose the term "Pauling-London-Pople effect". However, we emphasize that reservations concerning the rationalization of the effect of course in no way imply reservations concerning the remarkably reliable predictions based on the model [8, 45].

Turning finally to the question of the antisymmetric component of the shielding tensor, the selection rules and the consequences of the axial vector character of the antisymmetry vector are discussed in Sections II, and illustrated in Table 1 and Figure 3 in Section V. For nuclear shielding, it is known that the antisymmetric shielding component can be quite large for nuclei at positions of strained electronic structure involving multiple bonds [28, 46]. However, it is seen that even for a highly symmetric molecule like benzene, with a completely unstrained electronic structure, the antisymmetric shielding component is non-trivial at locations of lower site symmetry, as witnessed by the ratio of the two intensity measures in Table 1 for the two C_s site locations, and by the response graphs in Figure 4.

VII. Concluding Remarks

We have focussed on the perspective of the induced magnetic shielding field as a molecular response property of interest in its own right, accessible through computational and graphical studies. For the benzene molecule used as an illustrative examples, most of our results echo wellknown aspects of nuclear shielding in aromatic systems. However, among the features brought out specifically by combining computations and graphics for the field at nucleus-free locations, are the radial dominance of the shielding responses for all except the paramagnetic pi-electron contributions, and the fact that the field intensity corresponding to the antisymmetric part of the shielding tensor is of non-trivial magnitude at low site symmetry locations around the molecule. Turning briefly to potentially useful practical applications of this approach, we note that the numerical and graphical representations of the field and its intensities outside the molecule, can provide information on the magnetic perturbations, including relaxations [29], in other molecules due to the presence of the system under consideration. The present use of a computational scheme based on localized electronic contributions furthermore allows similar ab-initio treatments of magnetic perturbations due to molecular fragments, mutatis mutandis the multipolar approaches of refs. [7, 11–13].

In retrospect, we believe that Professor Mulliken was perhaps somewhat overstating his case in the remark quoted in the Introduction, concerning the status of computational chemistry in 1965. However, considered as a prophetic remark, the quotation certainly applies today, and it is a pleasure to dedicate this account to Professor Yngve Öhrn acknowledging his many important contributions to this development. These contributions include not only his own work on theoretical and computational approaches to central questions in chemistry, but also his continued engagement in the Sanibel Conferences and in the International Journal of Quantum Chemistry.

Acknowledgements

The work has been supported by The Danish Research Councils (Grants 11-0924-1 and 9600856), and through allocation of computer time at Uni-C, Lundtofte, Denmark (Grants 5.21.05.06 and 9701136). We are grateful to Dr. Frank Jensen, University of Odense, for valuable suggestions and discussions.

References

[1] P. Lazzeretti, Adv. Chem. Phys. **75**, 507 (1989).

[2] J.Linderberg and Y.Öhrn. *Propagators in Quantum Chemistry.* Academic Press, 1973.

[3] W. Kutzelnigg, U. Fleisher and M. Schindler. In J.A. Tossell, editor, *Nuclear Magnetic Shielding and Molecular Structure.* Kluwer, Dordrect, 1993.

[4] P. Lazzeretti. In J.A. Tossell, editor, *Nuclear Magnetic Shielding and Molecular Structure.* Kluwer, Dordrect, 1993.

[5] R. W. Bader and T. A. Keith, J. Chem. Phys. **99**, 3683 (1993).

[6] Aa. E. Hansen and T. D. Bouman, J. Math. Chem. **10**, 221 (1992).

[7] P. J. Stiles, Chem. Phys. Lett. **30**, 259 (1975).

[8] J. A. Pople, W. G Schneider and H. J. Bernstein. *High-Resolution Nuclear Magnetic Resonance.* McGraw-Hill, 1958.

[9] C. W. Haigh and R. B. Mallion, Progress in NMR Spectr. **13**, 303 (1980).

[10] P. von R. Schleyer, C. Maerker, A. Dransfield, H. Jiao and N. J. R. van E. Hommes, J. Am. Chem. Soc. **118**, 6317 (1996).

[11] H. M. McConnell, J. Chem. Phys. **27**, 226 (1957).

[12] A. D. Buckingham and P. J. Stiles, Mol. Phys. **24**, 99 (1972).

[13] P. J. Stiles, Mol. Phys. **29**, 1271 (1975).

[14] Aa. E. Hansen, K.V. Mikkelsen and K. L. Bak, Magn. Reson. Rev. **1**, 133 (1997).

[15] R. S. Mulliken. *Nobel Lectures, Chemistry 1863-1970.* Eslevier Amsterdam, 1972.

[16] Aa. E. Hansen and T. D. Bouman, J. Chem. Phys. **82**, 5035 (1985).

[17] Aa. E. Hansen and T. D. Bouman. In J.A. Tossell, editor, *Nuclear Magnetic Shielding and Molecular Structure.* Kluwer, Dordrect, 1993.

[18] Aa.E.Hansen and M.Bilde. In D.M.Grant and R.K.Harris, editors, *Enclyclopaededia of Nuclear Magnetic Resonance.*Vol.7, p.4292 John Wiley and Sons, Chichester, 1996.

[19] M. Bilde and Aa. E. Hansen, Mol. Phys. **92**, 237 (1997).

[20] Aa. E. Hansen and E. N. Svendsen, Mol. Phys. **28**, 1061 (1974).

[21] Aa.E. Hansen, Mol. Phys. **34**, 1473 (1977).

[22] Aa. E. Hansen, Am. J. Phys. **39**, 653 (1971).

[23] W. Kolos, Adv. Quant. Chem. **5**, 99 (1970).

[24] N. F. Ramsey, Phys. Rev. **86**, 243 (1952).

[25] A.D. Buckingham and S. M. Malm, Mol. Phys. **22**, 1127 (1971).

[26] U. Haeberlen. *High Resolution NMR in Solids.* Academic Press, New York, 1976.

[27] M. Mehring. *Principles of High Resolution NMR in Solids.* Springer Verlag, Berlin-Heidelberg, 2 edition, 1984.

[28] Aa. E. Hansen and T. D. Bouman, J. Chem. Phys. **91**, 3552 (1989).

[29] F. A. L. Anet and D. J. O'Leary, Concepts in Mag. Res. **3**, 19 (1991) and **4**, 35 (1992).

[30] Note that in ref. [14], the sign in the definition of the shielding vector, $\mathbf{T}_B$ is a misprint; cfr. the present eq.(18) and ref.[28].

[31] J. Oddershede, P. Jørgensen and D.L. Yeager, Comp. Phys. Rep. **2**, 33 (1984).

[32] T. D. Bouman and Aa. E. Hansen, Chem. Phys. Lett. **175**, 292 (1990).

[33] T. D. Bouman and Aa. E. Hansen, Chem. Phys. Lett. **197**, 59 (1992).

[34] R. H. Höller and H. Lischka, Mol. Phys. **41**, 1017 (1980) and **41**, 1041 (1980).

[35] P. Lazzaretti, M. Malagoli and R. Zanasi, J. Molec. Struct. Theochem. **234**, 127 (1991).

[36] J.A. Tossell, editor. *Nuclear Magnetic Shielding and Molecular Structure.* Kluwer, Dordrect, 1993.

[37] K. Ruud, T. Helgaker, R. Kobayashi, P. Jørgensen, K. L. Bak and H. J. Aa. Jensen, J. Chem. Phys. **100**, 8178 (1994).

[38] J. Gauss and J. F. Stanton, J. Chem. Phys. **104**, 2574 (1996).

[39] J. H. van Vleck. *The Theory of Electric and Magnetic Susceptibilities.* Oxford University Press, 1932.

[40] L. Pauling, J. Chem. Phys. **4**, 673 (1936).

[41] Aa. E. Hansen, T. D. Bouman, K. L. Bak, R. Kirby, T. B. Pedersen and M. Ø. Jensen. *RPAC Linear Response Properties Program Package, Version 11.X.* University of Copenhagen, 1998.

[42] M.-J. Frisch, G. W. Trucks, H. B. Schleigel, P. M. W. Gill, B. G. Johnson, M. A. Robb, J. R. Cheeseman, T. A. Keith, G. A. Petersson, T. A. Montgomery, B. Raghavachari, M. A. Al-Laham, V. G. Zakrzewski, J. V. Ortiz, J. B. Foresmann, J. Cioslowski, B. B. Stefanov, A. Nanaykkara, M. Challacombre, C. Y. Peng, W. Ayala, P. Y. Chen, M. W. Wong, J. L. Andres, E. S. Replogle, R. Gomberts, R. L. Martin, D. J. Fox, J. S. Brinkley, D. J. Defrees, J. Baker, M. Stewart, J. J. P. Head-Gordon, C. Gonzalez and J. A. Pople. *Gaussian 94, Revision F.3.* Gaussian Inc., Pittsburgh, PA, 1995.

[43] J. M. Foster and S. F. Boys, Rev. Mod. Phys. **32**, 300 (1960).

[44] J. Mason, Solid State Nuc. Mag. Res. **2**, 285 (1993).

[45] F. Sondheimer, I. C. Calder, J. A. Elix, Y. Gaoni, P. G. Garratt, K. Grohman, G. Dimiao, J. Mayer, M. V. Sargent and R. Wolowsky, The Chemical Society Special Publication. **No21**, 75 (1967).

[46] F.A.L. Anet, D.J. O'Leary, C. G. Wade and R. J. Johnson, Chem. Phys. Lett. **171**, 401 (1990).

Time-Dependent Variational Principle in Density Functional Theory

B. Weiner

Department of Physics, Penn State University, DuBois, Pennsylvania 15801

S. B. Trickey

Quantum Theory Project, Departments of Physics and of Chemistry, University of Florida, Gainesville, Florida 32611

ADVANCES IN QUANTUM CHEMISTRY, VOLUME 35

0065-3276/99 $30.00

1. INTRODUCTION

It is with great pleasure that we dedicate this article to our colleague and friend, Professor Yngve Öhrn, on his 65th birthday. His penetrating insight and clarifying formulation of physical problems in careful quantum mechanical terms have inspired and aided all his collaborators and coworkers. Over the last four decades he has had an impact in many areas of quantum molecular mechanics and a great influence on many researchers in this area. Yngve Öhrn's and Jan Linderberg's seminal *little yellow book* "Propagators in Quantum Chemistry" (1) has been the resource for seemingly innumerable studies and research projects for the last 25 years. Its focus is Double Time Propagators, with particular reference to the Electron Propagator and the Polarization Propagator. Both are exceptionally useful in the mathematical description of nuclear, atomic, molecular and condensed matter system properties, as they describe the linear response of a system to a time-dependent perturbation. Yngve Öhrn and various collaborators introduced many successful decoupling schemes for the propagator equations and used them to determine approximations to the Electron Propagator and the Polarization Propagator (for example 2, 3).

A major problem in the first order approximation of the Polarization Propagator, [equivalent to the Random Phase Approximation (RPA) - also known as the linearized Time-Dependent Hartree Fock (TDHF) approximation], is the inconsistency of the ground state involved in the definition of this propagator. The adjoints of the perturbationaly corrected excitation operators produced by RPA do not *kill* this ground state, as they should if the theory were consistent. In the late 70's Yngve Ohrn and Jan Linderberg (4) showed that the nearest one could get to a consistent ground state was to use states of the Antisymmetrized Geminal Power (AGP's) form. In ensuing years Yngve Öhrn and coworkers applied this generalized RPA theory to calculation of the excitation spectra of many small molecules (5).

The solution to the consistent ground state problem turns out to be closely related to the theory of Coherent States and the theory of group representations. The blending of these two topics involve the construction of group related generalized phase spaces. His combined interests in Quantum Molecular Dynamics, Wave Packet Dynamics, and Coherent States thus led Yngve Öhrn naturally to the Time-Dependent Variational Principle as a means of obtaining approximate solutions to the Time-Dependent Schrödinger Equation for systems of electrons and nuclei. By applying that principle to coherent states of both electrons and nuclei, one is able to avoid the Born-Oppenheimer approximation so that the time evolution of electrons and nuclei can be treated efficiently without first finding electronic potential energy surfaces, then studying the coupled motion of nuclei on these surfaces. Since the middle 1980's Yngve Öhrn and collaborators have been developing and applying this Electron Nuclear Dynamics theory in a system of computer programs named ENDyne. The present implementation is at the level of TDHF (i.e. single determinants constructed of complex general spin orbitals) and classical nuclei (i.e. infinitely narrow nuclear wavepackets). The results obtained so far have been impressive as compared both with calculations using other methods and with experiment. The variables used are nuclear coordinates and momenta and complex parameters labeling the electronic state. The approximate solutions to the TDSE are characterized by the evolution paths of these variables as determined by the *generalized clas*:*cal* equations of motion obtained by using the TDVP.

In this volume dedicated to Yngve Öhrn we feel it is particularly appropriate to extend his ideas and merge them with the powerful practical and conceptual tools of Density Functional Theory (6). We extend the formalism used in the TDVP to mixed states and consider the states to be labeled by the densities of electronic space and spin coordinates. (In the treatment presented here we do not explicitly consider the nuclei but consider them to be fixed. Elsewhere we shall show that it is indeed straightforward to extend our treatment in the same way as Öhrn et al. and obtain equations that avoid the Born-Oppenheimer Approximation.) In this article we obtain a formulation of *exact* equations for the evolution of electronic space-spin densities, which are equivalent to the Heisenberg equation of motion for the electrons in the system. Using the observation that densities can be expressed as quadratic expansions of functions, we also obtain *exact* equations for these one-particle functions.

A brief summary of motivation may perhaps be helpful. On the whole, contemporary versions of Density Functional Theory are presented in two separate conceptual and logical frameworks, one each for time-independent and time-dependent DFT. Even within the constrained search formulation of time-independent DFT, there are several rather deep and interconnected questions. Commonly these are discussed in terms of symmetry (and symmetry breaking), occupation number distributions, and functional differentiability. Their resolution is important for both fundamental reasons and to provide pathways to more powerful and reliable DFT approximations. Implicit in them is the issue of mixed states. Because customary formulations of time-dependent DFT are separate, it is not evident how resolution of those issues in the time-independent case would carry over. Further, the importance of parameter-space metrics in the END work of Öhrn et al. versus the absence of such metrics in conventional time-dependent DFT suggests strongly that a formulation directly from the TDVP would be beneficial and clarifying, in that it would provide a rigorous basis for the use of dynamics in parameter space, including mixed states right from the start, and would provide a significantly enhanced foundation for constructing approximations.

2. MIXED STATE TIME-DEPENDENT VARIATIONAL PRINCIPLE

Conventional presentations of DFT start with pure states but sooner or later encounter mixed states and densities (ensemble densities is the usual formulation in the DFT literature) as well. These arise, for example in formation or breaking of chemical bonds and in treatments of so-called "static correlation" (situations in which several different one-electron configurations are nearly degenerate). Much of the DFT literature treats these problems by extension and generalization from pure state, closed shell system results. A more inclusively systematic treatment is preferable. Therefore, the first task is to obtain the Time-Dependent Variational Principle (TDVP) in a form which includes mixed states.

In a landmark publication in 1981, Kramer and Saraceno (7) [hereafter "K&S" to distinguish from Kohn-Sham, usually denoted as "KS" in the DFT literature, a usage we follow] showed how to use the Time-Dependent Variational Principle to construct generalized classical equations of motion for quantum mechanical state vectors in terms of labels that characterize those vectors parametrically. (Obviously, "classical" in this setting refers to the form of the equations, not the content.) This construction leads to equations that determine evolution paths in a parameter space. In the case of a parameterization that covers all of state space those equations are *entirely equivalent* to the exact Time-Dependent Equation Schrödinger Equation (TDSE) and to approximate TDSE's for parameterizations that label a subset of states in a continuous fashion. In their work they showed that in certain cases the parameters could be related to group coset spaces (8) and give rise to families of Coherent States (CS) (9-12), where the set of coherent states is generated by the action of a coset on a reference state. A familiar example is the set of single determinantal states, in which all the states are generated by the action of a coset of the group of one-particle unitary transformations acting on a given single determinantal reference state, leading to the Thouless parameterization of these states (13). In the case of group-generated CS's, the labels are complex variables that holomorphically (complex analytically) parameterize a set of N-particle states and the generalized dynamics takes place in the manifold of these states (which in general is nonlinear). In general not all N-particle states are produced by the action of the chosen group on the reference state. The resulting equations of motion correspond to the restriction of the TDSE to this submanifold. This formulation of approximate TDSE's has been examined at great length in many contexts (14-18).

The most general formulation is in terms of real parameters, as any complex parameterization can always expanded into real and imaginary parts, while the converse construction (complex parameters from combination of real ones) is not always possible. The real parameterization does not take one directly into the group and coherent state formalism. However in the context of DFT the real parameterization is particularly natural, since the pervasive perspective is that N-particle states are labeled by densities which are real-valued functions of space and spin variables.

The K&S treatment is in terms of pure N-particle states i.e. vectors in an N-particle Hilbert space. As noted, to encompass DFT in its most general form one must consider mixed N-particle states i.e. states described by N-particle Density Operators. Thus this section extends the K&S treatment to include mixed states by using the vector space structure of N-particle operator space and applying the K&S treatment to state "super vectors" in this operator vector space [If one can define a linear structure on a space of operators, then those operators can be viewed as *supervectors* and any map that maps a linear operator space into itself can be viewed as a *superoperator*. The terms *supervector* and *superoperator* were first introduced by Zwanzig (19).] Although most of the formal manipulations and expressions are identical in appearance (with those of K&S), the interpretation in terms of operators is quite different.

Note to the reader: the notation can be somewhat intricate, thus it is summarized in Appendix A.

2.1 States and Superoperators

The mixed and pure states of an N-particle fermion system can be described by positive and normalized operators, $\mathcal{S}_N$, which form a convex set contained in the space of Trace Class (Appendix B) operators $\mathcal{B}_1\left(\mathcal{H}^N\right)$ acting in the N-particle fermion Hilbert space $\mathcal{H}^N$.

$$\mathcal{S}_N = \left\{ D^N ; D^N \geq 0; Tr\left\{D^N\right\} = 1; D^N \in \mathcal{B}_1\left(\mathcal{H}^N\right) \right\} \qquad (2.1)$$

where $\mathcal{H}^N$ is defined to be the N-fold antisymmetrized tensor product of the one-particle Hilbert space $\mathcal{H}^1$.

Positive elements, X, of the vector space of operators $\mathcal{B}_1\left(\mathcal{H}^N\right)$ can always be expressed as a product of a Hilbert-Schmidt operator (Appendix B) with its adjoint as $X=QQ^\dagger$. [The Harriman (20) decomposition of the density into positive sums of products of orbitals is in fact a very special case of this relationship, which will be important later.] One needs to keep in mind that this factorization is not unique however: QU for any unitary U produces the same X. Moreover UQ produces the same X for U belonging to the invariance group of X (Appendix C). The space of Hilbert-Schmidt operators, $\mathcal{B}_2\left(\mathcal{H}^N\right)$, is a Hilbert space that has an inner product defined in terms of the trace operation

$$(X|Y) = Tr\{X^{\dagger}Y\} \tag{2.2}$$

This fact allows us to express the set $\mathcal{S}_N$ as the unit sphere in $\mathcal{B}_2(\mathcal{H}^N)$ as

$$\mathcal{S}_N = \{Q; (Q|Q) = 1; Q \in \mathcal{B}_2(\mathcal{H}^N)\} \tag{2.3}$$

The action of the Hamiltonian, H, can be expressed as a superoperator mapping the Hilbert space $\mathcal{B}_2(\mathcal{H}^N)$ into itself by

$$\hat{H}: Q \to HQ;\ Q \in \mathcal{B}_2(\mathcal{H}^N) \tag{2.4}$$

The superoperator $\hat{H}$ inherits the unboundness of H and its domain is defined as

$$Dom(\hat{H}) = \{Q, HQ \in \mathcal{B}_2(\mathcal{H}^N)\} \tag{2.5}$$

In the following we consider Hamiltonians, $H(t)$, that are explicitly time-dependent. The definitions Eq. (2.4)-(2.5) also hold without modification in these cases.

2.2 State Evolution Equations

Now consider operators Q that depend on real parameters $\mathbf{x}$, which can be considered as coordinates of points of a linear or nonlinear manifold $\mathcal{M}$ of operators Q contained in $\mathcal{B}_2(\mathcal{H}^N)$, [including $\mathcal{B}_2(\mathcal{H}^N)$ itself], i.e. the real coordinates $\mathbf{x}$ denote a point $Q \in \mathcal{M}$. Following K&S, generalized classical equations of motion for evolution paths $\mathbf{x}(t)$ can be obtained for these operators, which are given by

$$\left.\begin{aligned} \{x_i, \mathcal{E}\} &= \dot{x}_i \\ \mathcal{E}(\mathbf{x}) = \frac{\mathcal{H}(\mathbf{x})}{\mathcal{S}(\mathbf{x})} &= \frac{(Q(\mathbf{x})|\hat{H}Q(\mathbf{x}))}{(Q(\mathbf{x})|Q(\mathbf{x}))} \end{aligned}\right\} \tag{2.6}$$

The Poisson brackets are defined for functions $\mathrm{f}: \mathcal{M} \to \mathbf{C}$ by

$$\{\mathrm{f}, \mathrm{g}\} = \sum_{i,j} \frac{\partial \mathrm{f}}{\partial x_i} \xi_{ij} \frac{\partial \mathrm{g}}{\partial x_j} \tag{2.7}$$

where the antisymmetric matrix that defines a "phase space" structure for the manifold was shown by K&S to be given by

$$\begin{aligned} \xi_{ij} &= \left(\eta^{-1}\right)_{ij} \\ \eta_{ij} &= i\left\{\frac{\partial}{\partial x_i'}\frac{\partial}{\partial x_j} - \frac{\partial}{\partial x_i}\frac{\partial}{\partial x_j'}\right\} \ln\{(Q(\mathbf{x}')|Q(\mathbf{x}))\}\Big|_{\mathbf{x}'=\mathbf{x}} \end{aligned} \tag{2.8}$$

Note that for general parameterizations this metric matrix is *neither skew diagonal nor constant*; see below. The equations of motion expressed in Eq. (2.6) are obtained by using the Principle of Stationary Action, $\delta\mathcal{A}=0$, with Lagrangian

$$\mathcal{L}\left(Q(\mathbf{x}(t)),Q(\mathbf{x}(t))^{\dagger},t\right)=\mathrm{Re}\left\{\frac{\left(Q(\mathbf{x}(t))\left|\left[i\frac{\hat{\partial}}{\partial t}-\hat{H}(t)\right]Q(\mathbf{x}(t))\right.\right)}{\left(Q(\mathbf{x}(t))|Q(\mathbf{x}(t))\right)}\right\} \tag{2.9}$$

and Action

$$\mathcal{A}=\int_{t_i}^{t_f}\mathcal{L}\left(Q(\mathbf{x}(t)),Q(\mathbf{x}(t))^{\dagger},t\right)dt \tag{2.10}$$

where the end points $\left(x(t_i),x(t_f)\right)$ of the paths are held fixed. The "superoperator", $i\frac{\hat{\partial}}{\partial t}$, that produces i times the time rate of change of operators and is defined in a way analogous with $\hat{H}(t)$ in Eq. (2.4), is not self-adjoint on the space $\mathcal{B}_2\left(\mathcal{H}^N\right)$. Therefore the process of taking the real part of the inner product in Eq. (2.9) is significant.

The metric term Eq. (2.8) is important for all cases in which the manifold $\mathcal{M}$ has non-zero curvature and is thus nonlinear, e.g. in the cases of Time-Dependent Hartree-Fock (TDHF) and Time-Dependent Multi-Configurational Self-Consistent Field (TDMCSCF) calculations. In such situations the metric tensor ξ varies from point to point and has a nontrivial effect on the time evolution. It plays the role of a time-dependent force (somewhat like the location-dependent gravitational force which arises in general relativity from the curvature of space-time). In the case of flat i.e. linear manifolds, as are found in Time-Dependent Configuration Interaction (TDCI) calculations, the metric is constant and does not have a significant effect on the dynamics.

If the inverse in Eq. (2.8) does not exist then the metric is singular, in which case the parameterization of the manifold of states is redundant. That is, the parameters are not independent, or splitting of the manifold occurs, as in potential curve crossing in quantum molecular dynamics. In both cases, the causes of the singularity must be studied and revisions made to the coordinate charts on the manifold (i.e. the way the operators are parameterized) in order to proceed with calculations.

The form of the action principle given above was first applied to quantum mechanics to describe the time evolution of pure states (i.e.

wavefunctions) by Frenkel (21) and leads, in the case that the manifold $\mathcal{M}$ is all of $\mathcal{H}^N$, to the TDSE. We have extended it to more general operator manifolds $\mathcal{M} \subseteq \mathcal{B}_2\left(\mathcal{H}^N\right)$ in the manner outlined above, and for the case $\mathcal{M} = \mathcal{B}_2\left(\mathcal{H}^N\right)$, have shown (22) that the equations of motion (2.6) are equivalent to the Heisenberg Equations of Motion (HEM) for the N-particle state operators D^N:

$$i\dot{D}^N = \left[H, D^N\right];\ D^N = QQ^\dagger;\ Tr\left\{D^N\right\} = 1 \tag{2.11}$$

In this particular case the $\mathbf{x} \equiv (\mathbf{x}_1, \mathbf{x}_2)$ parameters label the Q operators by

$$Q(\mathbf{x}) = Q(\mathbf{x}_1, \mathbf{x}_2) = \sum_{1 \le k, l \le d} c_{kl} |\Phi_k\rangle\langle\Phi_l| \tag{2.12}$$

where $x_{1kl} = \operatorname{Re} c_{kl}$, $x_{2kl} = \operatorname{Im} c_{kl}$, $d = \begin{pmatrix} r \\ N \end{pmatrix}$, the points $\mathbf{x} \equiv (\mathbf{x}_1, \mathbf{x}_2)$ lie on the hypersphere

$$\sum_{k,l=1}^{d} \left(x_{1kl}^2 + x_{2kl}^2\right) = 1 \tag{2.13}$$

$\left\{\Phi_k; 1 \le k \le \begin{pmatrix} r \\ N \end{pmatrix}\right\}$ is a complete orthonormal basis for $\mathcal{H}^N$, and r is the dimension of one-particle space, which in the exact case is infinite.

The equations of motion (2.6) can be expressed compactly in matrix form as

$$\frac{d\mathbf{x}}{dt} = \eta^{-1} \frac{\partial \mathcal{E}}{\partial \mathbf{x}} \tag{2.14}$$

and solved by standard integration techniques (23).

The preceding evolution equations also can be used to generate equations for stationary states by setting the time derivative to zero giving the two equivalent forms (signified by the double arrow):

$$\{x_i, \mathcal{E}\} = 0;\ \text{for all } i \Leftrightarrow \eta^{-1} \frac{\partial \mathcal{E}}{\partial \mathbf{x}} = \mathbf{0} \tag{2.15}$$

3. TIME-INDEPENDENT DENSITY FUNCTIONAL THEORY

In order to apply the preceding development to states parameterized by densities we must first analyze the structure and properties of this type of parameterization.

3.1 Definition of the energy functional of the density

The relationship between N-particle states, in which we include mixed states, represented by N-particle operators as defined in Eq. (2.1), and the space-spin density $\rho(\mathbf{y})$ is not 1-1. Here and throughout the following development, $\mathbf{y}$ is the combined space-spin variable $(\mathbf{r},\sigma)$. These facts are the essence of the power and simplicity of the density functional method and at the same time the source of its conceptual complexity and intricacy. The power and simplicity come from being able to characterize an N-particle system by a real positive semi-definite function of a single 3D spatial variable and one 2D complex spin variable, while the conceptual complexity and intricacy come from developing an explicit understanding of how such a density determines a N-particle state.

To develop a framework in which to treat these topics in a rigorous manner for both time-dependent and independent systems, we adapt the *constrained search* of Levy (24) to generate a well-defined energy functional of the density, then follow a constrained optimization analysis of the problem as described, for example, in the book by Hestenes (25). These techniques allow one to specify the way paths of N-particle density operators can be defined in N-particle operator space such that they are labeled by densities in 1-1 fashion. The expectation of the Hamiltonian with respect to N-particle states on these paths then becomes a well defined functional of the density $\rho(\mathbf{y})$ and the ground state energy is the minimum value of this functional. The treatment is somewhat akin to that of Kryachko and Ludeña's "orbits"(26) but differs (in an essential fashion) in determining the paths by optimization criteria which assure that the resulting functionals have well-defined functional derivatives irrespective of the topology of the density. Kryachko and Ludeña, in contrast, identify orbits by employing the Bader density surface criterion (27), which means that those orbits are connected inescapably with the molecular point group symmetry. Even at the time-independent level, molecular structure and bonding often involve changes in the point group symmetry, so that inescapable connection seems to intertwine two issues better left apart. Dynamics simply makes such conceptual and procedural issues harder.

First we define the linear map that produces the densities from N-particle states. It is a map from the space of N-particle Trace Class operators into the space of complex valued absolute integrable functions of space-spin variables

$$\Xi_N^1 : \mathcal{B}_1\left(\mathcal{H}^N\right) \to L_1\left(\mathbf{R}^3 \times \mathbf{C}^2\right) \equiv L_1(\mathbf{Y}) \tag{3.1}$$

defined by

$$\zeta(\mathbf{y}) = \Xi_N^1(X)(\mathbf{y}) = Tr\left\{\Phi^\dagger(\mathbf{y})\Phi(\mathbf{y})X\right\} \in L_1(\mathbf{Y}). \tag{3.2}$$

The field operators used in the definition above are given by

$$\Phi(\mathbf{y}) = \sum_{1\le i\le r} \overline{\varphi}_i(\mathbf{y})\mathrm{a}_i \tag{3.3}$$

in terms of the discrete field operators, $\left\{\mathrm{a}_i, \mathrm{a}_i^\dagger\right\}$, which are defined using a basis of one-particle functions of space and spin $\{\varphi_i; 1\le i\le r\}$ and their action on the vacuum vector $|\phi\rangle$

$$\mathrm{a}_i^\dagger|\phi\rangle = |\varphi_i\rangle \tag{3.4}$$

The field operators satisfy the fermion anti-commutation relationships

$$\begin{aligned} \left[\Phi(\mathbf{y}), \Phi^\dagger(\mathbf{y}')\right]_+ &= \delta(\mathbf{y}-\mathbf{y}') \\ \left[\mathrm{a}_i, \mathrm{a}_j^\dagger\right]_+ &= \delta_{ij} \end{aligned} \tag{3.5}$$

The kernel of Ξ_N^1 is a linear subspace of $\mathcal{B}_1\left(\mathcal{H}^N\right)$, which we use to define an equivalence relationship on $\mathcal{B}_1\left(\mathcal{H}^N\right)$

$$X \sim Y \Leftrightarrow X - Y \in Ker\left\{\Xi_N^1\right\} \Leftrightarrow \Xi_N^1(X) = \Xi_N^1(Y) = \zeta \tag{3.6}$$

where the doubled arrow indicates equivalent statements. We denote these equivalence classes by $[\zeta]_N$ and note that they form a linear space, in quotient notation, $\mathcal{B}_1\left(\mathcal{H}^N\right)\Big/ Ker\Xi_N^1$. The map Ξ_N^1, when restricted to the convex set of N-particle states, $\mathcal{S}_N$, has values in the convex set, $\mathcal{P}_{1N}$, of positive functions in $L_1(\mathbf{Y})$ that integrate to the value N

$$\mathcal{P}_{1N} = \left\{\rho; \rho(\mathbf{y}) \ge 0, \int \rho(\mathbf{y})d\mathbf{y} = N\right\} \tag{3.7}$$

Harriman (20) has shown that this map is "onto" i.e. any element of $\mathcal{P}_{1N}$ comes from at least one element of $\mathcal{S}_N$. Note that this property does not rule out the possibility that an element of $\mathcal{P}_{1N}$ can also come from operators *not* in $\mathcal{S}_N$. This "onto" property should be compared to the case that arises in the N-representability problem (28) where not every positive two-particle operator comes from a state in $\mathcal{S}_N$ so the contraction map in that case does *not* have the "onto" property.

The energy functional defined by the Hamiltonian, H,

$$\begin{aligned} E_H &: \mathcal{S}_N \to \mathbf{R} \\ E_H\left(D^N\right) &= Tr\left\{HD^N\right\} \end{aligned} \tag{3.8}$$

is, however, not uniquely defined on the equivalence classes $[\rho]_N$, i.e. it is not defined on the space $\mathcal{B}_1(\mathcal{H}^N)/Ker\Xi_N^1$ as many different D^N's give the same ρ while producing different values $E_H(D^N)$. Thus E_H is not well-defined as it is multivalued on individual equivalence classes. In order to obtain an energy functional that is well-defined, we continue with the constrained search logic and define another functional in terms of the space-spin density as

$$F_H(\rho) = \underset{D^N \in \mathcal{S}_N(\rho)}{Min} E_H(D^N) = E_H(D_*^N(\rho)) \tag{3.9}$$

where $\mathcal{S}_N(\rho) = [\rho]_N \cap \mathcal{S}_N$ is the set of N-particle states that produce the same density and $D_*^N(\rho)$ is the minimizer in the set $\mathcal{S}_N(\rho)$. Note that we exclude the case of a non-unique minimizer; see Savin (29) for a related discussion. The minimization of Eq. (3.9) contains three types of constraints, normalization, positivity and fixed density. The normalization and positivity constraints can be handled by the factorization $D^N = \dfrac{QQ^\dagger}{Tr\{QQ^\dagger\}}$, which leads to the energy functional definition

$$\begin{aligned} F_H(\rho) &= \underset{QQ^\dagger \in \mathcal{S}_N(\rho)}{Min} \frac{Tr\{HQQ^\dagger\}}{Tr\{QQ^\dagger\}} = \underset{Q \in \langle\rho\rangle_N}{Min} E_H(Q) \\ &= \underset{Q \in \langle\rho\rangle_N}{Min} \frac{(Q|\hat{H}Q)}{(Q|Q)} = E_H(Q_*(\rho)) \end{aligned} \tag{3.10}$$

One can show (30) that densities are square integrable and thus belong to the Hilbert space $L_2(\mathbf{Y})$ of square integrable functions. This allows one to define the set, $\langle\rho\rangle_N$, of feasible Q's by a quadratic constraint function for a *fixed* ρ as

$$\begin{aligned} &g: \mathcal{B}_2(\mathcal{H}^N) \to L_2(\mathbf{Y}) \\ &g(Q,\rho) = 0 \\ &g(Q,\rho) = \Xi_N^1(QQ^\dagger) - \rho \end{aligned} \tag{3.11}$$

The variation in Eq. (3.10) can be carried out by using a Lagrangian function

$$\mathcal{L}(Q,\lambda,\rho) = E_H(Q) - \int_{\mathbf{Y}} \lambda(\mathbf{y}) g(\mathbf{y}) d\mathbf{y} \tag{3.12}$$

whose stationary points determine the constrained minima of $E_H(Q)$. By considering the sensitivity of the minimizer, $Q_*(\rho)$, to variations in ρ and checking that certain conditions on the first and second derivatives of $\mathcal{L}(Q,\lambda,\rho)$ are satisfied, Hestenes (25) showed that one can define a path of solutions, $Q_*(\rho)$, parameterized by ρ. On this path one can define an energy functional

$$F_H(\rho) = E_H(Q_*(\rho)) \tag{3.13}$$

and a Lagrange parameter functional, $\lambda(\rho)$, which can be identified with the functional derivative $\frac{\delta F_H}{\delta \rho}$ of $F_H(\rho)$ along that path. Note that the energy is a 1-1 functional of the density on this path and simultaneously that the functional derivative is defined *on this* particular path. It is possible define other paths in $\mathcal{B}_2(\mathcal{H}^N)$ on which the energy is also a 1-1 functional, but on those paths the following crucial fact will not be true

$$\begin{aligned} E_0 &= \underset{\rho \in D}{Min} \left\{ F_H(\rho) \right\} \\ D &= \left\{ \rho \mid \rho(\mathbf{y}) \geq 0;\ \int \rho(\mathbf{y}) d\mathbf{y} = N \right\} \end{aligned} \tag{3.14}$$

where E_0 is the ground state energy of the system. The paths $Q_*(\rho)$ clearly define paths in $\mathcal{B}_1(\mathcal{H}^N)$ by $D_*^N(\rho) = Q_*(\rho) Q_*(\rho)^\dagger$. The preceding construction of the energy functional is discussed in more detail in (30).

The explicit form of the functional F_H is of course unknown and in practical applications has to be approximated. In order to facilitate the creation of these approximations one decomposes F_H into a sum of other functionals that focuses all the unknowns into one component, the exchange-correlation functional, F_{XC}.

$$F_H(\rho) = F_C(\rho) + F_{XC}(\rho) + F_{eN}(\rho) + F_{Ext}(\rho) + F_T(\rho) \tag{3.15}$$

(with subscripts *C, XC, eN, Ext*, and *T* denoting Coulomb, exchange-correlation, electron-nuclear attraction, external, and kinetic energies respectively). It is crucial to remark that (3.15) is *not* the Kohn-Sham decomposition familiar in conventional presentations of DFT. There is no reference, model, nor auxiliary system involved in (3.15). From the construction presented above it is clear that in order to maintain consistency and to define functional derivatives properly all these functionals need to be defined on the same path in $\mathcal{B}_2(\mathcal{H}^N)$. These two observations lead to what may be unfamiliar definitions for the kinetic energy functional and the exchange-correlation functional, as follows:

$$F_C(\rho)=\frac{1}{2}\int\frac{\rho(\mathbf{y})\rho(\mathbf{y}')}{\|\mathbf{y}-\mathbf{y}'\|}d\mathbf{y}d\mathbf{y}'$$

$$F_T(\rho)=Tr\left\{-\frac{1}{2}\sum_i\nabla_i^2\left\{Q_*(\rho)Q_*(\rho)^\dagger\right\}\right\}\equiv Tr\left\{TQ_*(\rho)Q_*(\rho)^\dagger\right\}=E_T\left(D_*^N(\rho)\right)$$

$$F_{XC}(\rho)=Tr\left\{\left(\frac{1}{r_{12}}\right)Q_*(\rho)Q_*(\rho)^\dagger\right\}-F_C(\rho)=E_{XC}\left(D_*^N(\rho)\right) \tag{3.16}$$

$$F_{eN}(\rho)=-\int\rho(\mathbf{y})\sum_{i\mu}\frac{Z_\mu}{\|\mathbf{r}_i-\mathbf{R}_\mu\|}d\mathbf{y}$$

$$F_{Ext}(\rho)=\int\rho(\mathbf{y})\sum_i V(\mathbf{r}_i)d\mathbf{y}$$

Note in particular that the exchange-correlation functional that emerges here does not involve the kinetic energy. From the perspective of the DFT literature, (3.16) is a formulation of the Hohenberg-Kohn functional that is constructed to ensure that the functional derivatives required for variational minimization actually exist. We return to these issues in Sect. 3.3. Also note that in the time-dependent case the external potential $V(\mathbf{r}_i)$ is often considered to be explicitly time-dependent and further, that if nuclear motions also are taken into account the eN term is also time-dependent.

3.2 Factoring the Energy Functional through the First Order Reduced Density Operator.

An appealing way to apply the constraint expressed in Eq. (3.14) is to make connection with Natural Orbitals (31), in particular, to express ρ as a functional of the occupation numbers, ***n***, and Natural General Spin Orbitals (NGSO's), $\{\psi_i\}$, of the First Order Reduced Density Operator (FORDO) associated with the N-particle state appearing in the energy expression Eq. (3.8). In order to introduce the variables ***n*** and $\{\psi_i\}$ in a well-defined manner, the constrained search process Eq. (3.9) needs to be factored into two stages. The first search is over all N-particle states that produce the same FORDO and hence produces an energy functional of the FORDO. The second search is over all FORDO's that correspond to a fixed density, thus producing an energy functional of the density. This sequential process constructs paths in N-particle state space that are labeled by FORDO's and paths in the set of FORDO's that are labeled by densities. On these paths there is 1-1 correspondence among N-particle states, FORDO's and densities.

FORDO's are determined by their occupation numbers and their NGSO's, a relationship that is only unique up to unitary transformations that mix NGSO's with the same occupation numbers. However one can parameterize this association to make it unique. Hence on the paths determined by the constrained energy functional, one has a 1-1 correspondence between $\{n_i, \psi_i\}$ and densities, and densities thus can be viewed as a functional of $\{n_i, \psi_i\}$. This construction leads to variational equations for the ground state energy in terms of occupation numbers and NGSO's.

The FORDO is defined by a linear contraction map, C_N^1, given by the following

$$\begin{aligned} &C_N^1 : \mathcal{B}_1\left(\mathcal{H}^N\right) \to \mathcal{B}_1\left(\mathcal{H}^1\right) \\ &D^1 = C_N^1\left(D^N\right) \\ &D_{ij}^1 = Tr\left\{a_j^\dagger a_i D^N\right\} \\ &D^1 = \sum_{i,j=1}^{r} D_{ij}^1 a_i^\dagger a_j \end{aligned} \tag{3.17}$$

The energy functional, $O_H\left(D^1\right)$, of the FORDO is defined by

$$O_H\left(D^1\right) = \underset{D^N \in \mathcal{S}_N\left(D^1\right)}{Min} E_H\left(D^N\right) = E_H\left(D_\#^N\left(D^1\right)\right) \tag{3.18}$$

in terms of the energy E_H on the path $D_\#^N\left(D^1\right)$, where $\mathcal{S}_N\left(D^1\right) = \left[D^1\right]_N \cap \mathcal{S}_N$ is the set of N-particle states that contract to the same FORDO D^1. (Note that non-positive operators also contract to D^1 thus the content of $[D^1]_N$ is not limited to states.) The equivalence classes $\left[X\right]_N$; $X \in \mathcal{B}_1\left(\mathcal{H}^1\right)$ are defined in an analogous manner to $\left[\rho\right]_N$, by replacing Ξ_N^1 by C_N^1, $L_1(\mathbf{Y})$ by $\mathcal{B}_1\left(\mathcal{H}^1\right)$ in Eqs. (3.1), (3.2), (3.6) and considering off-diagonal values in Eq. (3.2). The energy functional F_H from Eq. (3.9) then can be expressed as

$$F_H(\rho) = \underset{D^1 \in \mathcal{S}_{N1}(\rho)}{Min} O_H\left(D^1\right) = O_H\left(D_\#^1(\rho)\right) \tag{3.19}$$

in terms of O_H defined on the path $D_\#^1(\rho)$. Here $\mathcal{S}_{N1}(\rho) = \left[\rho\right]_1 \cap \mathcal{S}_{N1}$ is the set of FORDO's that produce the same density, $\mathcal{S}_{N1}$ is the set of N-representable, FORDO's and $\left[\rho\right]_1$ the set of one-particle Trace Class operators that map to the density ρ. The equivalence classes $\left[\rho\right]_1$ in one-particle Trace Class operator space $\mathcal{B}_1(\mathcal{H}^1)$ are defined analogously to $[\rho]_N$, which are in N-particle operator space. The path $D_*^N(\rho)$ can then be expressed as the composition

$$D_*^N(\rho) = D_\#^N\left(D_\#^1(\rho)\right) = \left(D_\#^N \circ D_\#^1\right)(\rho) \tag{3.20}$$

A FORDO D^1 always can be expressed in natural form as

$$D^1 = \sum_{1\le i\le r} n_i |\psi_i\rangle\langle\psi_i| = \sum_{1\le i\le r} |v_i\rangle\langle v_i|$$
$$\langle\psi_i|\psi_j\rangle = \delta_{ij};\ \ \langle v_i|v_j\rangle = n_i\delta_{ij};\ 0\le n_i\le 1 \tag{3.21}$$
$$\sum_{1\le i\le r} n_i = N$$

where we have introduced the occupation-number-normalized NGSO's $\{v_i\}$. On the path $D_\#(\rho)$ there is a 1-1 correspondence

$$\rho \leftrightarrow D_\#^1(\rho) \leftrightarrow \{v_i(\rho); 1\le i\le r\} \equiv \mathbf{v}(\rho) \tag{3.22}$$

As this relationship is 1-1, the density ρ can be expressed as a function of $\mathbf{v}$, i.e. $\rho = \rho(\mathbf{v})$, and both ρ and $\mathbf{v}$ can be treated as equivalent, but different, variables for the argument of the energy functional F_H. The minimization of Eq. (3.14) that determines the ground state energy then becomes

$$E_0 = \underset{\mathbf{v}\in\mathcal{K}}{Min}\{F_H(\mathbf{v})\} \tag{3.23}$$

The feasible region $\mathcal{K}$ is defined by the constraints

$$h_N(\mathbf{v}) = \sum_{i=1}^{r} |v_i|^2 - N = 0$$
$$h_{ii}(\mathbf{v}) = |v_i|^2 - 1 \le 0 \tag{3.24}$$
$$h_{ij}(\mathbf{v}) = \langle v_i|v_j\rangle = 0;\ i\ne j$$

and the constrained minimum of Eq. (3.23) can be obtained from the Lagrangian

$$\mathcal{L}_F(\mathbf{v}) = F_H(\mathbf{v}) - \sum_{1\le i,j\le r} \lambda_{ij}h_{ij} - \mu h_N \tag{3.25}$$

Noting that

$$\rho(\mathbf{y}) = \sum_{1\le i\le r} \langle\mathbf{y}|v_i\rangle\langle v_i|\mathbf{y}\rangle \tag{3.26}$$

one has, via the chain rule,

$$\frac{\delta F_H}{\delta \overline{v}_i} = \frac{\delta F_H}{\delta\rho}\frac{\delta\rho}{\delta\overline{v}_i} = \frac{\delta F_H}{\delta\rho} v_i \tag{3.27}$$

It is useful to note that Eq. (3.27) defines the action of a local operator $\frac{\delta F_H}{\delta\rho}:\mathcal{H}^1 \to \mathcal{H}^1$ by

$$\frac{\delta F_H}{\delta \overline{v}_i}(\mathbf{y}) = \frac{\delta F_H}{\delta \overline{v}_i(\mathbf{y})} = \frac{\delta F_H}{\delta\rho(\mathbf{y})} v_i(\mathbf{y}) = \left[\frac{\delta F_H}{\delta\rho} v_i\right](\mathbf{y}) \tag{3.28}$$

From Eq. (3.25) and the change of variables in Eq. (3.27), we obtain Euler equations $\frac{\delta \mathcal{L}_F}{\delta \overline{v}_i} = \frac{\delta \mathcal{L}_F}{\delta v_i} = 0$, which can be expressed as the generalized eigenvalue problem

$$\left(\frac{\delta F_H}{\delta\rho} - \mu\right) v_i = \sum_{1\le j\le r} \lambda_{ij} v_j \tag{3.29}$$

and its complex conjugate. The component potential functionals can be obtained by taking the functional derivatives of Eq. (3.16) leading to exact, but in some cases, unknown expressions for $\frac{\delta F_C}{\delta\rho}, \frac{\delta F_T}{\delta\rho}, \frac{\delta F_{XC}}{\delta\rho}, \frac{\delta F_{eN}}{\delta\rho}$ and $\frac{\delta F_{ext}}{\delta\rho}$. These derivatives are all evaluated on the same path $D_*^N(\rho) = \left(D_\#^N \circ D_\#^1\right)(\rho)$, which again, leads to the definitions of $\frac{\delta F_T}{\delta\rho}$ and $\frac{\delta F_{XC}}{\delta\rho}$ as local potentials in the one-particle Eq. (3.29) for the density.

3.3 Relation with the Conventional Kohn-Sham Procedure.

The reader should note that no restrictions were placed on the form of the density expansion Eq. (3.26); in particular there is no limit on the number of terms. As already noted, therefore Eqs. (3.29) are not conventional Kohn-Sham equations. Rather they are an exact one-particle form of the Hohenberg-Kohn variation procedure and use Hohenberg-Kohn potentials in the definition of the effective one-particle Hamiltonian $\left(\frac{\delta F_H}{\delta\rho} - \mu\right)$. They have some kinship with the generalized Kohn-Sham equations treated, for example, by Levy and Perdew (32) but there is still a key difference. Unlike Kohn-Sham procedures, in (3.16) no auxiliary state has been introduced to provide a partitioning and regrouping of the terms.

At least for the case of a non-degenerate ground state of a closed shell system, it is possible to delineate the standard Kohn-Sham procedure quite sharply. (The caveat is directed toward issues of degeneracy at the Fermi level, fractional occupation, continuous non-integer electron number, and the like. In many but of course not all works, these aspects of the theory seem to be intertwined in an unanalyzed way with incompatible assumptions about single determinantal KS states.) For that specific case, the standard KS auxiliary state is a single determinant of singly occupied orbitals, which is an Independent Particle State (IPS) (any conventionally doubly occupied orbitals simply occur twice). The form is appealing because it incorporates Pauli exclusion explicitly, and is explicitly N-representable while being easy to manipulate.

With these motivational remarks, we now recover standard KS theory (in the particular instance just defined) from Eq. (3.16) for this specific case but with an important new constraint. Let $\mathcal{S}_{N1IP}(\rho)$ be the set of all Independent Particle FORDO's corresponding to ρ with precisely N non-zero terms:

$$D_{IP}^{1} = \sum_{1 \le i \le N} |\varphi_i\rangle\langle\varphi_i| \qquad \langle\varphi_i|\varphi_j\rangle = \delta_{ij} \tag{3.30}$$

By Harriman's theorem already cited there is always at least one such FORDO for each legitimate ρ. We now define the KS kinetic energy functional, which for a fixed number of particles is system independent, as

$$T_{KS}(\rho) = \underset{D_{IP}^{1} \in \mathcal{S}_{N1IP}(\rho)}{Min} Tr\{TD_{IP}^{1}\} = E_T\left(D_{IP\lozenge}^{N}(\rho)\right) \tag{3.31}$$

where E_T is defined in Eq. (3.16) and $D_{IP\lozenge}^{N}(\rho)$ is the unique Independent Particle N-particle state that corresponds to the minimizer $D_{IP\lozenge}^{1}(\rho)$ of this constrained optimization, which if the conditions described in (25) are satisfied determines paths $D_{IP\lozenge}^{1}(\rho)$ and $D_{IP\lozenge}^{N}(\rho)$ in $\mathcal{B}_1(\mathcal{H}^1)$ and $\mathcal{B}_1(\mathcal{H}^N)$ respectively. It should be noted that the paths $D_{IP\lozenge}^{1}(\rho)$ and $D_{IP\lozenge}^{N}(\rho)$ are in general *very* different from the paths $D_*^1(\rho)$, $D_\#^1(\rho)$ and $D_*^N(\rho)$. The definitions in Eqs. (3.16) can then be transformed to standard KS form by regrouping and defining difference functionals between exact and independent particle paths in the following manner (note that the HK kinetic energy and XC terms involve a system dependent path $D_*^N(\rho)$ in N-particle state space i.e. $F_T(\rho) = E_T\left(D_*^N(\rho)\right)$ and $F_{XC}(\rho) = E_{XC}\left(D_*^N(\rho)\right)$):

$$\tilde{F}_T(\rho) = T_{KS}(\rho) - F_T(\rho) \equiv E_T\left(D_{IP\lozenge}^{N}(\rho)\right) - E_T\left(D_*^N(\rho)\right) \tag{3.32}$$

then

$$F_T(\rho) = T_{KS}(\rho) - \tilde{F}_T(\rho) \tag{3.33}$$

The KS XC potential term can then be defined by

$$F_{XC,KS}(\rho) = F_{XC}(\rho) - \tilde{F}_T(\rho) \tag{3.34}$$

leading to the identification

$$F_T(\rho)+F_{XC}(\rho)=T_S(\rho)+F_{XC,KS}(\rho) \tag{3.35}$$

By construction, KS functionals are well-defined, (but note that the XC KS term is defined with the help of two distinct paths) and give well-defined functional derivatives, so their variation proceeds as in the preceding section, leading superficially to the standard KS equations.

There is, however, an important distinction that seems to have been missed in most if not all of the DFT literature (29,33). In essentially all presentations of the standard KS procedure, the functions corresponding to $\{\varphi_i\}$ are restricted to the reals, usually implicitly. In fact this restriction cannot be true in general without either a violation of the constraints to the proper path or, alternatively, forcing the extremum of the constrained functional to lie above the actual energy minimum. (To illustrate the point, Harriman's construction of single-determinants associated explicitly with a specified density relied upon complex orbitals.) The proof is simple. Harriman's theorem provides at least one determinant for each feasible density but with the orbitals restricted only to $L_2(\mathbf{Y})$. Fukutome (34) has shown, however, that all possible single determinants with orbitals from $L_2(\mathbf{Y})$ separate into eight distinct classes according to spin and time reversal symmetries. Therefore all possible densities can be so classified (the densities associated with each class have a unique topology). Since some of the classes have orbital forms which are manifestly complex, it follows that to include all feasible densities in the paths and at the same time search them with a single KS determinant, the determinant must in general have complex orbitals.

There are two immediate consequences of this result. First, is a previously unappreciated ambiguity in the so-called adiabatic connection formulation of $F_{XC,KS}[\rho]$. In that treatment, the functional is found from a Pauli coupling constant integral which usually is said to connect from "the non-interacting ground state", i.e. the KS determinant, to the fully interacting ground state. If however, the KS determinant is restricted to real orbitals, then in general that coupling constant integral will not be connecting to the ground state of the non-interacting system but only an upper bound to its ground state. Secondly, in general, the exact KS potential will not be a pure real function, contrary to the unstated assumption in essentially all of the literature. Alternatively, if one insists on real orbitals, then the single KS determinant must be given up and replaced by a suitably chosen and characterized multi-determinantal auxiliary function. Taken together, these previously unnoticed aspects of KS theory also provide a significant opportunity for improvement in practical approximations, a topic that we address in Section 5.

In the preceding discussion we have expanded the density in terms of $N \leq M \leq r$ functions that belong to the one-particle Hilbert space $\mathcal{H}^1$ such that their norms are *less than or equal to one* and the trace of the density is equal to N. All these expansions could in principle be exact; there is no need for $M = r = \infty$, as is clearly demonstrated in the KS procedure, where $M = N$. If $M < \infty$ and $r = \infty$, then new forms of auxiliary states, i.e. different from single determinantal ones, are implicitly introduced.

Another class of expansions is also possible, but in these the functions cannot be interpreted as belonging to the Hilbert space of one-particle states, even though they are functions of one space and one spin variable and do belong to a Hilbert space. In such expansions the norms of the functions are *less than or equal to N* and $1 \leq M \leq \infty$, while the trace of the density is equal to N. In the extreme case of $M = 1$ one can even express the density as $\rho = \omega\overline{\omega}$, where $\omega = \rho^{\frac{1}{2}}$. This factorization leads to the Pauli potential (35); we shall discuss it in detail elsewhere (30). For each value of M, while $r = \infty$ one could choose a different partitioning of the XC and kinetic energy in a similar fashion to Eq. (3.35). Such choices would be closely related to the generalized KS schemes already mentioned.

4. TIME-DEPENDENT DFT.

4.1 Characterization of the Lagrangian as a function of the density

The definition of the Lagrangian in Eq. (2.9) in terms of the paths x(t) needs to be modified when time-dependent densities $\rho(t)$ are considered as the variables x(t) in Eq. (2.9), as $\mathcal{L}$ is not a well-defined functional of $\rho(t)$. It is necessary to proceed somewhat parallel with the determination of paths in the preceding section in order to surmount this difficulty. In the context of K&S therefore, the Lagrangian which results will have no new content; rather, the analysis is restructured to make the functional dependence on the density precise and well-defined. First define an intermediate Lagrangian

$$\mathcal{L}_1\left(Q(t), Q(t)^\dagger, \rho(t), \lambda(t)\right) = \mathcal{L}\left(Q(t), Q(t)^\dagger, t\right) - \int_{\mathbf{Y}} \lambda(t, \mathbf{y}) g(t, \mathbf{y}) d\mathbf{y} \qquad (4.1)$$

where $\mathcal{L}\left(Q(t), Q(t)^\dagger, t\right)$ is the Lagrangian defined in Eq. (2.9), $\lambda(t)$ a time-dependent Lagrange multiplier function and the constraint function $g(t)$ is defined for a *fixed* time-dependent ρ as

$$g(t, \mathbf{y}) = \Xi_N^1\left(Q(t)Q(t)^\dagger\right)(\mathbf{y}) - \rho(t, \mathbf{y}) = 0 \qquad (4.2)$$

The equality in the preceding constraint is in the sense of the $L_1(\mathbf{Y})$ norm for fixed t. The actual Lagrangian controlling the dynamics then is defined as

$$\begin{aligned}\mathcal{L}(\rho(t)) &= \underset{Q(t)\in\mathcal{B}_2(\mathcal{H}^N)}{Min}\left\{\mathcal{L}_1\left(Q(t),Q(t)^{\dagger},\rho(t),\lambda(t)\right)\right\} \\ &= \mathcal{L}_1\left(Q_*(\rho(t)),Q_*(\rho(t))^{\dagger},\lambda_*(\rho(t))\right)\end{aligned} \tag{4.3}$$

where the paths $Q_*(\rho(t)) \in \mathcal{B}_2(\mathcal{H}^N)$ are defined by the constrained minimization in the same manner as in the time-independent case. This Lagrangian is of the same form as in (7) modified only by constrained search considerations in order to get a well-defined functional of the time-dependent density.

4.2 Exact Equations of Motion in terms of the Density

With the Lagrangian in hand, the principle of stationary action

$$\delta\mathcal{A} = \int_{t_i}^{t_f} \delta\mathcal{L}(\rho(t))dt = 0 \tag{4.4}$$

leads to the equation of motion

$$\{\rho,\mathcal{E}\} = \dot{\rho} \tag{4.5}$$

where the equivalent classical Hamiltonian is given by

$$\mathcal{E}(\rho(t)) = \frac{\mathcal{H}(\rho(t))}{\mathcal{S}(\rho(t))} = \frac{\left(Q_*(\rho(t))|\hat{H}Q_*(\rho(t))\right)}{\left(Q_*(\rho(t))|Q_*(\rho(t))\right)} \tag{4.6}$$

and the inner products in Eq. (4.6) are as defined in Eq. (2.2). The Poisson brackets are defined in a fashion similar to Eq. (2.7)

$$\{\mathrm{f}(t),\mathrm{g}(t)\} = \int_{\mathbf{Y}\times\mathbf{Y}} \frac{\delta \mathrm{f}(\rho(t))}{\delta\rho(\mathbf{y})}\xi(t,\mathbf{y},\mathbf{y}')\frac{\delta g(\rho(t))}{\delta\rho(\mathbf{y}')}d\mathbf{y}d\mathbf{y}' \tag{4.7}$$

but now with a metric integral kernel, which depends on $\rho(t) \in L_1(\mathbf{Y})$ that is defined by $\xi(\mathbf{y}',\mathbf{y}) = (\eta^{-1})(\mathbf{y}',\mathbf{y})$, where

$$\eta(\mathbf{y}',\mathbf{y}) = i\left\{\frac{\delta}{\delta\tilde{\rho}(\mathbf{y}')}\frac{\delta}{\delta\rho(\mathbf{y})} - \frac{\delta}{\delta\rho(\mathbf{y}')}\frac{\delta}{\delta\tilde{\rho}(\mathbf{y})}\right\}\ln\left(Q_*(\tilde{\rho})|Q_*(\rho)\right)\Big|_{\tilde{\rho}=\rho} \tag{4.8}$$

The time-dependent Eq. (4.5) in terms of the density is *exact* and *equivalent to* the full Heisenberg equation of motion when no approximations or models are invoked. It is thus worthwhile to display it in more detail

$$\frac{d\rho(t)}{dt} = \int_{\mathbf{Y}\times\mathbf{Y}} \frac{\delta\rho}{\delta\rho(\mathbf{y})}\xi(t,\mathbf{y},\mathbf{y}')\frac{\delta\mathcal{E}(\rho)}{\delta\rho(\mathbf{y}')}d\mathbf{y}d\mathbf{y}' \tag{4.9}$$

The first term in this integral is a delta function and produces pointwise equations

$$\frac{d\rho(t,\mathbf{y})}{dt} = \int_{\mathbf{Y}} \xi(t,\mathbf{y},\mathbf{y}') \frac{\delta \mathcal{E}(\rho)}{\delta \rho(\mathbf{y}')} d\mathbf{y}' \tag{4.10}$$

4.3 Exact Equations of Motion in terms of one-particle Functions

The time-dependent density can be expressed as a sum of products of unnormalized, time-dependent NGSO's analogously with Eq. (3.26)

$$\rho(t,\mathbf{y}) = \sum_{1\le i\le r} \langle \mathbf{y} | v_i(t) \rangle \langle v_i(t) | \mathbf{y} \rangle \tag{4.11}$$

and one can set up a 1-1 correspondence between $\rho(t)$ and $v(t) \equiv \{v_i(t); 1 \le i \le r\}$. This correspondence allows us to express the Lagrangian as a functional $\mathcal{L}(v,\bar{v})$. Note however that unlike functionals used in the Time-dependent Hartree Fock approximation (14), this Lagrangian *is not* complex analytic in the variables $(v,\bar{v})$ separately.

The equation of motion Eq. (4.5) and Eq. (4.9) can be expressed in terms of the variables $v(t) \equiv \{v_i(t); 1 \le i \le r\}$

$$\{v_i, \mathcal{E}\} = \dot{v}_i \equiv \{\bar{v}_i, \mathcal{E}\} = \dot{\bar{v}}_i \tag{4.12}$$

In more detail these relations are

$$\begin{aligned} \frac{dv_i(t)}{dt} &= \sum_{1\le j\le r} \xi_{ij}(v(t)) \frac{\delta \mathcal{E}(v(t))}{\delta v_j} \\ \frac{d\bar{v}_i(t)}{dt} &= \sum_{1\le j\le r} \bar{\xi}_{ij}(v(t)) \frac{\delta \mathcal{E}(v(t))}{\delta \bar{v}_j} \end{aligned} \tag{4.13}$$

where

$$\eta_{ij}(v(t)) = i \left\{ \frac{\delta}{\delta \tilde{v}_i} \frac{\delta}{\delta v_j} - \frac{\delta}{\delta v_i} \frac{\delta}{\delta \tilde{v}_j} \right\} \ln\left(Q_*(\tilde{v}) | Q_*(v)\right) \Big|_{\tilde{v}=v} \tag{4.14}$$

and $\xi_{ij} = \left(\eta^{-1}\right)_{ij}$. Using Eq. (4.11) the components $\nabla_{v_i} \mathcal{E}$ of the gradient of the equivalent *classical energy* can be expanded as

$$\frac{\delta \mathcal{E}(v(t))}{\delta v_j} = \int_{\mathbf{Y}} \frac{\delta \mathcal{E}(v(t))}{\delta \rho(\mathbf{y})} \frac{\delta \rho(\mathbf{y})}{\delta v_j} d\mathbf{y} = \int_{\mathbf{Y}} \frac{\delta \mathcal{E}(v(t))}{\delta \rho(\mathbf{y})} \bar{v}_j(\mathbf{y}) d\mathbf{y} \equiv \frac{\delta \mathcal{E}(v)}{\delta \rho} \bar{v}_j \tag{4.15}$$

and

$$\frac{\delta \mathcal{E}(v(t))}{\delta v_j} = \frac{1}{\mathcal{S}(v(t))} \left\{ \frac{\delta \mathcal{H}(v(t))}{\delta v_j} - \mathcal{E}(v(t)) \frac{\delta \mathcal{S}(v(t))}{\delta v_j} \right\} \tag{4.16}$$

Using Eqs (4.11), (4.13), (4.15) and (4.16) one then can obtain time-dependent equations for the unnormalized NGSO's ν

$$\frac{d\bar{\nu}_i}{dt} = \frac{1}{\mathcal{S}(\nu)} \sum_{1 \le j \le r} \xi_{ij}(\nu) \left\{ \frac{\delta \mathcal{H}(\nu)}{\delta \rho} - \mathcal{E}(\nu) \frac{\delta \mathcal{S}(\nu)}{\delta \rho} \right\} \nu_j \tag{4.17}$$

Note that the time-dependent version of the constraints of Eq. (3.24)

$$\begin{aligned} h_N(\nu(t)) &= \sum_{i=1}^{r} |\nu_i(t)|^2 - N = 0 \\ h_{ii}(\nu(t)) &= |\nu_i(t)|^2 - 1 \le 0 \\ h_{ij}(\nu(t)) &= \langle \nu_i(t) | \nu_j(t) \rangle = 0; \; i \ne j \end{aligned} \tag{4.18}$$

need to be maintained, which necessitates the use of time-dependent Lagrange parameters in a modified evolution equation based on Eq. (4.17).

Again we point out the rather remarkable fact that the coupled Eqs. (4.17) are *exact* and are equivalent to the full HEM if no other constraints, other than those of Eq. (4.18), are placed on ν, $\mathcal{H}$, nor $\mathcal{S}$. These coupled equations describe evolution paths in $\mathcal{H}^1 \equiv L_2(\mathbf{Y})$ and the local potentials $\frac{\delta \mathcal{H}}{\delta \rho}$ and $\frac{\delta \mathcal{S}}{\delta \rho}$ act as time-dependent operators mapping $\mathcal{H}^1 \to \mathcal{H}^1$. The curvature tensor $\xi_{ij}(\nu)$ can be obtained either from the expression in Eq. (4.14) or from Eq. (4.8) by using the coordinate transformation Eq. (4.11) that introduces matrix elements of the operator defined by the kernel

$$\eta(\mathbf{y}', \mathbf{y}) = i \left\{ \frac{\delta}{\delta \tilde{\rho}(\mathbf{y}')} \frac{\delta}{\delta \rho(\mathbf{y})} - \frac{\delta}{\delta \rho(\mathbf{y}')} \frac{\delta}{\delta \tilde{\rho}(\mathbf{y})} \right\} \ln\left(Q_*(\tilde{\rho}) | Q_*(\rho) \right) \Big|_{\tilde{\rho} = \rho} \tag{4.19}$$

between NGSO's $\{\nu_i(t); 1 \le i \le r\}$.

The local potential $\frac{\delta \mathcal{H}}{\delta \rho}$ can be expanded in terms of the component potentials Eq. (3.16), now considered as time-dependent, leading to a time-dependent exchange-correlation potential

$$\frac{\delta F_{XC}(\rho(t))}{\delta \rho} = \frac{\delta}{\delta \rho} \left\{ Tr\left\{ \left(\frac{1}{r_{12}} \right) Q_*(\rho(t)) Q_*(\rho(t))^\dagger \right\} - F_C(\rho(t)) \right\} \tag{4.20}$$

where the path $Q_*(\rho(t))$ is defined by Eq. (4.3).

4.4 Other Forms of Time-Dependent Density Functional Theory

Time-Dependent Density Functional theory (TDDFT) has been considered with increasing interest since the late 1970's and many papers have been published on the subject. The treatments presented by Runge and Gross (36) and Gross and Kohn (37) are widely cited in the discussion of the evolution of pure states. The evolution of mixed states has been considered extensively by Rajagopal et al. (38), but that treatment differs in many aspects from the form given here.

In essentially all of the prior formulations of TDDFT a complex Lagrangian is used, which would amount to using the full expectation value in Eq. (2.9), not just the real part as in our presentation. The form we use is natural for conservative systems and, if not invoked explicitly at the outset, emerges in some fashion when considering such systems. A discussion of the different forms of Frenkel's variational principle, although not in the context of DFT, can be found in (39).

Another place where we diverge from other developments of TDDFT is in the use of the *metric term* Eq. (2.9). These terms arise in a non-trivial manner as the paths $Q_*(\rho)$ are manifestly nonlinear functionals of ρ and thus have significant affects on the evolution of the density. Regarding the Time-Dependent KS form of the theory as used by e.g. Theilhaber (40), it has been suggested (14) that the metric terms should cancel as in TDHF. However we do not concur with this suggestion as the overlap functional that appears in the Lagrangian that produces the one-particle equations depends on the paths $D_*^N(\rho), Q_*(\rho)$ in $\mathcal{B}_1(\mathcal{H}^N)$ and $\mathcal{B}_2(\mathcal{H}^N)$ respectively, not on independent particle paths. Independent particle paths are determined by these equations via the auxiliary KS single determinantal state, but *it is not* those auxiliary states that appear in the Lagrangian nor determine the generalized phase space metric. Our form of TDKS equations would modify Eqs. (17)-(18) to refer to only N one-particle functions, (thus each function must be normalized to 1), and use the kinetic and XC functionals $T_S(\rho)$ and $F_{XC,KS}(\rho)$ from Eq. (3.35) to generate the potentials in the evolution equations.

One could view the occurrence of the metric terms in the equations of motion as an annoying complication, but we hold a more positive view. First they assure that whatever the choice of parameters to be used as dynamical variables, that choice will not introduce unphysical artifacts. Second, the metric terms are another component of the theory with potential for providing guiding principles for development of XC models. Those terms also allow the mathematical origin of physical affects to be assigned.

The mixed state TDDFT of Rajagopal et al. (38) differs from our formulation in the aspects mentioned above and in the nature of the operator space where the *supervectors* reside. A particularly notable distinction is in the use of the factorization $D = QQ^{\dagger}$ of the state density operator that leads to unconstrained variation over the space of Hilbert-Schmidt operators, rather than to a constrained variation of the space of Trace-Class operators.

For a review of TDDFT the reader should consult (36) and Görling (41). In the latter work TDKS is developed and a fairly exhaustive list of TDDFT references is given.

4.5 Time-Independent NGSO Equations from Time-Dependent Theory

The critical points of the equivalent classical Hamiltonian $\mathcal{E}$ occur at stationary state energies of the quantum Hamiltonian H and correspond to stationary states in both the quantum and generalized classical pictures. These points are characterized by the constrained generalized eigenvalue equation obtained by setting the time variation to zero in Eq. (4.17)

$$\begin{aligned} \frac{d\bar{v}_i}{dt} = 0 = \frac{1}{\mathcal{S}(v)} \sum_{1\le j\le r} \xi_{ij}(v)\left\{\frac{\delta\mathcal{H}(v)}{\delta\rho} - \mathcal{E}(v)\frac{\delta\mathcal{S}(v)}{\delta\rho}\right\}v_j \\ \Rightarrow \left\{\frac{\delta\mathcal{H}(v)}{\delta\rho} - \mathcal{E}(v)\frac{\delta\mathcal{S}(v)}{\delta\rho}\right\}v_j = 0;\ 1\le j\le r \end{aligned} \tag{4.21}$$

The ground state energy corresponds to the lowest value of $\mathcal{E}$ that satisfies Eq. (4.21) in a self-consistent manner. The variables v belong to the nonlinear manifold defined by the constraints expressed in Eq. (3.24). Often it is possible to find intrinsic coordinates for this manifold and convert the problem into an unconstrained one. Notice in particular that the N-particle state energy, $\mathcal{E}$, appears explicitly in Eq. (4.21), thus this equation can be used to determine the ground state energy, which is the lowest self-consistent root of Eq. (4.21) or excited state energies given by higher self-consistent roots. Techniques commonly used in geometry optimization to obtain saddle points could used be effectively in this latter context.

5. APPROXIMATIONS

5.1 Model Hamiltonians

In the preceding sections we have introduced effective one-particle equations, both time-dependent and time-independent, for one-particle functions that determine the density through the expansion

$$\rho(t,\mathbf{y}) = \sum_{1\le i\le M} \langle \mathbf{y}|v_i(t)\rangle\langle v_i(t)|\mathbf{y}\rangle;\ 1\le M\le r \tag{5.1}$$

The time-independent case corresponds to fixed time $t=0$. The only constraints on this expansion are that the Hilbert space norm of the orthogonal functions $\{v_i\}$ should be less than or equal to one, *if the functions are to be interpreted as one-particle states* and less than or equal to N otherwise, and the integral of ρ over space and spin variables should equal the number of particles. The effective one-particle Hamiltonian that determines these functions in Eqs. (3.29) and (4.21) in general depends on the density and its derivatives and is composed of terms that make up the Hohenberg-Kohn energy functional. The equations we have displayed are exact and lead to the exact solutions of the quantum mechanical equations. However finding exact solutions is not possible in general because (a) the XC potential terms is not known, (b) the metric tensor is not known in the time-dependent case, and (c) it is not feasible to solve the equations if the dimension of the space of functions is infinite and $M=\infty$ (i.e. the standard form of the equations that we have presented).

Approximations thus must be introduced that involve modeling both the XC potential and the metric tensor, and a truncation of the space within which to choose the unknown functions $\{v_i\}$ to finite dimension $r<\infty$. The modeling is based on the restricted ansatz chosen for the form of states used to determine paths that approximate $D_*^N(\rho), D_\#^1(\rho)$ and $D_\#^N\left(D^1\right)$. It can be carried out, for example, by postulating and fitting functional forms involving the density, its derivatives and fitting parameters to match the properties of high quality CI calculations. If the expansion size $M\neq r$ then, in analogy to the KS case, the form of the exact potentials will be different and thus the chosen functional forms in the modeling/fitting procedure will have different properties.

Traditionally the expansion Eq. (5.1) used in the KS procedure has been in terms of $\frac{N}{2}$ real functions of space variables only. In order to allow some spin-polarized solutions this treatment is extended to allow N real space functions half associated with alpha spin and half with beta spin. Generalizations of the KS procedure would allow functions of a more general form and expansions with a greater number of functions than N in Eq. (5.1). Such generalizations would be based on auxiliary states other than a real restricted single determinant. Approximate forms of these *generalized* KS equations would correspond to the approximate forms of the time-independent one-particle equations discussed in this article in the case when $M\neq r$. More detail is found in (42).

Approximate time-dependent KS formulations differ more sharply from approximations to our time-dependent formulation than do the time-independent ones as they do not explicitly refer to a time-dependent metric term. In time-dependent KS, these terms either are added implicitly to the approximate XC potential or combined with other potentials.

5.2 Symmetry Constraints

All of the approximation procedures noted above lead to density-dependent effective one-particle Hamiltonians. In such approximation schemes it is possible to obtain better results by relaxing some physical symmetry constraints on the form of approximate solutions i.e. allowing symmetry-broken solutions. The types of symmetry-broken solutions that are possible by relaxing spin and time-reversal symmetry have been discussed at length in (42), where we applied Fukutomes' analysis of the Hartree-Fock solutions to DFT. An alternative perspective on that same analysis is that, if more general solutions were included in the formulation of an approximate Hamiltonian, the resulting solutions would not be symmetry-broken. In short, an approximate Hamiltonian which does not have the generality discussed in the preceding subsection may yield an energetically favorable solution by breaking symmetry. Such symmetry breaking can be construed as a reintroduction of the missing flexibility.

REFERENCES

1. Linderberg J., Öhrn Y., "Propagators in Quantum Chemistry"; Academic Press, London, 1973.
2. Öhrn Y., Born G., Adv. Quantum Chem., 1981, 13, 1.
3. Weiner B., Jensen H. J. Aa, Öhrn Y., J. Chem. Phys., 1983, 80, 2009.
4. Linderberg J., Öhrn Y., Int. J. Quantum Chem., 1977, 12, 161;
   Öhrn Y., Linderberg J., Int. J. Quantum Chem., 1979, 15, 343.
5. Ortiz J. V., Weiner B., Öhrn Y., Int. J. Quantum Chem., 1981, S15, 113;
   Jensen H. J. Aa, Weiner B., Öhrn Y., Int. J. Quantum Chem., 1982, S16, 615;
   Weiner B., Öhrn Y., J. Phys. Chem., 1987, 91, 563.
6. Hohenberg H., Kohn W., Phys. Rev. B, 1964, 136, 864.
   Kohn W., Sham L. J., Phys. Rev A, 1965, 140, 1133.
   Dreizler R. M., Gross E. K. U., “Density Functional Theory”; Springer Verlag, Heidelberg, 1990.
   Parr R.G., Yang W., “Density Functional Theory of Atoms and Molecules”; Oxford University Press, 1989.
7. Kramer P., Saraceno M., "Geometry of the Time-Dependent Variational Principle in Quantum Mechanics"; Springer, New York, 1981.
8. Zhang W., Feng D. H., Gilmore R., Rev. Mod. Phys. 1990, 62, 867.
9. Klauder J. R., Skagerstam B. S., "Coherent States. Applications in Physics and Mathematical Physics"; World Scientific, Singapore, 1985
10. Feng D. H., Klauder J. R., Strayer M. R., "Coherent States: Past, Present and Future"; World Scientific, Singapore, 1994.
11. Perelomov A. M., "Generalized Coherent States and their Applications"; Springer, New York, 1986.

12. Glauber R. J., Phys. Rev., 1963, 131, 2766.
13. Thouless D. J., "The Quantum Mechanics of Many-Body Systems"; Academic, New York, 1961.
14. Deumens E., Diz A., Longo R., Ohrn Y., Rev. Mod. Phys. 1994, 66, 917.
15. Mechtly B., Shaw P. B., Weiner B., Phys. Rev. A, 1989, 40, 7275.
16. Blaziot J.P., Orland H., Phys. Rev. C, 1981, 24, 1740.
17. Deumens E., Ohrn Y., Weiner B., J. Math. Phys., 1991, 32, 1166.
18. Weiner B., Deumens E., Ohrn Y., J. Math. Phys., 1994, 35, 1139.
19. Zwanzig, R., Physica, 1964, 1109.
20. Harriman J. Phys. Rev. A, 1981, 24, 680.
21. Frenkel J. "Wave Mechanics: Advanced General Theory"; Clarendon, Oxford 1934.
22. Weiner B., Trickey S.B., "Mixed State TDVP" unpublished.
23. Press W. H., Flannery B. P., Teukolsky S. A., Vetterling W. T., "Numerical Recipes"; Cambridge University 1986.
24. Levy M., Proc. Natl. Acad. Sci., 1979, 76, 6062; Bull. Am. Phys. Soc., 1979, 24, 626; J. Chem. Phys., 1979, 70, 1573.
25. Hestenes M. R., "Optimization Theory"; Wiley, New York 1975.
26. Kryachko E. S., Ludeña E. V., "Energy Density Functional Theory of Many-Electron Systems"; Kluwer, Dordrecht, 1990.
27. Bader R. F. W., "Atoms in Molecules - A Quantum Theory"; Oxford University Press, Oxford, 1990.
28. Coleman A. J., Rev. Mod. Phys., 1963, 35, 668.
29. Savin A., In "Recent Developments and Applications of Modern Density Functional Theory", Seminario J. M., Ed., Elsevier, 1996, 327.
30. Weiner B., Trickey S.B., "Paths in DFT", unpublished.
31. Löwdin P. O., 1955, Phys. Rev. 97, 1474.
32. Levy M., Perdew J P., In "Density Functional Methods in Physics", Dreizler R.M., da Providencia J., Eds. Plenum, New York, 1985, 11.
33. Theophilou A. K. 1998, In "Proceedings of 1997 Duke DFT Satellite Symposium", Levy M., Yang W., Trickey S.B., Eds. Int. J. Quantum Chem. (in press).
34. Fukutome H., Int. J. Quantum Chem., 1981, 20, 955.
35. March N.H., Phys. Lett., 1981, 84A, 319;
Levy M., Perdew J.P., Sahni V., Phys. Rev. A, 1984, 30, 2745;
Levy M. Ou-Yang H., Phys. Rev. A, 1988, 38, 625;
Holas A., March N.H., Phys. Rev. A, 1991, 44, 5521;
Nagy A., March N.H., Int. J. Quantum Chem., 1991, 39, 615;
Flores J.A., Keller J., Phys. Rev. A, 1992, 45, 6259.
36. Runge E., Gross E. K., Phys. Rev. Lett., 1884, 52, 997.
37. Gross E. K., Kohn W., Adv. Quantum Chem., 1990, 27, 255.
38. Rajagopal A. K., Bout F. A., Phys. Lett. A, 1994, 195, 312.
Rajagopal A. K., Phys. Lett. A, 1994, 193, 363.
Rajagopal A. K., Bout F. A., Phys. Rev E, 1994, 50, 721.

Rajagopal A. K., Bout F. A., Phys. Rev A, 1995, 51, 1883.
39. Christiansen O., Jorgensen P., Hattig C., Int. J. Quantum Chem. (in press).
40. Theilhaber J., Phys. Fluids B (Plasma Phys.), 1992, 4, 2044;
Phys. Rev. B, 1992, 46, 12990;
Int. J. Quantum Chem., Quantum Chem. Symp., 1994, 28, 611.
41. Görling A., Phys. Rev. A, 1997, 55, 2630.
42. Weiner B., Trickey S.B., 1998, In "Proceedings of 1997 Duke DFT Satellite Symposium", Levy M., Yang W., Trickey S.B., Eds. Int. J. Quantum Chem. (in press).

APPENDIX A: GLOSSARY OF MATHEMATICAL SYMBOLS

$\mathcal{S}_N$	The convex set of N-particle states
$\mathcal{S}_N(\rho)$	The set of N-particle states that produce the same density ρ
$\mathcal{S}_{N1}$	The of First Order Reduced Density Operators (FORDO's)
$\mathcal{S}_{N1}(\rho)$	The set of FORDO's that produce the same density ρ
$\mathcal{S}_{N1IP}(\rho)$	The set of Independent Particle FORDO's that produce the same density ρ
$\mathcal{S}_{Np}$	The set of p-particle Trace Class operators that are N -representable.
$\mathcal{B}_1(\mathcal{H}^N)$	The space of trace class operators acting in the Hilbert space $\mathcal{H}^N$
$\mathcal{H}^N$	The Hilbert space of pure N-particle fermion states. It is an N-fold antisymmetric tensor product of the Hilbert space of pure one-particle states.
D^N	N-particle state operator, if it is not a projector onto a 1D subspace it represents a mixed N-particle state
$\mathcal{B}_2(\mathcal{H}^N)$	The Hilbert space of Hilbert Schmidt operators acting in $\mathcal{H}^N$
$\hat{H}$	The Hamiltonian superoperator (the Hamiltonian represented as an operator acting on Hilbert Schmidt operators)
$(\mid)$	Inner product in $\mathcal{B}_2(\mathcal{H}^N)$
$\mathcal{M}$	Submanifold (in general, nonlinear) of operators in $\mathcal{B}_2(\mathcal{H}^N)$.
$\mathcal{E}(\mathbf{x})$	Equivalent classical Hamiltonian defined on $\mathcal{M}$; $\mathbf{x}$ is a coordinate system on $\mathcal{M}$ i.e. the points of $\mathcal{M}$ are states and the coordinates of these points are $\mathbf{x}$.

$\mathcal{S}(\mathbf{x})$	Normalization function of states in $\mathcal{M}$
$\{,\}$	Poisson Brackets defined on tangent spaces of $\mathcal{M}$
ξ	The symplectic metric tensor defined on the tangent spaces of $\mathcal{M}$
$\frac{\hat{\partial}}{\partial t}$	Superoperator acting on Hilbert-Schmidt operators producing their time derivative.
$\mathcal{L}$	Lagrangians
$\binom{r}{N}$	Combinatorial coefficient "r choose N"
$\rho(\mathbf{y})$	Charge density as a function of space and spin variables $(\mathbf{r},\sigma)$
$\mathbf{Y}$	$\mathbf{R}^3 \times \mathbf{C}^2$
$L_1(\mathbf{Y})$	Linear normed space of absolutely integrable complex-valued functions of 3 real and 2 complex variables. Arbitrary elements are denoted by ζ and densities by ρ
$L_2(\mathbf{Y})$	Hilbert space of square-integrable complex-valued functions of 3 real and 2 complex variables.
Ξ_N^1	Linear map from the space of bounded N-particle operators to the space of absolutely integrable complex-valued functions of the variables $\mathbf{y}$.
C_N^p	Contraction map from N-particle Trace Class Operators to p-particle Trace Class operators.
$\Phi(\mathbf{y})$	Continuous Fermi field annihilation operator that depends on the space-spin variable $\mathbf{y}$.
$\{a_i, a_i^\dagger\}$	Discrete Fermi field annihilation and creation operators
$\mathcal{B}_1(\mathcal{H}^N)/Ker\Xi_N^1$	Linear space of equivalence classes of Trace Class operators. The operators are equivalent if there difference lies in the kernel of Ξ_N^1
$\mathcal{P}_{1N}$	Positive cone of space-spin densities derived from N-particle states.
E_H	Linear energy functional based on the Hamiltonian H; it acts on the space of N-particle Trace Class operators.
E_T	Linear energy functional based on the Kinetic Energy operator.
E_{XC}	Linear energy functional based on the XC terms.
F_H	Nonlinear energy functional based on the Hamiltonian H; it acts on the space of absolutely integrable complex valued functions of the variables $\mathbf{y}$.

O_H	Nonlinear energy functional based on the Hamiltonian H; it acts on the space of one-particleTrace Class operators.
$D_*^N(\rho)$	Path of N-particle states; each state on the path corresponds to a density ρ and is the minimum energy state for that density.
$D_\#^N(D^1)$	Path of N-particle states; each state on the path corresponds to a fixed FORDO and is the minimum energy state for that FORDO.
$D_\#^1(\rho)$	Path of FORDO's; each FORDO on the path corresponds to a density ρ and is the FORDO that corresponds to the minimum energy state for that density.
$D_{IP\Diamond}^N(\rho)$ and $D_{IP\Diamond}^1(\rho)$	Paths of IP N-particle states and the unique FORDO's that correspond to these states.
$Q_*(\rho)$	Path of Hilbert Schmidt operators
$[\zeta]_N$	Equivalence classes of N-particle Trace Class operators that all map to the same function ζ.
$[\zeta]_1$	Equivalence classes of 1 particle Trace Class operators that all map to the same function ζ
$[X]_N$	The set of N-particle Trace Class operators that contract to the one-particle operator X.
$\langle\rho\rangle_N$	Set of N-particle Hilbert-Schmidt operators that produce N-particle states associated with the same space-spin density.
C	The complex numbers
CS	Coherent State
FORDO	First Order Reduced Density Operator
HK	Hohenberg-Kohn
K&S	Kramer and Saraceno
KS	Kohn-Sham
NGSO	Natural General Spin Orbitals
R	The real numbers
TDHF	Time-Dependent Hartree-Fock
TDVP	Time-Dependent Variational Principle

APPENDIX B: SPACES OF OPERATORS

Bounded Operators $\mathcal{B}(\mathcal{H})$

The set of bounded operators acting in a Hilbert space $\mathcal{H}$ form a normed linear space. The norm is given by the bound on the operator

$$\|X\| = \sup_{\psi} \frac{\langle \psi | X\psi \rangle}{\langle \psi | \psi \rangle}; \quad \psi \in \mathcal{H} \tag{B1}$$

Trace Class Operators $\mathcal{B}_1(\mathcal{H})$

The set of trace class operators form a subset of the set of bounded operators defined by

$$\mathcal{B}_1(\mathcal{H}) = \left\{ X; Tr\left\{ \left(X^\dagger X \right)^{\frac{1}{2}} \right\} < \infty \right\} \tag{B2}$$

This set also forms a normed linear space with norm defined by

$$\|X\|_1 = Tr\left\{ \left(X^\dagger X \right)^{\frac{1}{2}} \right\} \tag{B3}$$

Hilbert-Schmidt Operators $\mathcal{B}_2(\mathcal{H})$

The Hilbert-Schmidt operators are another subset of the set of Bounded Operators defined by

$$\mathcal{B}_2(\mathcal{H}) = \left\{ X;\ Tr\left\{ X^\dagger X \right\} < \infty \right\} \tag{B4}$$

This set forms a Hilbert space with an inner product defined by

$$\left(X | Y \right) = Tr\left\{ X^\dagger Y \right\} \tag{B5}$$

which defines the Hilbert-Schmidt norm

$$\|X\|_2 = \left(X | X \right)^{\frac{1}{2}} \tag{B6}$$

The set of Trace Class operators and Hilbert-Schmidt operators are not in general contained in each other, but they are connected in the following manner

$$X \in \mathcal{B}_2(\mathcal{H}) \Rightarrow X^\dagger X \in \mathcal{B}_1(\mathcal{H}) \tag{B7}$$

If the dimension of $\mathcal{H}$ is finite all of these spaces of operators are identical.

APPENDIX C: INVARIANCE GROUPS

If X is an operator then its invariance group *Inv*(X) is defined to be

$$Inv(X) = \left\{ U;\ UXU^\dagger = X;\ U^\dagger U = UU^\dagger = I \right\} \tag{C1}$$

If v is a vector then its invariance group *Inv*(v) is defined by

$$Inv(\mathrm{v}) = \left\{ U;\ U\mathrm{v} = \mathrm{v};\ U^\dagger U = UU^\dagger = I \right\} \tag{C2}$$

Quantum Control in Semiconductor Heterostructures

Jeffrey L. Krause
Quantum Theory Project
University of Florida
Gainesville, FL 32611-8435

Abstract

Specially tailored, ultrafast laser fields can be used to control the charge carriers in semiconductor heterostructures. Two examples are presented in this work. The first is a ten-well superlattice with an injection well and a detection well. The laser field that creates the maximum density in the detection well is computed using the techniques of optimal control theory. In the second example, a genetic algorithm is used to design a superlattice in which the overlap of the lowest five eigenfunctions is as large as possible. A subsequent round of optimization determines the laser field that drives an electronic wave packet to maximum overlap with a chosen target distribution, at a chosen target time.

1. Forward

It is a pleasure to contribute this paper to a volume celebrating the 65th birthday of Yngve Öhrn. Yngve is my official Mentor at the University of Florida, and he is an inspiration to me, both personally and professionally. Yngve has taught me that to do science you have to be tough, but also have a sense of humor.

2. Introduction

Quantum wells are produced by epitaxially growing alternating layers of an insulator and a semiconductor. In the resulting structures, the charge carriers (holes and electrons) are constrained in the direction of growth, and unconstrained in the orthogonal dimensions. Unlike molecular potentials, the constraining potentials in quantum wells can be designed to have nearly arbitrary shape. Application of a DC field causes the electron levels in the conduction band to move into tunneling resonance, while the hole levels in the valence band move out of resonance. Thus, varying the DC field allows control of the energy levels and density of states. When a heterostructure is

ADVANCES IN QUANTUM CHEMISTRY, VOLUME 35

0065-3276/99 $30.00

excited with an appropriate ultrafast laser pulse, an electronic wave packet is created in the conduction band, which tunnels back and forth across the structure [1]. As the wave packets oscillate they radiate, typically in the THz regime [2].

Advances in ultrafast laser technology, in conjunction with improved techniques for growing semiconductor heterostructures, have enabled detailed investigations of coherent electronic dynamics and carrier-carrier interactions in the solid state. High quality structures exhibit dephasing times well in excess of the 10 fs resolution possible with existing ultrafast light sources. Investigations of the coherent interaction of a laser field with the induced polarization in a heterostructure have led to the observation of a variety of coherent phenomena, including Bloch oscillations [3], heavy and light hole quantum beats [1], Rabi-flopping [4], and THz and far-infrared emission [5–7].

The ability to create and observe coherent dynamics in heterostructures offers the intriguing possibility to control the dynamics of the charge carriers. Recent experiments have shown that control in such systems is indeed possible. For example, phase-locked laser pulses can be used to coherently amplify or suppress THz radiation in a coupled quantum well [5]. The direction of a photocurrent can be controlled by exciting a structure with a laser field and its second harmonic, and then varying the phase difference between the two fields [8,9]. Phase-locked pulses tuned to excitonic resonances allow population control and coherent destruction of heavy hole wave packets [10]. Complex filters can be designed to enhance specific characteristics of the THz emission [11,12]. These experiments are impressive demonstrations of the ability to control the microscopic and macroscopic dynamics of solid-state systems.

The purpose of this work is to demonstrate that the techniques of quantum control, which were developed originally to study atoms and molecules, can be applied to the solid state. Previous work considered a simple example, the asymmetric double quantum well (ADQW). Results for this system showed that both the wave packet dynamics and the THz emission can be controlled with simple, experimentally feasible laser pulses. This work extends the previous results to superlattices and chirped superlattices. These systems are considerably more complicated, because their dynamic phase space is much larger. They also have potential applications as solid-state devices, such as ultrafast switches or detectors.

3. Theory

3.1. Semiconductor Dynamics

Calculating the exact response of a semiconductor heterostructure to an ultrafast laser pulse poses a daunting challenge. Fortunately, several approximate methods have been developed that encompass most of the dominant physical effects. In this work a model Hamiltonian approach is adopted to make contact with previous advances in quantum control theory. This method can be systematically improved to obtain agreement with existing experimental results. One of the main goals of this research is to evaluate the validity of the model, and to discover the conditions under which it can be reliably applied.

The wave function for confined carriers in a quantum well can be written as [13],

$$\varphi^{\alpha}_{n,\vec{k}}(\vec{r}) = e^{i\vec{k}\cdot\vec{r}} F^{\alpha}_{n}(z) u^{\alpha}(\vec{r}), \tag{1}$$

where $\alpha = c, h, l$, labels the charge carrier (electron, heavy-hole or light-hole, respectively), n labels the subband, u^{α} are the band edge bulk Bloch states, and $\vec{k}$ is the wave vector of the plane wave in the $x - y$ plane. The envelope functions, $F^{\alpha}_{n}(z)$, are solutions of a one-dimensional, effective-mass Schrödinger equation (in atomic units) [13–15],

$$\left[-\frac{1}{2m_{\alpha}}\frac{d^2}{dz^2} + V_{\alpha}(z) \pm E_0 z\right] F^{\alpha}_{n}(z) = E^{\alpha}_{n} F^{\alpha}_{n}(z), \tag{2}$$

where m_{α} is the effective mass along the z-direction and E_0 is the magnitude of the applied DC field. In the third term, the upper sign $(+)$ is used for the holes, and the lower sign $(-)$ is used for the electrons. $V_{\alpha}(z)$ is the confining potential along the direction of growth.

In the DC-biased structures considered here, the dynamics are dominated by electronic states in the conduction band [1]. A simplified version of the theory assumes that the excitation occurs only at zone center. This reduces the problem to an n-level system (where n is approximately equal to the number of wells in the structure), which can be solved using conventional first-order perturbation theory and wave-packet methods. A more advanced version of the theory includes all of the hole states and electron states subsumed by the bandwidth of the excitation laser, as well as the perpendicular k states. In this case, a density-matrix picture must be used, which requires a solution of the time-dependent Liouville equation. Substituting the Hamiltonian into the Liouville equation leads to a modified version of the optical Bloch equations [13,15]. These equations can be solved readily, if the k states are not coupled (i.e., in the absence of Coulomb interactions).

The simplified theory is adequate to obtain qualitative agreement with experiment [1,16]. Comparisons between the simplified and more advanced versions of the theory show excellent agreement for the dominant (electronic) contribution to the time-dependent dipole moment, except during the initial excitation, where the k states are coupled by the laser field [17]. The contributions to the dipole from the heavy holes and light holes are not included in the simplified approach. This causes no difficulty in the ADQW because the holes are trapped and do not make a major contribution to the dynamics [1]. This assumption may not be valid in the more general case of superlattices, as discussed below.

3.2. Quantum Control

The simplified theory allows the time-dependent wave function to be calculated rapidly for any specified laser field. However, controlling the dynamics of the charge carriers requires the answer to an inverse question [18–22]. That is, given a specific target or objective, what is the laser field that best drives the system to that objective? Several methods have been developed to address this question. This section sketches one method, valid in the weak response (perturbative) regime in which most experiments on semiconductors are performed.

The theory begins by constructing a target operator $\hat{A}$ that specifies the desired outcome of the experiment. In general, $\hat{A}$ is a projection operator onto a set of observables. The objective is to maximize the target yield, or the expectation value of $\hat{A}$ at a chosen target time, t_f [23,24],

$$A(t_f) = \left\langle \psi(t_f) \mid \hat{A} \mid \psi(t_f) \right\rangle . \tag{3}$$

To prevent the optimization procedure from discovering trivial, or non-physical solutions, the yield must be optimized with respect to a set of constraints. These constraints can take many forms, including details of the experimental apparatus and the physical system [23–30].

If the target is a pure electronic state (or coherent superposition state) in the conduction band, $|\Phi_c\rangle$, the target operator can be written as

$$\hat{A}_c = |\Phi_c\rangle \langle \Phi_c| . \tag{4}$$

Then, in the perturbative regime, a control kernel can be constructed as [23,24],

$$M_c(t_2, t_1) = \langle \psi_c^0(t_2) \mid \Phi_c\rangle \langle \Phi_c \mid \psi_c^0(t_1 + t_2)\rangle , \tag{5}$$

where $|\psi_c^0(t)\rangle$ is the time-evolved wave packet in the absence of the field (i.e., the promoted state). The optimal fields are then computed as solutions of an eigenequation by discretizing a symmetrized (Hermitian) version of the control kernel onto a numerical grid, over the allotted time interval [23,24],

$$\int_{t_0}^{t_f} d\tau' M_c^S(\tau,\tau')E(\tau') = \lambda E(\tau). \tag{6}$$

In this equation, the eigenvectors λ are the target yields, and the eigenvectors, $E(t)$, are the optimal fields. The eigenvector associated with the largest eigenvalue is the globally optimal field, in the weak response limit.

In some cases, calculating the optimal fields explicitly is inconvenient, either for computational reasons, or because the quantity to be optimized cannot be expressed in terms of a simple control functional. In such situations alternative procedures can be applied. One method is to express the laser field $E(t)$ as a function of a small number of parameters, and then vary the parameters to maximize the yield. For example, the laser field can be written

$$E(t) = A_0 e^{-(t-\bar{t})^2/\Gamma^2} e^{-i\phi(t)}, \tag{7}$$

where A_0 is the amplitude, $\bar{t}$ is the center time and $\Gamma\sqrt{\ln 16}$ is the temporal width (full-width at half maximum). Then the time-dependent phase, $\phi(t)$, can be expanded in a Taylor series as

$$\phi(t) = \phi + \omega(t-\bar{t}) + \frac{1}{2}b(t-\bar{t})^2 + \cdots, \tag{8}$$

where ϕ is the (irrelevant) phase constant, ω is the center frequency, and b is the linear chirp. In the perturbative limit, A_0 is also irrelevant, since the yield scales linearly with the amplitude of the field. The four remaining parameters, $\bar{t}$, Γ, ω and b, are sufficient to characterize the laser pulse for a large variety of experimentals.

To determine the optimal parameters, traditional methods, such as conjugate gradient and simplex are often not adequate, because they tend to get trapped in local minima. To overcome this difficulty, higher-order methods, such as the genetic algorithm (GA) can be employed [31,32]. The GA is a general purpose functional minimization procedure that requires as input an evaluation, or test function to express how well a particular laser pulse achieves the target. Tests have shown that several thousand evaluations of the test function may be required to determine the parameters of the optimal fields [17]. This presents no difficulty in the simple, pure-state model discussed above.

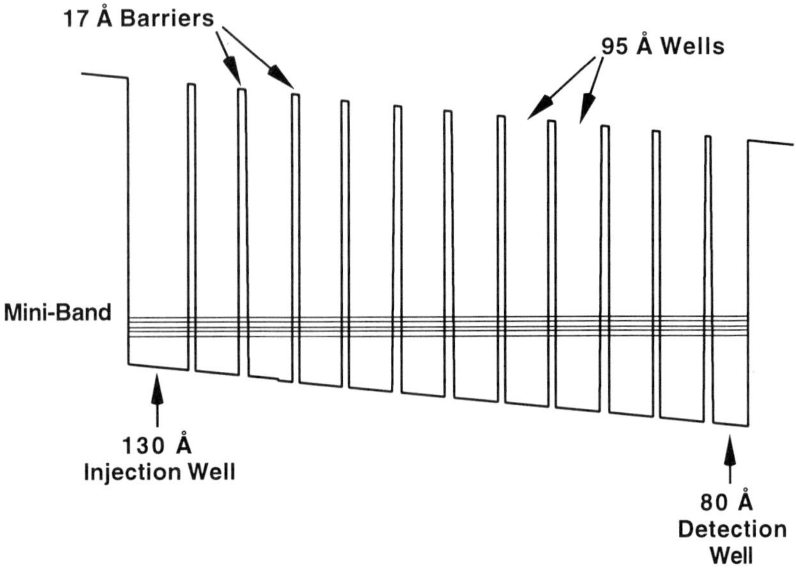

Figure 1. DC-biased, ten-well superlattice. A shaped laser field excites a wave packet localized initially in the injection well. The objective is to created the maximum density possible in the detection well, at a chosen target time.

4. Results

As a first example of the control of charge carriers in semiconductor heterostructures, consider the superlattice depicted schematically in Fig. 1. The structure consists of a series of ten GaAs wells with widths of 95 Å, separated by AlGaAs barriers of 17 Å. At one end of the well is a 130 Å injection well, and at the other end is a 80 Å detection well. A DC bias field forms a mini-band in which the twelve lowest electronic levels move into near tunneling resonance. Excitation with a laser to the injection well creates an electronic wave packet that tunnels back and forth through the structure. The object in this case is to produce, at a chosen target time, as much density as possible in the detection well. Note that this is a challenging objective, because the detection well is not in strong tunneling resonance with the ten equally-spaced wells.

The globally optimal laser field for this example is presented in Fig. 2. The field is relatively simple with structure at early times, followed by a large peak with a nearly Gaussian profile. Note that the control formalism enforces no specific structure on the field *a priori*. That is, the form of the field is totally unconstrained during the allotted time interval, so simple solutions are not guaranteed. Also shown in Fig. 2 is the locally optimal

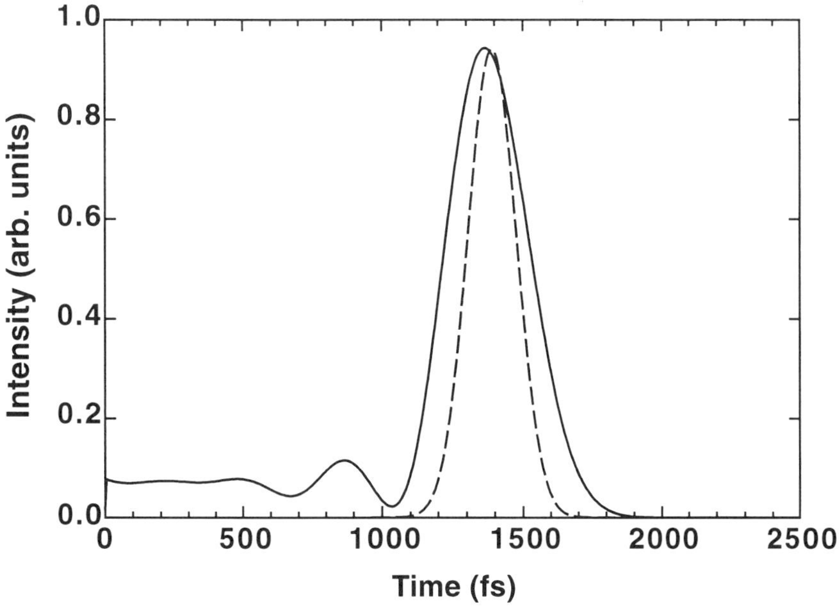

Figure 2. Optimal laser fields for the control scenario in Fig. 1. The solid line is the globally optimal laser field. The dashed line is the locally optimal Gaussian field.

field discovered by the genetic algorithm, as parameterized in Eqs. 7 and 8. The locally optimal field, in this case, performs nearly as well as the globally optimal field, indicating that the complicated structure at early times in the globally optimal field is not critical to attaining the control objective.

The square of the wave packet produced by the globally optimal field at the target time is shown in Fig. 3. Analysis of both the globally optimal field and the locally-optimal Gaussian field indicates the presence of a small negative chirp, which creates a large density in the wells adjacent to the detection well at the target time. This allows some density to leak through to the the target well. If the chirp is removed from the pulse, the density in the target well diminishes by about a factor of six.

As this first example indicates, experiments involving superlattices have an additional complication compared to the ADQW treated previously. In the ADQW, essentially all quantities of interest can be calculated analytically. The well widths and barriers widths are designed such that the electronic levels move into tunneling resonance at a value of the DC field sufficient to immobilize the holes. This is not guaranteed to be the case in superlattices, where the electronic levels display a complicated pattern of repulsion and anticrossing as the DC field is varied. In addition, as shown

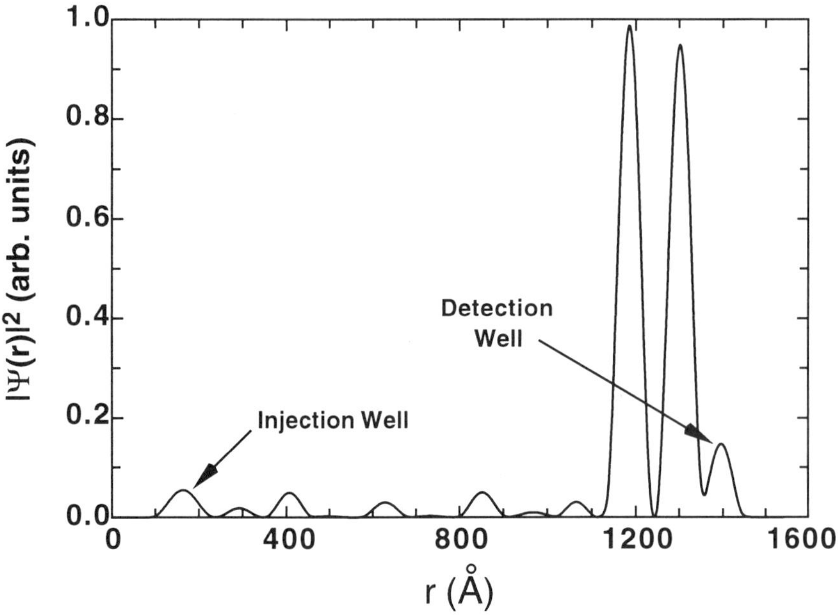

Figure 3. Wave packet produced by the globally optimal field in Fig. 2 at the target time.

in the 10-well superlattice, there is no guarantee that a specified control objective is possible in a given structure. Thus, the superlattice presents a double optimization problem. An initial round of optimization is required to design the structure, and then a second round is required to optimize the excitation field.

An example of the results of this double optimization procedure is presented in Fig. 4. In this case, a 5-well structure is designed to optimize tunneling. To do this, the magnitude of the DC field is fixed (to prevent the system from discovering a solution in which the bias field is zero, and the wells are equally spaced) and the well widths and barriers are allowed to vary freely over a prescribed range (to prevent the system from discovering a solution consisting of just a single well). The evaluation function is chosen to be

$$\eta = \prod_{i=1}^{5} P_i, \tag{9}$$

where

$$P_i = \sum_{j=1}^{5} |\phi_{ij}|^2 . \tag{10}$$

In this equation, $|\phi_{ij}|^2$ is the density of the ith eigenfunction in the jth

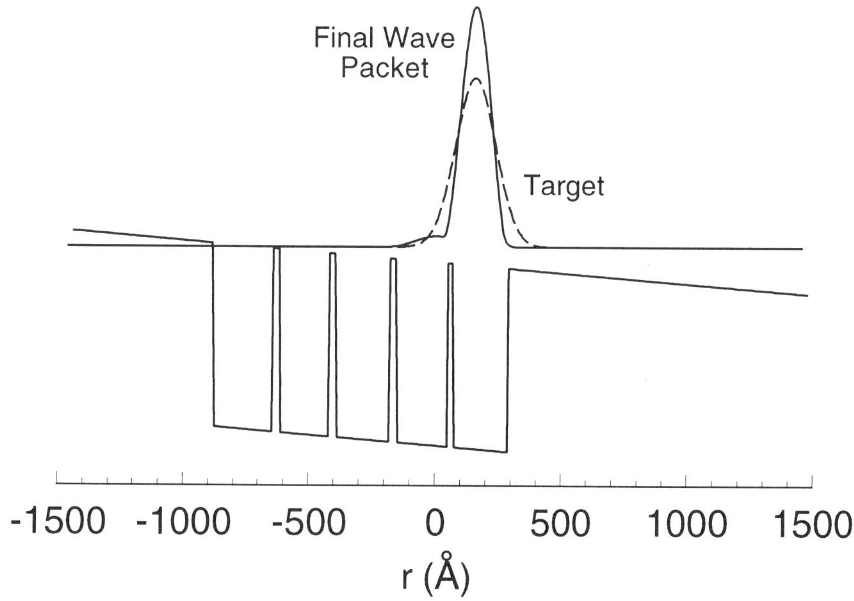

Figure 4. Control of wave packet dynamics in an optimal 5-well structure. The parameters of the structure are designed to maximize tunneling. The wave packet at the target time (solid line) has a large overlap with the target (dashed line).

well. This function assures that the eigenfunctions are as delocalized as possible, subject to the constraints on the well widths and the bias field. The optimal structure discovered by the GA in this case is shown in Fig. 4. The first four wells are slightly chirped (that is, they decrease slightly in width from well one to well four), and the first and fifth wells are about the same width. This may be an attempt to reduce the effects of an avoided crossing between two eigenvalues. For the second round of control, a target is chosen as a Gaussian distribution localized in the fifth well. As shown in the figure, the wave packet produced with the globally optimal field has large overlap with the target at the target time, indicating that control in this example is quite effective.

5. Discussion

The theory discussed in this paper treats the biased superlattices as one-dimensional systems in a single particle envelope approximation in which the electrons and holes act independently. Scattering mechanisms, which cause a loss of coherence, have not yet been included in the formalism. Loss of coherence represents a significant obstacle to quantum control in semiconductors. Phase breaking collisions result in a loss of wave packet coherence and place an upper limit on the ability to manipulate coherent charge motion in these systems (phonons are not important in the current experiments because they are performed at cryogenic temperatures). One of the key issues to be addressed in future work is to what extent (if any) dephasing can be overcome by a suitably chosen control field. Note, though, that for certain applications and target states, it may not be necessary to minimize dephasing effects. For example, control of incoherent charge transport in devices might use time-dependent pulse trains to switch tunneling resonances on or off [33].

As the number of eigenstates available for coherent coupling increases, the dynamical behavior of the system becomes considerably more complex, and issues such as Coulomb interactions become more important. For example, over how many wells can the wave packet survive, if the holes remain locked in place? If the holes become mobile, how will that affect the wave packet and, correspondingly, its controllability? The contribution of excitons to the experimental signal must also be included [34], as well as the effects of the superposition of hole states created during the excitation process. These questions are currently under active investigation.

In summary, this work has shown that superlattices are promising systems for investigation of quantum control in the solid state. The examples presented here show that the dynamics of charge carriers can be controlled using relatively simple, experimentally laser fields. Superlattices are ideal candidates for quantum control precisely because their complexity does not allow for simple, intuition-guided experiments, and because their dynamics are largely unknown.

6. Acknowledgments

Acknowledgment is made to the Donors of The Petroleum Research Fund, administered by the American Chemical Society, for partial support of this research. J.L.K. is a Cottrell Scholar of the Research Corporation. The author would also like to thank David H. Reitze and Christopher J. Stanton for inspiring conversations.

References

[1] Leo, K., Shah, J., Göbel, E. O., Damen, T. C., Schmitt-Rink, S., Schäfer, W., and Köhler, K. *Phys. Rev. Lett.* **1991**, *66*, 201.

[2] Roskos, H. G., Nuss, M. C., Shah, J., Leo, K., Miller, D. A. B., Fox, A. M., Schmitt-Rink, S., and Köhler, K. *Phys. Rev. Lett.* **1992**, *68*, 2216.

[3] Dekorsy, T., Ott, R., Kurz, H., and Köhler, K. *Phys. Rev. B* **1995**, *51*, 17275.

[4] Cundiff, S. T., Knorr, A., Feldmann, J., Koch, S. W., Göbel, E. O., and Nickel, H. *Phys. Rev. Lett.* **1994**, *73*, 1178.

[5] Planken, P. C. M., Brener, I., Nuss, M. C., Luo, M. S. C., and Chuang, S. L. *Phys. Rev. B.* **1993**, *48*, 4903.

[6] Brener, I., Planken, P. C. M., Nuss, M. C., Luo, M. S. C., Chuang, S. L., Pfeiffer, L., Leaird, D. E., and Weiner, A. M. *J. Opt. Soc. Am. B* **1994**, *11*, 2457.

[7] Bonvalet, A., Nagle, J., Berger, V., Migus, A., Martin, J.-L., and Joffe, M. *Phys. Rev. Lett.* **1996**, *76*, 4392.

[8] Baranova, B. A., Chudinov, A. N., and Zel'dovich, B. Y. *Opt. Comm.* **1990**, *79*, 116.

[9] Dupont, E., Corkum, P. B., Liu, H. C., Buchanan, M., and Wasilewski, Z. R. *Phys. Rev. Lett.* **1995**, *74*, 3596.

[10] Heberle, A. P., Baumberg, J. J., and Köhler, K. *Phys. Rev. Lett.* **1995**, *75*, 2598.

[11] Weiner, A. M. *J. Opt. Soc. Am. B* **1994**, *11*, 2480.

[12] Liu, Y., Park, S.-G., and Weiner, A. M. *IEEE J. Sel. Top. Quant.* **1996**, *2*, 709.

[13] Kuznetsov, A. V. and Stanton, C. J. *Phys. Rev. B* **1993**, *48*, 10828.

[14] Bastard, G. *Wave Mechanics Applied to Semiconductor Heterostructures*; Halsted Press: New York, NY, 1988.

[15] Kuznetsov, A. V., Sanders, G. D., and Stanton, C. J. *Phys. Rev. B* **1995**, *52*, 12045.

[16] Luo, M. S. C., Chuang, S. L., Planken, P. C. M., Brener, I., and Nuss, M. C. *Phys. Rev. B.* **1993**, *48*, 11043.

[17] Krause, J. L., Reitze, D. H., Sanders, G. D., Kuznetsov, A. V., and Stanton, C. J. *J. Phys. B* **1998**, *57*, 9024.

[18] Warren, W. S., Rabitz, H., and Dahleh, M. *Science* **1993**, *259*, 1581.

[19] Tannor, D. J. and Rice, S. A. *Adv. Chem. Phys.* **1988**, *70*, 441.

[20] Shapiro, M. and Brumer, P. *Int. Rev. Phys. Chem.* **1994**, *13*, 187.

[21] Neuhauser, D. and Rabitz, H. *Acc. Chem. Res.* **1993**, *26*, 496.

[22] Krause, J. L., Whitnell, R. M., Wilson, K. R., and Yan, Y. J. In *Femtosecond Chemistry*; Manz, J. and Wöste, L., Eds.; VCH: Weinheim, 1995; pp 743–779.

[23] Yan, Y. J., Gillilan, R. E., Whitnell, R. M., Wilson, K. R., and Mukamel, S. *J. Phys. Chem.* **1993**, *97*, 2320.

[24] Krause, J. L., Whitnell, R. M., Wilson, K. R., Yan, Y. J., and Mukamel, S. *J. Chem. Phys.* **1993**, *99*, 6562.

[25] Kosloff, R., Rice, S. A., Gaspard, P., Tersigni, S., and Tannor, D. J. *Chem. Phys.* **1989**, *139*, 201.

[26] Rabitz, H. and Shi, S. *Adv. Molec. Vib. Coll. Dyn.* **1991**, *1A*, 187.

[27] Shi, S. and Rabitz, H. *Chem. Phys.* **1989**, *139*, 185.

[28] Tersigni, S. H., Gaspard, P., and Rice, S. A. *J. Chem. Phys.* **1990**, *93*, 1670.

[29] Shapiro, M. and Brumer, P. *Chem. Phys. Lett.* **1993**, *208*, 193.

[30] Somlói, J., Kazakov, V. A., and Tannor, D. J. *Chem. Phys.* **1993**, *172*, 85.

[31] Judson, R. S. and Rabitz, H. *Phys. Rev. Lett.* **1992**, *68*, 1500.

[32] Grefenstette, J. D. GENESIS v5.0, Copyright (c), 1990.

[33] Allen, S. J., Bhattacharya, U., Campman, K., Drexler, H., Gossard, A., Keay, B. J., Maranowski, K., Medieros-Ribeiro, G., Rodwell, M., Scott, J. S., Unterrainer, C., Wanke, M., and Zeuner, S. *Physica B* **1996**, *227*, 367.

[34] Fox, A. M., Miller, D. A. B., Livescu, G., Cunningham, J. E., Henry, J. E., and Jan, W. Y. *Phys. Rev. B.* **1990**, *42*, 1841.

Selective Photodynamic Control of Chemical Reactions : A Rayleigh-Ritz Variational Approach

Vandana K.† and Manoj K. Mishra

Department of Chemistry

Indian Institute of Technology Bombay,

Powai, Mumbai 400 076, India

अज्ञानतिमिरान्धस्य ज्ञानाञ्जनशलाकया ।
चक्षुरुन्मीलितं येन तस्मै श्रीगुरवे नमः ॥

Salutations to the teacher who removes the darkness of ignorance by opening the eyes to the light of knowledge.

Dedicated to my (MKM) teacher and my (VK) teacher's teacher Yngve Öhrn on his 65 th birthday.

† CSIR Senior Research Fellow

ADVANCES IN QUANTUM CHEMISTRY, VOLUME 35

0065-3276/99 $30.00

Contents

1 Introduction

Easy availability of ultrafast high intensity lasers has fuelled the dream of their use as molecular scissors to cleave selected bonds (1-3). Theoretical approaches to laser assisted control of chemical reactions have kept pace and demonstrated remarkable success (4,5) with experimental results (6-9) buttressing the theoretical claims. The different established theoretical approaches to control have been reviewed recently (10). While the focus of these theoretical approaches has been on field design, the photodissociation yield has also been found to be extremely sensitive to the initial vibrational state from which photolysis is induced and results for H_2^+ (11), HI (12,13), HCl (14) and HOD (2,3,15,16) reveal a crucial role for the initial state of the system in product selectivity and enhancement. This critical dependence on initial vibrational state indicates that a suitably optimized linear superposition of the field free vibrational states may be another route to selective control of photodissociation.

Towards this end, a scheme to establish the optimal linear mix of the field free vibrational eigenstates for the given photolysis pulse and chosen photodissociation objective has been pursued in our group (17-20) whereby, the emphasis is shifted from control through design of an appropriate field, to control through the design of an optimal linear combination of the field free vibrational eigenstates for the chosen photolysis pulse and preferred photodissociation objective. This Field Optimized Initial STate (FOIST) based approach to control, shifts the focus from control via field design to the design of an optimal initial state for the chosen field and is computationally simple to implement. Applications of our FOIST based selective control to HI (17,

19) and IBr (18,19) have shown considerable enhancement in selectivity and product yield.

It is our purpose in this review to present a brief summary of some of the main results from the applications of our FOIST based approach to selective control of photodissociation. The formal and computational considerations of this method are summarised in section 2 and in section 3, we discuss some representative results from our applications. Some concluding remarks summarising the main results and avenues for further research are collected in section 4.

2 Method

For molecules possessing a dipole moment $\vec{\mu}$, the effect of the radiation field $\vec{\epsilon}(t)$ may be obtained by solving the time dependent Schrödinger equation,

$$i\hbar\frac{\partial}{\partial t}\psi = \hat{H}(t)\psi, \qquad \hat{H}(t) = \hat{H}_0 - \vec{\mu}\cdot\vec{\epsilon}(t) \tag{1}$$

with H_0 being the field free Hamiltonian. The solution ψ at some time T can be expressed as $\psi(T) = \hat{U}(T,0)\psi(0)$, where $\hat{U}(T,0)$ is the (not necessarily unitary) propagator and $\psi(0)$ is the wavefunction for the initial state of the molecule to which the photodissociation pulse $\vec{\epsilon}(t)$ is applied.

2.1 Product yield as the time integrated flux

Defining the time integrated flux operator $\hat{F}$ as

$$\hat{F} = \int_0^T dt\, \hat{U}^\dagger(t,0)\; \hat{\jmath}\; \hat{U}(t,0), \qquad \hat{\jmath} = \frac{1}{2m}[\hat{p}\,\delta(r-r_d) + \delta(r-r_d)\,\hat{p}] \tag{2}$$

with m as the reduced mass, $\hat{p}$ the momentum operator along the reaction coordinate and r_d the grid point in the asymptotic region where the flux is evaluated. In the case of more than one possible dissociation channel, the operator $\hat{j}$ is channel specific and in a discrete representation of the electronic curves on a spatial grid, r_d denotes the grid point appropriate to the desired channel. The time-integrated flux,

$$\int_0^T dt\ \langle\hat{j}\rangle_t = \int_0^T dt\ \langle\psi(t)|\ \hat{j}\ |\psi(t)\rangle = \int_0^T dt\langle\psi(0)|\hat{U}^\dagger(t,0)\ \hat{j}\ \hat{U}(t,0)|\psi(0)\rangle \quad (3)$$

$$or \quad \int_0^T dt\ \langle\hat{j}\rangle_t = \langle\psi(0)|\hat{F}|\psi(0)\rangle, \quad (4)$$

represents the product yield and may be controlled by altering the field dependent part $\hat{F}$ or the field free initial state $\psi(0)$.

2.2 FOIST based selective control

Earlier control schemes (1,4,5,10) have attempted control over photodissociation entirely through field manipulation by altering $\hat{F}$ for a fixed $\psi(0)$. In our FOIST scheme (17-19), control over product yield is sought through preparation of the initial wavefunction $\psi(0)$ as a coherent superposition of vibrational eigenstates of the ground electronic state for the chosen photolysis pulse (for the short femtosecond pulses to be considered here, rotational motion is ignored). Of course, the field itself may also be altered which will change the nature of the optimal $\psi(0)$. By expanding $\psi(0)$ in a basis of $(M+1)$ field free vibrational eigenfunctions,

$$\psi(0) = \sum_{m=0}^{M} c_m\ \phi_m, \quad (5)$$

flux maximization is reduced to the familiar Rayleigh - Ritz variational optimization (17) of $\{c_m\}$ through diagonalization of an $(M+1 \times M+1)$ matrix

F whose elements are

$$F_{kl} = \langle \phi_k | \hat{F} | \phi_l \rangle \approx \Delta t \sum_{n=0}^{N_t} \langle \psi_k(n\Delta t) | \ \hat{j} \ | \psi_l(n\Delta t) \rangle, \tag{6}$$

with Δt as the step size for the numerical time propagation and $N_t \Delta t = T$.

The largest eigenvalue f_{max} of **F** is the maximum product yield (flux) and the corresponding eigenvector $\{c_m^{max}\}$ is the set of coefficients which define the optimal initial wavefunction

$$\psi^{max}(0) = \sum_{m=0}^{M} c_m^{max} \ \phi_m \tag{7}$$

constituting the superposition that will provide the maximum achievable product yield f_{max} out of the particular channel specified by $\hat{F}$ for the chosen field $\vec{\epsilon}(t)$. As in other variational calculations, the larger the basis set size, "better" the results. However, due to the difficulties in the simultaneous overtone excitation of many vibrational levels, the basis set expansion in Eq. 5 is restricted to only the ground plus the first two excited vibrational levels leading to a 3 $\times$ 3 **F** matrix which is computationally trivial to diagonalize.

2.3 Time dependent wave packet analysis of FOIST

In the time dependent wave packet analysis (21) the Schrödinger equation

$$H_{ex}\chi_\ell(0) = i\hbar \frac{\partial \chi_\ell(0)}{\partial t}. \tag{8}$$

is solved with $\chi_\ell = \mu_{0\ell}\phi(0)$ as the initial wavefunction on the ℓ-th excited state and $\mu_{0\ell}$ is the transition dipole moment between the ground (0) and the ℓ-th excited state. The field free vibrational ground state $\phi(0)$ in our case is the optimal linear combination $\psi^{max}(0)$. The time-evolution of the promoted wavefunction $\chi_\ell(0)$ is governed by,

$$\chi_\ell(t) = exp(-iH_{ex}t/\hbar)\chi_\ell(0), \tag{9}$$

and the overlap $\langle\chi_\ell(0)|\chi_\ell(t)\rangle$ of the time evolving wavefunction $\chi_\ell(t)$ with the initial $\chi_\ell(0)$ is the autocorrelation function. To ensure correct branching ratio (22), the autocorrelation function is evaluated after a sufficiently large time interval τ such that the norm of the wavefunction on different curves has stabilised and the system population is completely out of the curve crossing region. The Fourier transform of the autocorrelation function yields frequency dependent partial absorption cross-section

$$\sigma(\omega) = C\omega \int_{-\infty}^{+\infty} e^{-i(\omega+E_0)t}\langle\chi_\ell(\tau)|\chi_\ell(\tau+t)\rangle dt, \tag{10}$$

where C is a constant (21), ω is the frequency of the incident radiation and E_0 is the energy corresponding to the initial state. The branching ratio is given by the ratio of the sum of partial photoabsorption cross-sections for the two channels.

2.4 Numerical Considerations

Photodissociation of IBr (18,19,23-27) has been studied extensively and is our representative system of choice. The results to be presented are for the photolysis of IBr employing a multicolor cw field of the form $\epsilon(t) = A\sum_{p=0}^{2}\cos(\omega-\omega_{p,0})t$ where A is the amplitude, ω the photodissociation frequency and $\omega_{p,0} = (E_p - E_0)/\hbar$ is the Bohr frequency for transition between the p_{th} and the ground (0_{th}) vibrational energy levels. The vibrational eigenvalues E_p and corresponding eigenfunctions ϕ_p of the electronic ground state are computed using the Fourier Grid Hamiltonian (FGH) method (28). The optimal linear combination maximizing flux out of the lower I + Br channel (to be labelled as 1) is denoted by ψ_1^{max} and ψ_2^{max} represents the combination which maximizes flux out of the upper channel (labelled 2) with I + Br* as the dissociation product. The split-operator fast Fourier transform with Pauli matrix propagation algorithm (29) was utilized to integrate the time dependent Schrödinger equation. The total spatial grid of 14 a.u. starting from 2.0 a.u. was divided into 1024 equally spaced points. The flux was monitored at 14.75 a.u. beyond which "ramp" type optical barrier of height 0.01 a.u. was set up till the end of the grid to prevent any unphysical reflection of the wave packet. In the following section we present some illustrative results from the application of our FOIST scheme to IBr photodissociation.

3 Selective control of IBr photodissociation

The potential energy curves (Fig. 1), the non-adiabatic coupling, transition dipole moments and other system parameters are same as those used in our previous work (18,19,23,27). The excited states 1:$B(O^+)$ and 2:$B(^3\Pi_0^+)$ are non-adiabatically coupled and their potential energy curves cross at R = 6.08 a.u. The ground 0:$X(^1\Sigma_0^+)$ state is optically coupled to both the 1:$B(O^+)$ and the 2:$B(^3\Pi_0^+)$ states with the transition dipole moment $\mu_{01} = 0.25\mu_{02}$. The results to be presented are for the cw field $\epsilon(t) = A\sum_{p=0}^{2}\cos(\omega - \omega_{p,0})t$ described earlier. However, for IBr, we have shown (18) that similar selectivity and yield may be obtained using Gaussian pulses too.

3.1 Flux maximization using FOIST

The variation of maximized flux out of channels 1 and 2 as a function of field amplitude at a laser frequency $\omega = 0.087$ a.u. (19094 cm^{-1}) using the optimal combinations ψ_1^{max} and ψ_2^{max} of the ground, first and second vibrational levels (filled symbols) is compared with maximum from any one of the v=0, v=1 and v=2 vibrational eigenstates (open symbols) in Fig 2. Since the optimal FOIST combination comprises of v=0, 1 and 2 vibrational states, the optimized flux is compared with the maximum attainable if total initial population was to be in any one of these and not just with the yield from v=0. The circles represent flux out of the I + Br, and triangles out of the I + Br* channel. For the broad range of field strengths considered in Fig. 2, the sum of flux out of I + Br and I + Br* channels adds to 1 which means there is total dissociation. At lower amplitudes, yield out of the I + Br* channel is greater and at larger amplitudes the yield is predominantly out the of the I + Br channel.

The results of Fig. 2 therefore show that for the chosen frequency, almost total dissociation is attainable for a range of field amplitudes and we can get more I + Br or I + Br* (the desired selectivity) with simple cw fields employed in the FOIST scheme. We may therefore infer that considerable dissociation and selectivity (almost any branching ratio) may be achieved using the scheme (18,19) presented here. The extent of optimization however varies with field amplitude and at the two ends of the field range, the FOIST results do not provide much improvement over the maximum available from one of the pure vibrational states in the optimization manifold. It should however be noted that FOIST alone can provide which of the v=0, 1 or 2 is optimal for the chosen objective.

3.2 Spatial attributes of the optimal initial state

The selective flux maximization from the FOIST scheme shown in Fig. 2 is achieved by altering the spatial profile of the initial state to be subjected to the photolysis pulse and since changes in flux are due to the flow of probability density, it is useful to examine the attributes of the probability density profiles from the field optimized initial states.

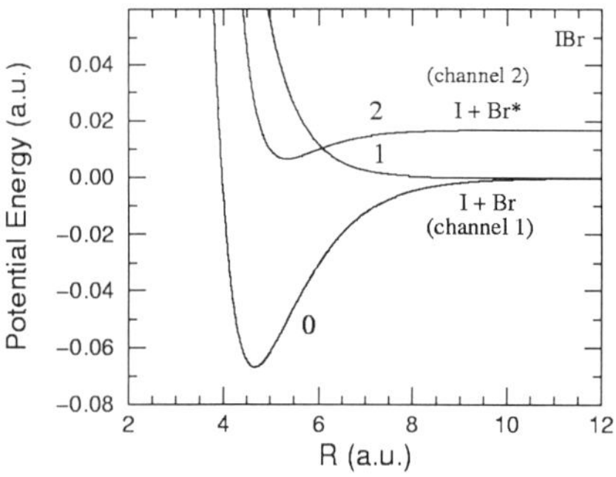

Fig. 1 Potential energy curves of IBr.

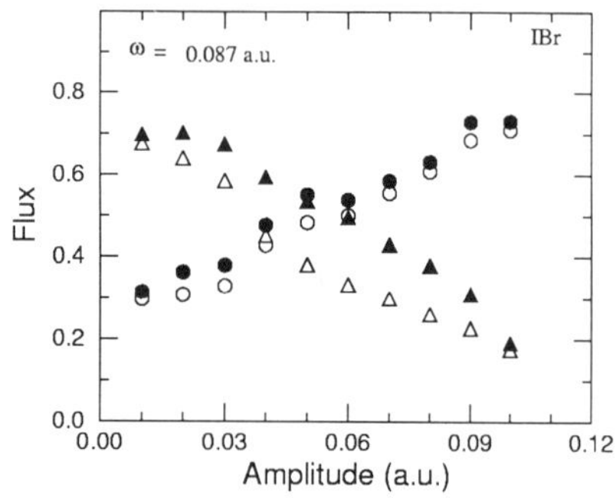

Fig. 2 Flux out of channels 1 and 2 as a function of field amplitude.

The probability density plots for the first three vibrational states ϕ_0, ϕ_1 and ϕ_2 of the IBr molecule are plotted in Fig. 3 and the probability density profile from the optimal superpositions ψ_1^{max} and ψ_2^{max} which maximize flux out of I + Br and I + Br* channels respectively for a field frequency $\omega = 0.087$ a.u. (19094 cm^{-1} / 524 nm) and amplitude $A = 0.03$ a.u. are displayed in Fig. 4. The chosen field amplitude is the lowest for which almost 100% dissociation (18) occurs using a pulse length of 480 fs (20,000 a.u.), followed by further propagation without the field for another 387 fs (16,000 a.u.).

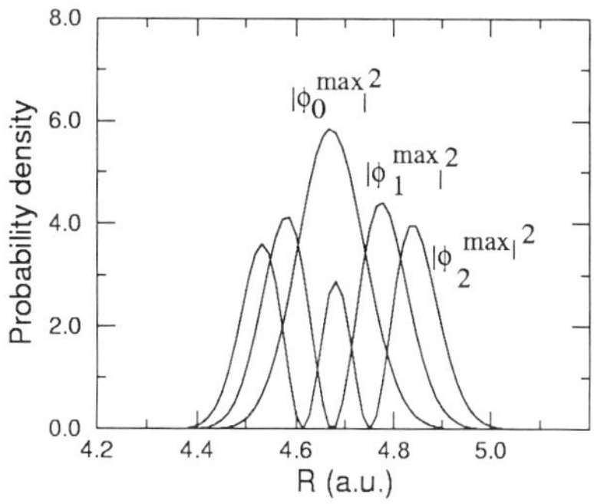

Fig. 3 Probability density for the vibrational eigenstates ϕ_0, ϕ_1 and ϕ_2

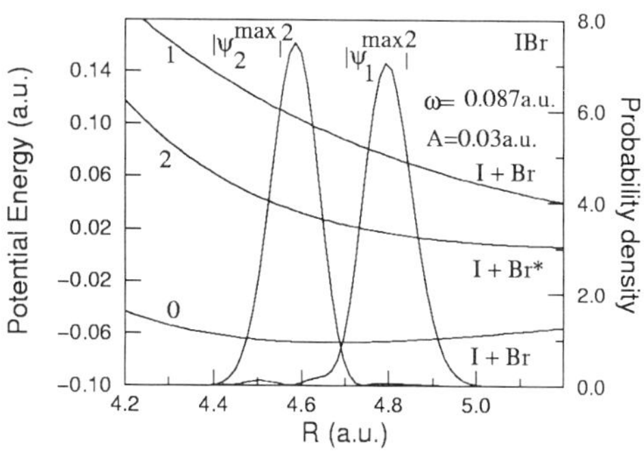

Fig. 4 Probability density for the optimal superpositions ψ_1^{max} and ψ_2^{max}

The probability density for the ground vibrational level ϕ_0 peaks at 4.66 a.u., that for ψ_1^{max} at 4.8 a.u. and for ψ_2^{max} at 4.58 a.u. The optimal wavefunction ψ_2^{max} seems to maximize flux out of channel 2 (I + Br*) by localizing the probability density to the left of that given by the ϕ_0 wavefunction and

ψ_1^{max} maximizes flux out of channel 1 (I + Br) by localizing probability density to the right of the ϕ_0 probability density peak. Though the results presented here are for a single representative frequency, this trend of ψ_2^{max} peaking to the left of ϕ_0 and ψ_1^{max} to its right persists throughout the frequency range specified earlier. The probability density profiles for ψ_1^{max} and ψ_2^{max} are mutually exclusive and much more compact as compared to those from ϕ_0, ϕ_1 or ϕ_2. Furthermore, the probability density profiles from pure eigenstates ϕ_0, ϕ_1 and ϕ_2 (Fig. 3) subsume the spatial attributes of both ψ_1^{max} and ψ_2^{max} (Fig. 4) which explains why selective photodissociation cannot ensue from the use of only one of these molecular eigenstates as the initial state. This is achieved in the cases studied by us through dominant component coming from v=0 for ψ_2^{max} and from v=1 at lower amplitudes and v=2 at higher amplitudes for ψ_1^{max}.

This need for a suitable mixing of vibrational states for selective control of photodissociation is also seen in the optimal control theory based calculations (23) on IBr where the additional frequency components of the optimal field separated from each other by IBr ground state vibrational spacings, large expectation value for the internuclear distance on the electronic ground state corresponding to vibrational stretch for highly excited vibrational levels of IBr and extremely intense fields required to achieve this in the very beginning of the control procedure point to the same central role of initial mixing of vibrational states in achieving selective control. In our FOIST scheme, ψ_1^{max} and ψ_2^{max} represent the premixing of vibrational states required for selective control with an additional advantage that the photolysis pulse may be chosen beforehand for practical convenience.

3.3 The dynamical evolution of the optimal initial state

In the FOIST based selective control, it is the ψ_1^{max} or the ψ_2^{max} which are transported to the excited electronic states. In the case of IBr, the $B(^3\Pi_0^+)$ state is coupled four times more strongly with the ground state ($\mu_{01} = 0.25\mu_{02}$) as compared to the $B(O^+)$ state. Also, the $B(O^+)$ state is far off resonance within the frequency band considered by us and the amplitude transferred to the $B(^3\Pi_0^+)$ level will therefore be much larger and dominate the photodissociation outcome. The broad mechanistic details may therefore be inferred from arguments employing the dynamics ensuing from the evolution of the wavepacket on the $B(^3\Pi_0^+)$ curve alone. We may therefore deduce from the structural features of $|\ \psi_2^{max}\ |^2$ (Fig. 4) that for any given frequency, transitions from the initial state represented by ψ_2^{max} will occur to the energetically higher or steeper, more repulsive region of the excited $B(^3\Pi_0^+)$ potential energy curve. The excited molecule described by ψ_2^{max} will therefore traverse the $B(^3\Pi_0^+)$ - $B(O^+)$ crossing with greater velocity and exit out of the excited I + Br* channel compared to where the molecule is represented by ψ_1^{max} which transfers it to a relatively smoother region of the $B(^3\Pi_0^+)$ state potential energy curve thereby facilitating a slower adiabatic exit out of the lower I + Br channel. Selective maximization is effected through localization of the probability density at internuclear distances which enable Franck - Condon transitions to appropriate region of the excited states. Transfer of the wavefunction to the more steep, repulsive part of the excited potential energy curves favours high velocity diabatic exit into the higher channel. Localization away from the repulsive wall favours slow adiabatic exit into the lower channel.

This interpretation of the control mechanism utilizing ψ_1^{max} and ψ_2^{max}

as the initial states is consistent with the analysis of the frequency dependence of IBr photodissociation as a function of the molecular radial velocity using the Landau - Zener theory presented by Devries et al where increase in photodissociation yield out of the I+Br* channel with increase in frequency is well correlated with an increase in radial velocity at the crossing point (24). From Fig. 4, it is obvious that $| \psi_2^{max} |^2$ and $| \psi_1^{max} |^2$ are localized in a mutually exclusive manner and that the optimization leads to a significant altering of the spatial profiles to excite the molecule to the region most suited for directing flux out of the desired channel. A more detailed substantiation of these assertions has been attempted elsewhere (19).

3.4 Absorption spectrum and branching ratio

The mechanism of the FOIST based selective control of IBr photodissociation has been further probed by the use of ψ_1^{max} and ψ_2^{max} in the TDWP calculation of IBr absorption spectrum (Fig. 5) and branching ratio (Fig. 6).

The absorption cross-section obtained using ψ_2^{max} as the initial condition is smooth with only a negligible interference pattern around 600 nm. The absorption spectrum from ψ_1^{max} as the initial condition contains a series of sharp peaks characteristic of predissociation dynamics (25) in the higher wavelength region due to a slower approach to the $B(^3\Pi_0^+)$ - $B(O^+)$ crossing. The longer wavelength region is dominated by predissociation in the vicinity of the energy values around the $B(^3\Pi_{0^+})$ - $B(O^+)$ crossing leading to the complicated interference pattern seen in the ψ_1^{max} absorption profile. In contrast, the smoothness of the absorption spectrum from ψ_2^{max} as the initial condition stems from the initial placement of the ψ_2^{max} wavepacket on the steeper part of the excited curve which facilitates faster diabatic exit with little time for interference. The absorption spectrum from ϕ_0 (Fig. 5) as the initial condition has also been plotted and compares well with other calculated and experimental (25,26) absorption spectra for IBr. The absorption spectra peak at the wavelengths corresponding to the vertical Franck - Condon transition energies with ψ_1^{max}, ψ_2^{max} and ϕ_0 as the initial states.

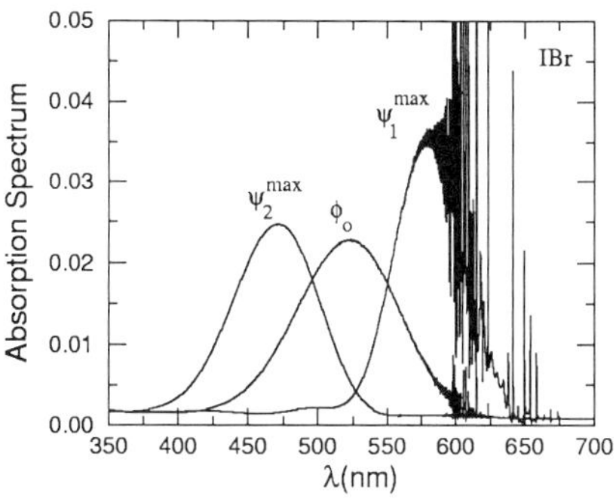

Fig. 5 Total absorption spectrum

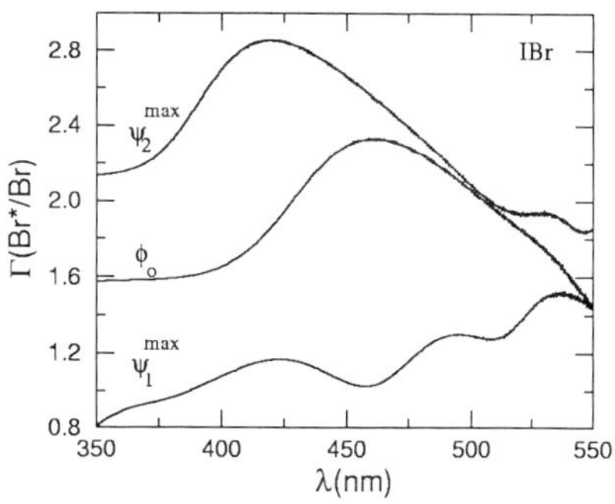

Fig. 6 Branching ratio $\Gamma(Br^*/Br)$

The branching ratio Γ(Br* / Br) with ϕ_0, ψ_1^{max} and ψ_2^{max} as the initial conditions are plotted in Fig. 6. At all energy values, Γ(Br* / Br) is much larger in magnitude with ψ_2^{max} as the initial condition compared to that with ϕ_0 or ψ_1^{max} as the initial condition and Γ(Br* / Br) is uniformly smaller with

ψ_1^{max} as the initial condition as compared to ψ_2^{max} or ϕ_0 as the initial condition. This does seem to suggest that preparation of the initial state for a suitable photolysis pulse using our FOIST scheme can provide selective control through constructing appropriate linear combination of vibrational states.

4 Concluding remarks

Results from the selective control of IBr photodissociation using the optimal superpositions selected by the Rayleigh - Ritz variational procedure for maximization of flux out of the desired channel for the chosen field confirms the utility of the FOIST approach advocated here. The results reveal that the selective maximization is effected through localization of the probability density at internuclear distances which enable Franck - Condon transitions to appropriate region of the excited states. Transfer of the wavefunction to the more steep, repulsive part of the excited potential energy curves favours high velocity diabatic exit into the higher channel. Localization away from the repulsive wall favours slow adiabatic exit into the lower channel. The nascent mechanistic notions linking selectivity to appropriate modification of the initial state have been further examined by an analysis of the resulting absorption spectrum and branching ratios and this central role for modified spatial profile in selective control provides a new possibility for experimental exploration.

The experimental realization of the optimal initial states is however a completely uncharted area at this time. In an earlier paper (17), we have presented the formulae to obtain field parameters required to achieve these field optimized initial states and the optimal control (30) approach may also be easily and profitably employed to attain this FOIST comprising of only three vibrational levels. The parametric equations of motion for Time dependent Schrödinger equation (31) also provide an economic map of vibrational population as a function of field parameters and may be harnessed for this task. We however believe that while the theoretical tools are useful, the central result from our investigations (17-20,23,27) is that instead of putting the entire onus of selective control on a theoretically designed laser pulse which may not be easy to realize in practise, the approach where different vibrational population mixes are experimentally obtained and subjected to readily attainable photolysis pulses leading to an empirical experimental correlation between selectivity attained for diverse photolysis pulses and initial vibrational population mix used, represents a more promising and desirable alternative. Our results, we hope will spur such experimental tests and a concerted partnership between field and initial state shaping is required to better realize the chemical dream (1,2) of using lasers as molecular scissors and tweezers to control chemical reactions.

It is our hope that the approach advocated here will merit experimental attention, where, instead of attempting selective control by using an active field manipulating a passive molecule in the ground vibrational state, experiments will be planned to use a variety of population mixes as the initial state. Should a pattern of the kind where altered spatial probability density profiles of the type studied here are experimentally confirmed to lead to selective control, generation of these profiles can be reduced to finding a suitable linear combination of known vibrational eigenfunctions without requiring any time dependent quantum mechanical calculations whatsoever. In an extremely optimistic scenario, since the profile of the standard simple harmonic oscillator eigenfunctions is well known, a lot can be done by mere inspection as well. Also, we have utilized three vibrational states, but there could be frequency ranges where only two vibrational states play a dominant role and further simplification may be obtained by examining flux as a function of initial vibrational states for an even easier shaping of the optimal linear combination.

In conclusion, the Rayleigh - Ritz variational maximization of flux by generating an optimal spatial profile for the initial wavefunction offers a new and flexible alternative for laser assisted selective control of chemical reactions. It is our hope that the FOIST based approach presented here will attract requisite experimentation and a concerted partnership between field and initial state shaping advocated here will assist in keeping the dream of controlling chemical reactions by modifying the underlying quantal dynamics, alive, and attractive for further pursuit. A detailed study of the variational properties of the time integrated flux operator underlying our FOIST scheme to correlate field attributes with the resulting optimal mix of field free vibrational eigenstates in ψ_1^{max} and ψ_2^{max} will provide additional mechanistic insights and an effort along these lines is underway in our group.

It is difficult not to be sentimental when I(MKM) recall my graduate studentship under Yngve Öhrn. I feel privileged to have had him as my supervisor and it is with utmost pleasure that I take this opportunity to record my deep gratitude to him for his fine example combining high scientific achievement with superlative human qualities.

5 Acknowledgements

VK acknowledges the support from CSIR, India (SRF 2-14/94(II)/E.U.II). MKM acknowledges financial support from the Board of Research in Nuclear Sciences (BRNS 37/19/97-R & D-II) of the Department of Atomic Energy, India.

6 References

1. Warren, W. A., Rabitz, H. A., and Dahleh, M., *Science*, **259**, 1581 (1993).
2. Crim, F. F., *Annu. Rev. Phys. Chem.*, **44**, 397 (1993).
3. Crim, F. F., *Science*, **249**, 1387 (1990).
4. Brumer, P., and Shapiro, M., *Annu. Rev. Phys. Chem.*, **43**, 257 (1992).
5. Tannor, D. J., and Rice, S. A., *Adv. Chem. Phys.*, **70**, 441 (1988).
6. Zhu, L., Kleiman, V., Li, X., Lu, S. P., Trentelman, K., and Gordon, R. J., *Science*, **270**, 77 (1995).
7. Park, S. M., Lu, S. P., Xie, Y., and Gordon, R. J., *J. Chem. Phys.*, **94**, 8622 (1991).
8. Wang, X., Bersohn, R., Takahashi, K., Kawasaki, M., Kim, HL., *J. Chem. Phys.*, **105**, 2992 (1993).
9. Baumert, T., Gerber, G., *Isr. J. Chem.*, **34**, 103 (1994).

10. Gordon, R. J., and Rice, S. A., *Ann. Rev. Phys. Chem.*, **48**, 601 (1997).

11. Chu, S. I., *J. Chem. Phys.*, **94**, 7901 (1991).

12. Levy, I., and Shapiro, M., *J. Chem. Phys.*, **89**, 2900 (1988).

13. Kalyanaraman, C., and Sathyamurthy, N., *Chem. Phys. Letts.*, **209**, 52 (1993).

14. Gersonde, H., Hennig, S., and Gabriel, H., *J. Chem. Phys.*, **101**, 9558 (1994).

15. Cohen, Y., Bar, I., and Rosenwaks, S., *J. Chem. Phys.*, **102**, 3612 (1995).

16. Amstrup, B., and Henriksen, N. E., *J. Chem. Phys.*, **97**, 8285 (1992)

17. Gross, P., Gupta, A. K., Bairagi, D. B., and Mishra, M. K., *J. Chem. Phys.*, **104**, 7045 (1996).

18. Bairagi, D. B., Gross, P., and Mishra, M. K., *J. Phys. Chem. A*, **101**, 759 (1997).

19. K, Vandana., and Mishra, M. K., *J. Chem. Phys.*, (in press)

20. K, Vandana., Bairagi, D. B., Gross, P. and Mishra, M. K., *Pramana - J. of Physics*, **50**, 521 (1998)

21. Heller, E. J., *Acc. Chem. Res.*, **14**, 368 (1981); Lee, S. Y., and Heller, E. J., *J. Chem. Phys.*, **71**, 4777 (1979)

22. Das, S., and Tannor, D. J., *J. Chem. Phys.*, **91**, 2324 (1989).

23. Gross, P., Bairagi, D. B., Mishra, M. K., and Rabitz, H. A., *Chem. Phys. Letts.*, **223**, 263 (1994).

24. M. S. Devries, N. J. A. van Veen and A. E. DeVries, Chem. Phys. Letts. **56**, 15 (1978).

25. Guo, H., *J. Chem. Phys.*, **99**, 1685 (1993).

26. Bony, H., Shapiro, M., and Yogev, A., *Chem. Phys. Letts.*, **107**, 603 (1984).

27. K, Vandana., Chakrabarti, N., Sathyamurthy, N., and Mishra, M. K., *Chem. Phys. Letts.*, **288**, 545 (1998)

28. Marston, M, M., Balint-Kurti, G. G., *J. Chem. Phys.*, **91**, 3571 (1989).

29. Kosloff, R., *J. Phys. Chem.*, **92**, 2087 (1988); Feit, M. D., Fleck, J. A., and Steiger, A., *J. Comput. Phys.*, **48**, 412 (1982); Gross, P., Neuhauser, D., and Rabitz, H. A., *J. Chem. Phys.*, **99**, 1752 (1993).

30. Shi, S., Woody, A., and Rabitz, H. A., *J. Chem. Phys.*, **88**, 6870 (1988); Shi, S., Rabitz, H., *Comput. Phys. Commun.*, **63**, 71 (1991).

31. Gupta, A. K., Gross, P., Bairagi, D. B., and Mishra, M. K., *Chem. Phys. Letts.*, **257**, 658 (1996).

Semi-Classical Pictures of Non-Adiabatic Induced Electron Ejection in Molecular Anions

Jack Simons

Henry Eyring Center for Theoretical Chemistry
Chemistry Department
University of Utah
Salt Lake City, Utah 84112

ADVANCES IN QUANTUM CHEMISTRY, VOLUME 35

0065-3276/99 $30.00

ABSTRACT

Molecular anions that possess excess internal vibrational and/or rotational energy can eject their "extra" electron through a radiationless transition event involving non Born-Oppenheimer coupling. In such processes, there is an interplay between the nuclear and electronic motions that allows energy to be transferred from the former to the latter and that permits momentum and/or angular momentum to also be transferred in a manner that preserves total energy, momentum, and angular momentum. There are well established quantum mechanical expressions for the rates of this kind of radiationless process, and these expressions have been used successfully to compute electron ejection rates. In this paper, we recast the quantum rate equation into more physically clear forms by making use of semi-classical approximations that have proven useful in rewriting the quantum expressions for rates of other processes (e.g., photon absorption) in a more classical manner. It is hoped that by achieving alternative and clearer interpretations of the electron ejection rate equation, it will be possible to more readily predict when such rates will be significant.

1. INTRODUCTION

1.1 Relation to Experiments

There exist a series of beautiful spectroscopy experiments that have been carried out over a number of years in the Lineberger (1), Brauman (2), and Beauchamp (3) laboratories in which electronically stable negative molecular ions prepared in excited vibrational-rotational states are observed to eject their "extra" electron. For the anions considered in those experiments, it is unlikely that the anion and neutral-molecule potential energy surfaces undergo crossings at geometries accessed by their vibrational motions in these experiments, so it is believed that the mechanism of electron ejection must involve vibration-rotation to electronic energy flow. That is, the couplings between nuclear motions and electronic motions known as non Born-Oppenheimer (BO) couplings have been postulated to cause the electron ejection rather than curve crossings in which the anion's energy surface intersects that of the neutral at some geometries.

In earlier works (4), we and others (5) have formulated and computed such non BO coupling strengths for several of the anion systems that have been studied experimentally including:

1. Dipole-bound anions (5a, 4f) in which the extra electron is attracted primarily by the dipole force field of the polar molecule and for which rotation-to-electronic coupling is most important in inducing electron ejection.

2. NH^- ($X^2\Pi$) for which (4d) vibration of the N-H bond couples only weakly to the non-bonding $2p_\pi$ orbital and for which rotation-to-electronic coupling can be dominant in causing electron ejection for high rotational levels.

3. Enolate anions (4e) that have been "heated" by infrared multiple photon absorption for which torsional motion about the H_2C-C bond, which destabilizes the π orbital containing the extra electron, is the mode contributing most to vibration-to-electronic energy transfer and thus to ejection.

Our calculations have been successful in interpreting trends that are seen in the experimentally observed rates of electron ejection. However, until now, we have not had a clear physical picture of the energy and momentum (or angular momentum) balancing events that accompany such non BO processes. It is the purpose of this paper to enhance our understanding of these events by recasting the rate equations in ways that are more classical in nature (and hence hopefully more physically clear). This is done by

1. starting with the rigorous state-to-state quantum expression for non BO transition rates (4g),

2. including what is known from past experience (4) about the magnitudes and geometry dependencies of the electronic non BO matrix elements arising in these rate expressions, to

3. make the simplest reasonable semi-classical approximation to the nuclear motion (7) (i.e., vibration-rotation) and its coupling to the electronic motions. By so doing, we are able to arrive at expressions for rates of electron ejection that, in our opinion, offer better physical insight into these radiationless processes and thus offer the potential for predicting when such rates will be significant in other systems.

1.2 State-to-State Quantum Rate Expression

Within the Born-Oppenheimer approximation to molecular structure, the electronic Schrödinger equation

$$h_e\,(r|Q)\ \psi_K(r|Q) = E_k(Q)\ \psi_k(r|Q) \tag{1}$$

is solved to obtain electronic wavefunctions $\psi_k(r|Q)$, which are functions of the molecule's electronic coordinates (collectively denoted r) and atomic coordinates (denoted Q), and the corresponding electronic energies $E_k(Q)$, which are functions of the Q coordinates. The electronic Hamiltonian

$$h_e(r|Q) = \Sigma_i\ \{-\hbar^2/2m_e\ \nabla_i^2 + 1/2\ \Sigma_{j \neq i}\ e^2/r_{i,j}\ - \Sigma_a\, Z_a\, e^2\ /r_{i,a}$$

$$+\ 1/2\ \Sigma_{a \neq b}\ Z_a\, Z_b\ e^2/R_{a,b}\ \} \tag{2}$$

contains, respectively, the sum of the kinetic energies of the electrons, the electron-electron repulsion, the electron-nuclear Coulomb attraction, and the nuclear-nuclear repulsion energy. In h_e, second-order differential operators involving the coordinates of the electrons appear, but the coordinates of the atomic centers appear only parametrically in the various Coulomb potentials. Hence, the solutions $\{\psi_k$ and $E_k\}$ depend only parametrically on the nuclear positions.

Given the solutions to the electronic Schrödinger equation, the solutions of the full Schrödinger equation (i.e., the equation in which all nuclei and electrons are moving)

$$H(r|Q)\,\Psi(r,Q) = E\,\Psi(r,Q) \tag{3}$$

are expressed as sums over the (complete set of functions of the electronic coordinates r) electronic functions $\{\psi_k(r|Q)\}$

$$\Psi(r,Q) = \Sigma_k\,\psi_k(r|Q)\,\chi_k(Q), \tag{4}$$

with the "expansion coefficients" $\chi_k(Q)$ carrying the remaining Q-dependence. When substituted into the full Schrödinger equation, this expansion of Ψ gives equations which are to be solved for these $\{\chi_k\}$ functions:

$$(H - E)\Sigma_k\,\psi_k(r|Q)\,\chi_k(Q) = 0. \tag{5}$$

Using the fact that the full Hamiltonian H is h_e plus the kinetic energy operator for nuclear motion T

$$H = h_e + T = h_e + \Sigma_a(-\hbar^2/2m_a\,\nabla_a^2), \tag{6}$$

and premultiplying the above Schrödinger equation by ψ_n and integrating over the electronic coordinates gives the set of coupled equations that need to be solved for the $\{\chi_k\}$:

$$\begin{aligned}&\Sigma_k \int \psi_n^*(r|Q)\,\{h_e + T - E\}\,\psi_k(r|Q)\,\chi_k(Q)\,dr\\ &= \{E_n(Q) - E\}\,\chi_n(Q) + T\,\chi_n(Q)\\ &+ \Sigma_a\Sigma_k\,\{\int \psi_n^*(r|Q)\,(-i\hbar\partial\psi_k/\partial R_a)(-i\hbar\partial\chi_k/\partial R_a)/m_a\,dr\\ &+ \int \psi_n^*(r|Q)\,(-\hbar^2\partial^2\,\psi_k/\partial R_a^2)/2m_a\,dr\,\chi_k\} = 0.\end{aligned} \tag{7}$$

The expression

$$\{E_n(Q) \ -E\} \ \chi_n(Q) + T \ \chi_n(Q) = 0 \tag{8a}$$

is, within the Born-Oppenheimer model, the equation governing the nuclear motion functions $\{\chi_{k\,L}(Q)\}$ in the absence of the so-called non Born-Oppenheimer (non BO) coupling terms (i.e., the latter two terms in Eq.(7)). Within this model, the vibration-rotation functions $\{\chi_k(Q)\}$ of each specific electronic state labeled k are found by solving the vibration-rotation Schrödinger equation

$$\{T + E_k(Q)\} \ \chi_{k,L}(Q) = \varepsilon_{k,L} \ \chi_{k,L}(Q). \tag{8b}$$

There are a complete set of functions of Q (i.e., the $\{\chi_{k,L}\}$ for each electronic state k.

In the theory of radiationless transitions as covered in this paper (6,4g), the two non BO terms are treated as perturbations (not externally applied, but arising as imperfections within this model of molecular structure) that can induce transitions between unperturbed states each of which is taken to be a specific Born-Oppenheimer product state:

$$\Psi_{k,L}(r,Q) \cong \psi_k(r|Q) \ \chi_{k,L}(Q). \tag{9}$$

It is reasonably well established that the non BO coupling term involving second derivatives of the electronic wavefunction contributes less to the coupling than does the term $(-i\hbar\partial\psi_k/\partial R_a)\ (-i\hbar\partial\chi_k/\partial R_a)/m_a$ having first derivatives of the electronic and vibration-rotation functions. Hence, it is only the latter terms that will be discussed further in this paper.

With this background, it should not be surprising that it has been shown that the rate R (sec^{-1}) at which transitions from a Born-Oppenheimer initial state $\Psi_i = \psi_i \chi_i$ to a final state $\Psi_f = \psi_f \chi_f$ is given, via first-order perturbation theory, as:

$$R = (2\pi/\hbar) \int |\langle\chi_i| \langle\psi_i |P| \psi_f\rangle(P/\mu)\chi_f\rangle|^2 \delta(\varepsilon_f + E - \varepsilon_i)\, \rho(E)\, dE. \tag{10}$$

Here, $\varepsilon_{i,f}$ are the vibration-rotation energies of the initial (anion) and final (neutral) states, and E denotes the kinetic energy carried away by the ejected electron (e.g., the initial state corresponds to an anion and the final state to a neutral molecule plus an ejected electron). The density of translational energy states of the ejected electron is $\rho(E) = 4\pi m_e L^3 (2meE)^{1/2}/h^2$. We have used the short-hand notation involving P P/μ to symbolize the multidimensional derivative operators that arise in the non BO couplings as discussed above:

$$(P\psi_f)(P/\mu\chi_f) = \Sigma_a (-i\hbar\partial\psi_f/\partial R_a)(-i\hbar\partial\chi_f/\partial R_a)/m_a\,, \tag{11}$$

where R_a runs over the Cartesian coordinates (X_a, Y_a, Z_a) of the ath atom whose mass is m_a. In Eq. (10), the product ρ dE is unitless, $\delta(\varepsilon_f + E - \varepsilon_i)$ has units of $sec^2/(gm\ cm^2)$, the square of the P matrix element has units of $(gm\ cm/sec)^2$, the square of the P/μ matrix element has units of $(cm/sec)^2$, and $2\pi/\hbar$ has units of $(sec/(gm\ cm^2))$. Hence the product has units of sec^{-1} .

1.3 The Electronic Non BO Matrix Elements

The integrals over the anion and neutral-plus-free-electron electronic states

$$m_{i,f} = \langle\psi_f |P| \psi_i\rangle \tag{12}$$

are known to be large in magnitude only under special circumstances:

1. The orbital of the anion from which an electron is ejected to form the state ψ_f of the neutral (usually the anion's highest occupied molecular orbital (HOMO)) must be strongly modulated or affected by movement of the molecule in one or more directions (Q). That is $\partial\psi_i/\partial Q$, which appears in $P\psi_i$, must be significant.
2. The state-to-state energy gap $\varepsilon_i - \varepsilon_f$, which is equal to the energy E of the ejected electron, must not be too large; otherwise, the oscillations in the ejected

electron's wavefunction ψ_f will be so rapid as to render overlap with $\partial\psi_i/\partial Q$ negligible.

Moreover, symmetry can cause $m_{i,f} = <\psi_f |P| \psi_i>$ to vanish. In particular, if the direct product of the symmetry of ψ_i and of $\partial/\partial Q$ do not match that of ψ_f, then $m_{i,f}$ will vanish.

Let us consider these conditions in more detail before proceeding further because they form the basis for approximations that are introduced later. The derivatives or responses of the anion's orbitals to nuclear motions $\partial\psi_i/\partial Q$ arise from two sources:

1. The orbital's LCAO-MO coefficients depend on the positions of the atoms (or, equivalently, on bond lengths and internal angles). For example, the π^* orbital of an olefin anion that contains the "extra" electron is affected by stretching or twisting the C-C bond involving this orbital because the LCAO-MO coefficients depend on the bond length and twist angle. As the bond stretches or twists, the π^* orbital's LCAO-MO coefficients vary, as a result of which the orbital's energy, radial extent, and other properties also vary.
2. The atomic orbitals (AO) themselves respond to the motions of the atomic centers. For example, vibration of the $X^2\Pi$ NH^- anion's N-H bond induces d_π character into the $2p_\pi$ orbital containing the extra electron as shown in Fig. 1. Alternatively, rotation of this anion's N-H bond axis causes the $2p_\pi$ HOMO to acquire some $2p_\sigma$ character (see Fig. 1 again). Such AO responses can be evaluated using the same analytical derivative methods that have made computation of potential energy gradients and Hessians powerful tools in quantum chemistry.

$d\Psi/dR = (\Psi(R+\delta) - \Psi(R))/\delta$
causes $2p_x$ to acquire d character

$d\Psi/d\theta = (\Psi(\theta+\delta) - \Psi(\theta))/\delta$
causes $2p_x$ to acquire $2p_z$ character

Figure 1. Orbital response of NH^- 's $2p_\pi$ orbital to (a) vibratation of the N-H bond (left) and (b) rotation of the N-H bond (right).

Another view of how the LCAO-MO coefficients vary with geometry can be achieved by differentiating $h_e \psi_i = E_i \psi_i$ with respect to Q (an arbitrary molecular motion) and then premultiplying by the anion-plus-free-electron function ψ_f and integrating over the electronic coordinates r to obtain:

$$<\psi_f|\partial h_e/\partial Q|\psi_i>/(E_i - E_f - E) = <\psi_f|\partial/\partial Q \psi_i>. \tag{13}$$

In this form, one sees that the response of the anion's electronic state, when projected against the neutral-plus-free-electron state to which it will decay, will be enhanced at geometries where the anion and neutral potential surfaces approach closely (so the denominator in Eq. (13) is small). Enhancement is also effected when the initial and final states have a strong matrix element of the "force operator" $\partial h_e/\partial Q$. The latter is effectively a one-electron operator involving derivatives of the electron-nuclear Coulomb attraction potential $\Sigma_i \Sigma_a Z_a e^2/r_{ii,a}$, so the matrix element $<\psi_f|\partial h_e/\partial Q|\psi_i>$ can be visualized as $<\phi_f|\partial h_e/\partial Q|\phi_i>$, where ϕ_i is the anion's HOMO and ϕ_f is the continuum orbital of the ejected electron. At geometries where the anion-neutral energy surfaces are far removed, the denominator in Eq.(13) will attenuate the coupling. If the state-to-state energy difference $\varepsilon_i - \varepsilon_f = E$ accompanying the electron ejection is large, the integral $<\phi_f|\partial h_e/\partial Q|\phi_i>$ will be small because the continuum orbital ϕ_f will be highly oscillatory and thus will not overlap well with $(\partial h_e/\partial Q)\phi_i$.

In summary, for non BO coupling to be significant (4) the anion's HOMO must be strongly modulated by a motion (vibration or rotation) of the molecule's nuclear framework and the state-to-state energy gap must not be too large as to render the HOMO-to-continuum-orbital overlap insignificant. For the HOMO to be strongly modulated, it is helpful if the anion and neutral energy surfaces approach closely (n.b., this is not the same as requiring that the state-to-state energy gap ε_i - ε_f be small) at some accessible geometries.

It should be emphasized that it is necessary but not sufficient for $E_f(Q)$ - $E_i(Q)$ to be small over an appreciable range of geometries; this only guarantees that the denominator in Eq. (13) is small. It is also necessary that $E_f(Q)$ - $E_i(Q)$ decrease at a significant rate as the point of closest approach is reached; this is why we say the surfaces must **approach** closely. Viewed another way, if $E_f(Q)$ - $E_i(Q)$ were small yet unvarying over some range of geometries (Q), then the HOMO's electron binding energy (and thus radial extent) would remain unchanged over this range of geometries. In such a case, movement along Q would not modulate the HOMO, and thus $\partial\psi_i/\partial Q$ would vanish. Let us consider a few examples to further illustrate.

1.4 A Few Examples

In Fig. 2 are depicted anion and neutral potential curves that are qualitatively illustrative of (1b,4d) the $X^2\Pi$ NH^- case mentioned earlier. In this anion, the HOMO is a non-bonding $2p_\pi$ orbital localized almost entirely on the N atom. As such, its LCAO-MO coefficients are not strongly affected by motion of the N-H bond (because it is a non-bonding orbital). Moreover, the anion and neutral surfaces have nearly identical R_e and ω_e values, and similar D_e values, as a result of which these two surfaces are nearly parallel to one another and are separated by ca. 0.4 eV or more than 3000 cm^{-1}. It has been seen experimentally that excitation of NH$^-$ to the low rotational states of the v=1 vibrational level, which lies above v=0 NH neutral, results in very slow (e.g., cs. 10^8 sec^{-1}) electron ejection, corresponding to ca. one million vibrational periods before detachment occurs. However, excitation to high rotational levels (e.g., J = 40) of v=1 produces much more rapid electron ejection (ca. 10^9 -10^{10} sec^{-1}). These data have been interpreted as saying that vibrational coupling is weak because of the non-bonding nature of the 2p$_\pi$ MO, while rotational coupling becomes significant for high J.

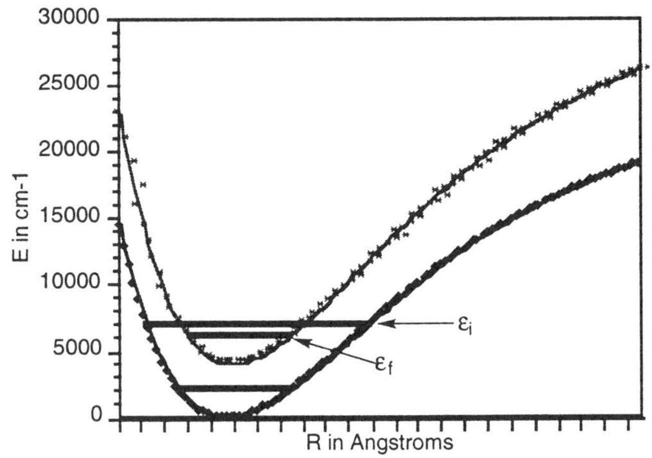

Figure 2. Anion (lower) and neutral (upper) potential energy surfaces illustrative of NH$^-$ where the surface spacing does not vary strongly along R.

Fig. 3 shows a hypothetical case similar to the NH$^-$ situation but for which the anion and neutral curves approach closely at longer bond lengths. In this case, one would expect larger rates of detachment than in NH$^-$ because

1. The state-to-state gap ε_i -ε_f is small for the two states labeled in Fig. 3.

2. The anion and neutral curves <u>approach</u> one another at R values that are accessible to the vibrational wavefunctions of the two states shown in Fig. 3, thus allowing strong modulation of the HOMO.

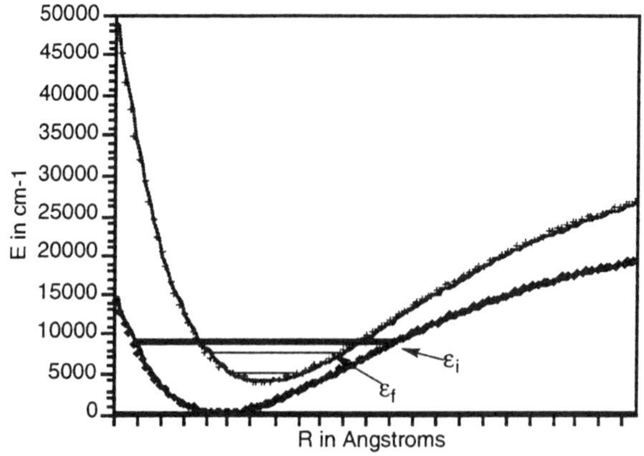

Figure 3. Anion (lower) and neutral (upper) potential energy surfaces illustrative of cases where the surface spacing varies strongly along R and becomes small at some R.

In Fig. 4 are shown anion and neutral potential curves, as functions of the "twist" angle of the H_2C-C bond in a typical enolate anion (2,4e) such as acetaldehyde enolate H_2CCHO^-. Angles near θ =0 correspond to geometries where the p_π orbital of the H_2C moiety is delocalized over the two p_π orbitals of the neighboring C and O atoms, thus forming a delocalized π HOMO.

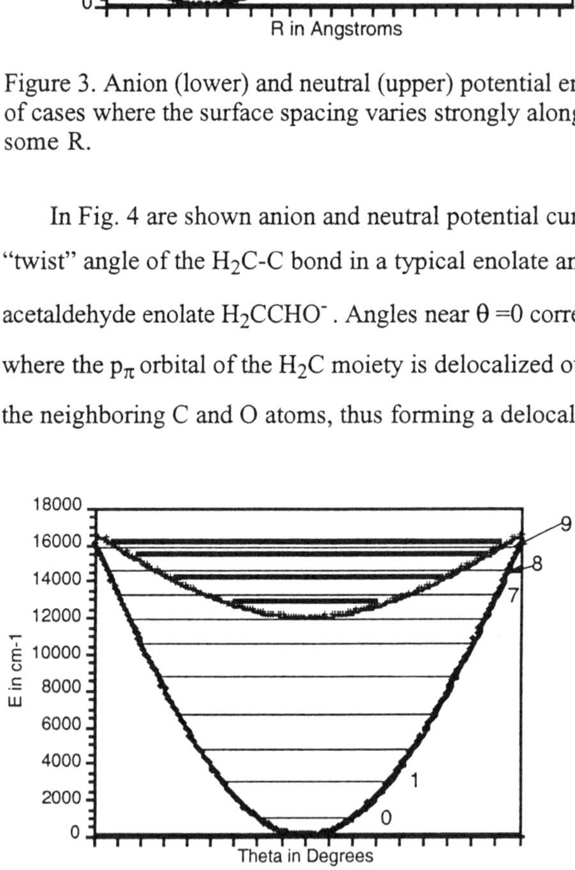

Figure 4. Anion (lower) and neutral (upper) potential energy surfaces illustrative of enolate cases where the surface spacing varies strongly along the H_2C-C torsion angle θ and becomes very small near $\theta = 90°$.

At angles near $\theta = 90°$, the p_π orbital of the H_2C group is no longer stabilized by delocalization; so the HOMO's energy is much higher (as a result of which the anion-neutral surfaces approach closely). In this case, excitation of, for example, v=7 in the H_2C-C torsional mode of the anion might be expected to produce electron ejection because v=7 of the anion lies above v = 0 of the neutral. However, over the range of θ values accessible to both the v=7 vibrational function of the anion and the v=0 function of the neutral, the anion-neutral potential energy gap is quite large (i.e., $E_{ff}(Q) - E_i(Q)$ is large even though $\varepsilon_i - \varepsilon_f$ is small). In contrast, excitation of v=9 of the anion could produce more rapid electron ejection (to v=2 of the neutral, but not to v=0 of the neutral) because for the v=9 $\rightarrow$ v=2 transition there are angles accessed by both v=9 anion and v=2 neutral vibrational functions for which $E_{ff}(Q) - E_i(Q)$ is small and changing; moreover, the state-to-state gap $\varepsilon_i - \varepsilon_f$ is also small in this case.

The purpose of these examples and of considering the nature of the electronic non BO matrix element was to prepare for critical approximations that are to be introduced. In particular,

1. We will focus on transitions for which $\varepsilon_i - \varepsilon_f$ is small.

2. We will focus on molecular deformations that most strongly modulate the anion's HOMO, so

3. we will focus on geometries Q near which the anion-neutral surface spacing is small and changing.

2. TIME CORRELATION FUNCTION EXPRESSION FOR RATES

Before dealing further with the non Born-Oppenheimer case, it is useful to recall how one can cast other rate expressions, such as the rate of photon absorption (7) accompanying an electronic transition in a molecule, in terms of a

Fourier transform of a time dependent function that involves dynamical motions on the initial and final electronic states' potential energy surfaces.

2.1 The Optical Spectroscopy Case

2.1.1 From Wentzel-Fermi Golden Rule to the Time Domain

The expression for the rate R (sec^{-1}) of photon absorption due to coupling V between a molecule's electronic and nuclear charges and an electromagnetic field is given through first order in perturbation theory by the well known Wentzel Fermi "golden rule" formula (7,8):

$$R = (2\pi/\hbar)|\langle\psi_i \chi_i |V| \psi_f \chi_f\rangle|^2 \delta(\varepsilon_f - \varepsilon_i - \hbar\omega). \tag{14}$$

Here, $\psi_{i,f}$ and $\chi_{i,f}$ are the initial and final state electronic and vibration-rotation state wavefunctions, respectively, and $\varepsilon_{i,f}$ are the respective state energies which are connected via a photon of energy $\hbar\omega$. For a particular electronic transition (i.e., a specific choice for ψ_i and ψ_f and for a specific choice of <u>initial</u> vibration-rotation state, it is possible to obtain an expression for the total rate R_T of transitions from this particular initial state into all vibration-rotation states of the final electronic state. This is done by first using the Fourier representation of the Dirac δ function:

$$\delta(\varepsilon_f - \varepsilon_i - \hbar\omega) = (1/2\pi\hbar) \int \exp[it(\varepsilon_f - \varepsilon_i - \hbar\omega)/\hbar]\, dt \tag{15}$$

and then summing over the indices labeling the final vibration-rotation states χ_f:

$$R_T = (2\pi/\hbar)\Sigma_f \ (1/2\pi\hbar) \int \exp[it(\varepsilon_f - \varepsilon_i - \hbar\omega)/\hbar] \langle\psi_i \chi_i |V| \psi_f \chi_f\rangle \langle\psi_i \chi_i |V| \psi_f \chi_f\rangle^* \, dt. \tag{16}$$

Next, one introduces the electronic transition matrix element (which may be an electric dipole matrix element, but need not be so restricted for the development presented here)

$$\mu_{i,f} = \langle\psi_f|V|\psi_i\rangle \tag{17}$$

and uses the facts that the $\psi_{i,f}$ are eigenfunctions of the electronic Hamiltonian h_e and that the $\chi_{i,f}$ are eigenfunctions of the vibration-rotation Hamiltonia T + $V_{i,f}(Q)$ belonging to the two electronic states having potential energies $V_{i,f}(Q)$ and vibration-rotation kinetic energy T (both of which are functions of the molecule's atomic position coordinates collectively denoted Q)

$$h_e\,\psi_{i,f} = V_{i,f}(Q)\,\psi_{i,f} \tag{18a}$$

$$[T + V_{i,f}(Q)\,]\,\chi_{i,f} = \varepsilon_{i,f}\,\chi_{i,f}\,. \tag{18b}$$

These identities then allow R_T to be rewritten as

$$R_T = (2\pi/\hbar)\Sigma_f \;\; (1/2\pi\hbar)\int \exp[-it\,\omega]$$

$$\langle\chi_i\,|\exp(-ith_i/\hbar)\;\mu_{i,f}{}^*|\,\chi_f\rangle\,\langle\chi_f\,\exp(ith_f/\hbar)|\;\mu_{i,f}\,|\,\chi_i\rangle\;dt. \tag{19}$$

In this form, the completeness of the $\{\chi_f\}$

$$\Sigma_f\,|\chi_f\rangle\langle\chi_f| = 1 \tag{20}$$

can be used to eliminate the sum over the vibration-rotation states belonging to the final electronic state and thus express R_T in the following manner:

$$R_T = \;(2\pi/\hbar)\;(1/2\pi\hbar)\int \exp[-it\,\omega]$$

$$\langle\, \exp(ith_i/\hbar)\chi_i\,|\;\mu_{i,f}{}^*\;\exp(ith_f/\hbar)\;\mu_{i,f}\,|\,\chi_i\rangle\;dt. \tag{21}$$

The above expression is often visualized (and computed) (9) in terms of the Fourier transform of the overlap of two time-propagated wavefunctions:

(a) One function F_1 is equal to the initial vibration-rotation function $|\chi_i\rangle$ upon which the electronic transition perturbation $\mu_{i,f}$ acts, after which the resultant product function is propagated for a time t on the final-state's potential energy surface by using the propagator $\exp(ith_f/\hbar)$.

(b) The second function F_2 is equal to the initial vibration-rotation function $|\chi_i\rangle$ which is propagated for time t on the <u>initial</u>-state's potential energy surface using the propagator $\exp(ith_i/\hbar)$ (which, of course simply produces $\exp(it\varepsilon_i/\hbar)$ times $|\chi_i\rangle$) , after which the electronic transition perturbation $\mu_{i,f}$ is allowed to act.

The overlap $\langle F_2 | F_1\rangle$ is then Fourier transformed at the energy $\hbar\omega$ of the photon to obtain the rate R_T of absorption of photons of frequency ω.

Before returning to the non-BO rate expression, it is important to note that, in this spectroscopy case, the perturbation (i.e., the photon's vector potential) appears explicitly only in the $\mu_{i,f}$ matrix element because this external field is purely an electronic operator. In contrast, in the non-BO case, the perturbation involves a product of momentum operators, one acting on the electronic wavefunction and the second acting on the vibration/rotation wavefunction because the non-BO perturbation involves an explicit exchange of momentum between the electrons and the nuclei. As a result, one has matrix elements of the form $\langle\chi_i| \langle\psi_i |P| \psi_f\rangle(P/\mu)\chi_f\rangle$ in the non-BO case where one finds $\langle\chi_i| \langle\psi_i |V| \psi_f\rangle|\chi_f\rangle$ in the spectroscopy case. A primary difference is that derivatives of the vibration/rotation functions appear in the former case (in $(P/\mu)\chi$) where only χ appears in the latter.

2.1.2 The Semi-Classical Approximation to R_T

There are various approximations (7) to the above expression for the absorption rate R_T that offer further insight into the photon absorption process and form a basis for comparison to the non Born-Oppenheimer rate expression. The most classical (and hence, least quantum) approximation is to ignore the fact that the kinetic energy operator T does not commute with the potentials $V_{i,f}$ and thus to write

$$\exp(ith_{i,f}/\hbar) \cong \exp(itT/\hbar)\exp(itV_{i,f}/\hbar). \quad (22)$$

Inserting this into the above equation for R_T and also assuming that T also commutes with $\mu_{i,f}$ (or that $\mu_{i,f}$ does not depend significantly on geometry Q), gives

$$R_T = (2\pi/\hbar)\,(1/2\pi\hbar) \int \exp[-it\,\omega]$$
$$< \chi_i \exp(itV_i/\hbar) |\, \mu_{i,f}{}^* \exp(itV_f/\hbar) |\, \mu_{i,f} \,|\, \chi_i > dt. \quad (23)$$

Then, carrying out the Fourier integral over time gives,

$$(1/2\pi\hbar) \int \exp[it(V_f - V_i - \hbar\omega)/\hbar]\,dt = \delta(V_f - V_i - \hbar\omega). \quad (24)$$

This delta function can be used in the expression for R_T to constrain the multidimensional integral over vibration-rotation coordinates (denoted Q) to those specific values which obey the energy conservation condition

$$\hbar\omega = V_f(Q) - V_i(Q), \quad (25)$$

thereby yielding

$$R_T = (2\pi/\hbar) < \chi_i \,|\, \mu_{i,f}{}^* \,\delta(V_f - V_i - \hbar\omega)\, \mu_{i,f} \,|\, \chi_i > . \quad (26)$$

This semi-classical result can be interpreted as saying that R_T is given as the norm of the function $\mu_{i,f}\chi_i$, consisting of the perturbation $\mu_{i,f}$ acting on the initial vibration-rotation state, <u>constrained</u> to those regions of space which obey the condition $\hbar\omega = V_f(Q) - V_i(Q)$. This condition is equivalent to constraining the integration to those regions within which the change in classical kinetic energy in moving from the initial-state surface V_i to the final-state surface V_f is zero. One can visualize such geometries as those at which the upper potential energy surface $V_f(Q)$ is intersected by the lower surface V_i (Q) once the lower surface is shifted to higher energy by an amount $\hbar\omega$. In Fig. 5 is shown such an intersection between a $V_f(Q)$ surface and a lower-state $V_i(Q)$ surface that has been shifted upward in energy. In this example, transitions between the initial level ε_i and the final level ε_f, whose difference $\varepsilon_f - \varepsilon_i = \hbar\omega$ determines the energy shift, would occur at R values near where the two surfaces cross.

2.1.3 Relation to Landau-Zener Surface Hopping Rates

It is instructive to examine further the approximate semi-classical form for R_T shown above because, when viewed as a rate of transition between two intersecting energy surfaces, one anticipates that connection can be made with the well known Landau-Zener theory (10). For a non-linear molecule with N atoms, the potentials $V_{i,f}(Q)$ depend on 3N-6 internal degrees of freedom (for a linear molecule, $V_{i,f}$ depend on 3N-5 internal coordinates). The subspace S

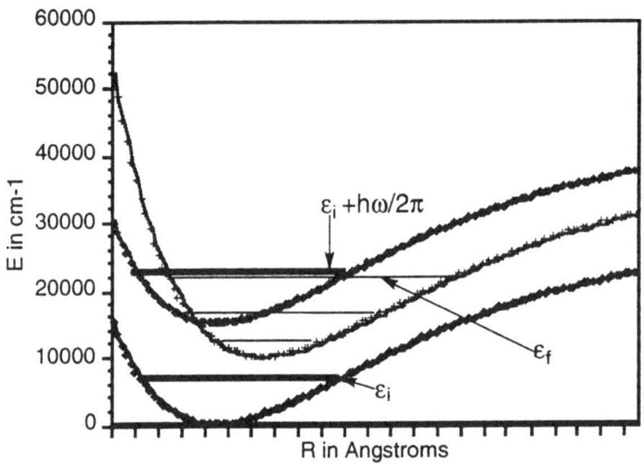

Figure 5. Anion (lowest) and neutral (second lowest at large R) potential energy surfaces arising in the photon absorption case. The curve that is highest in energy at large R is the anion surface that has been shifted upward in energy by the photon's energy $\hbar\omega$ (which has been chosen to make ε_f equal to ε_i plus the photon's energy).

within which $V_f(Q) = V_i(Q) + \hbar\omega$ will be of dimension 3N-7 for the non-linear molecule case. Any geometrical arrangement of the molecule can be described by specifying an orientation Ω (i.e., three orientation or Euler angles for a non-linear molecule), a 3N-7 component vector **s** lying within the subspace S and a distance d along the unit vector **n** that is normal to S at **s**:

$$\mathbf{Q} = \{\Omega\ , \mathbf{R}\ \},\ \text{where}\ \mathbf{R} = \mathbf{s} + d\ \mathbf{n}. \tag{27}$$

Points lying within S are characterized by d=0.

The integral appearing in R_T can be rewritten (7b) as an integral over orientations, an integral over the 3N-7 dimensional subspace S, and a one-dimensional integral over d:

$$R_T = (2\pi/\hbar)\int d\Omega \int dS \int dd\ |\chi_i(\mathbf{s},d,\Omega)|^2\ |\mu_{i,f}\ (\mathbf{s},d,\Omega)|^2\ \delta(V_f(\mathbf{s},d) - V_i(\mathbf{s},d) - \hbar\omega). \tag{28}$$

Expanding $V_f(\mathbf{s},d) - V_i(\mathbf{s},d) - \hbar\omega$ in powers of d about the point d=0 and realizing that $V_f(\mathbf{s},0) - V_i(\mathbf{s},0) - \hbar\omega = 0$, allows the delta function to be written as a delta function for the variable d:

$$\delta(V_f(\mathbf{s},d) - V_i(\mathbf{s},d) - \hbar\omega) = \delta(V_f(\mathbf{s},0) - V_i(\mathbf{s},0) - \hbar\omega + [F_f(\mathbf{s}) - F_i(\mathbf{s})]\ d\ ) = |\ F_f(\mathbf{s}) - F_i(\mathbf{s})|^{-1}\ \delta(d). \tag{29}$$

Here, $F_{i,f}(\mathbf{s})$ are the gradients of the respective potentials $V_{i,f}$ along the direction **n** normal to S evaluated at the point **s**,d=0; these gradients, of course, are the negatives of the classical forces normal to S experienced on the $V_{i,f}$ surfaces. With this expression for the delta function, the rate R_T can be expressed as an integral over orientations and over coordinates totally within the space S:

$$R_T = (2\pi/\hbar)\int d\Omega \int dS\ \ |\chi_i(\mathbf{s},0,\Omega)|^2\ |\mu_{i,f}\ (\mathbf{s},0,\Omega)|^2\ |\ F_f(\mathbf{s}) - F_i(\mathbf{s})|^{-1}\ . \tag{30}$$

The Landau-Zener expression for the probability P of moving from $V_i(Q)$ + $h\omega$ to $V_f(Q)$ under the influence of the perturbation $\mu_{i,f}$ $(\mathbf{s},0,\Omega)$ is

$$P = 2\pi \ |\mu_{i,f} (\mathbf{s},0,\Omega)|^2 \ [\ |\ F_f(\mathbf{s}) - F_i(\mathbf{s})|\ \hbar v_d]^{-1}, \tag{31}$$

where v_d is the speed at which flux passes through the intersection of the two surfaces (n.b., any velocity components lying within S do not cause flux to move between the surfaces since, within S, the surfaces are degenerate). If this probability is multiplied by the probability of the molecule residing within S (i.e., by $|\chi_i(\mathbf{s},0,\Omega)|^2$) , then by the rate at which the molecule moves from d=0 to d=δ (i.e., by v_d/δ), and if one then integrates over all orientations and all S and over the small range d=0 to d=δ, one obtains:

$$\int dS\ d\Omega\ \ |\chi_i(\mathbf{s},0,\Omega)|^2 (v_d/\delta)\ P\ \delta$$

$$= (2\pi/\hbar)\int d\Omega \int dS\ \ |\chi_i(\mathbf{s},0,\Omega)|^2\ |\mu_{i,f} (\mathbf{s},0,\Omega)|^2\ \ |\ F_f(\mathbf{s}) - F_i(\mathbf{s})|^{-1} = R_T. \tag{32}$$

That is, the semi-classical approximation to the photon absorption rate is equivalent to a Landau-Zener treatment of the probability of hopping from V_i + $\hbar\omega$ to V_f induced by the electronic coupling perturbation $\mu_{i,f}$ $(\mathbf{s},0,\Omega)$.

2.2 The Non Born-Oppenheimer Case

2.2.1 From Wentzel-Fermi Golden Rule to the Time Domain

Let us now consider how similar the expression for rates of radiationless transitions induced by non Born-Oppenheimer couplings can be made to the expressions given above for photon absorption rates. We begin with the corresponding (6,4g) Wentzel-Fermi "golden rule" expression given in Eq. (10) for the transition rate between electronic states $\psi_{i,f}$ and corresponding vibration-rotation states $\chi_{i,f}$ appropriate to the non BO case:

$$R = (2\pi/\hbar) \int |<\chi_i| <\psi_i |P| \psi_f >(P/\mu)\chi_f>|^2 \delta(\varepsilon_f + E - \varepsilon_i) \rho(E) dE. \quad (10)$$

We recall that $\varepsilon_{i,f}$ are the vibration-rotation energies of the molecule in the anion and neutral molecule states, E denotes the kinetic energy carried away by the ejected electron, and the density of translational energy states of the ejected electron is $\rho(E)$. Also recall that we use the short hand notation to symbolize the multidimensional derivative operators that arise in non BO couplings and that embody the momentum-exchange between the vibration/rotation and electronic degrees of freedom:

$$(P\psi_f)(P/\mu\chi_f) = \Sigma_a (-i\hbar\partial\psi_f/\partial R_a)(-i\hbar\partial\chi_f/\partial R_a)/m_a, \quad (11)$$

where R_a is one of the Cartesian coordinates (X_a, Y_a, Z_a) of the ath atom whose mass is m_a. These 3N coordinates span the same space as the three center of mass coordinates plus the Ω , **s**, and d coordinates used earlier in detailing the semi-classical photon absorption rate expression.

In the event that some subset $\{Q_i\}$ of internal vibration or rotation coordinates have been identified as inducing the radiationless transition, $(P\psi_f)(P/\mu\chi_f)$ would represent $\Sigma_j (-i\hbar\partial\psi_f/\partial Q_j)(-i\hbar\partial\chi_f/\partial Q_j)/(\mu_j)$, where μ_φ is the reduced mass associated with the coordinate Q_j. As indicated in the discussion of Sec. I. C, it is usually straightforward to identify which distortional modes need to be considered by noting which modes most strongly <u>modulate the anion's HOMO</u>. So, for the remainder of this work, we will assume that such active modes have been identified as a result of which the sum $\Sigma_j (-i\hbar\partial\psi_f/\partial Q_j)(-i\hbar\partial\chi_f/\partial Q_j)/(\mu_j)$ will include only these modes. The integration over all of the other coordinates contained in the matrix element $<\chi_i| <\psi_i |P| \psi_f >(P/\mu)\chi_f>$ in Eq. (10) can then be carried out (assuming the electronic element $<\psi_i |P| \psi_f >$ to not depend significantly on these coordinates) to produce an effective Franck-Condon like factor (FC):

$$<\chi_i| <\psi_i |P| \psi_f >(P/\mu)\chi_f>$$

$$= \Pi_{j=\text{inactive}} \int dQ_j <\chi_{i,j} |\chi_{f,j}> \Pi_{j=\text{active}} \int dQ_j <\chi_{i,j}| <\psi_i |P| \psi_f >(P/\mu)\chi_{f,j}>$$

$$= FC\ \Pi_{j=\text{active}} \int dQ_j <\chi_{i,j}| <\psi_i |P| \psi_f >(P/\mu)\chi_{f,j}>. \quad (32)$$

Since, by assumption, the anion and neutral molecule do not differ significantly in their geometries (and vibrational frequencies) along the coordinates contributing to the FC factor (otherwise, the anion-neutral energy gap would depend substantially on these modes), the FC factor is probably close to unity in magnitude. Hence, for the remainder of this paper, we will focus only on the active-mode part of this expression, and will do so assuming only one such mode is operative (i.e., we treat one active mode at a time).

Introducing the electronic coupling matrix element

$$m_{i,f} = <\psi_f |P| \psi_i>, \quad (33)$$

which plays a role analogous to the $\mu_{i,f}$ of photon absorption theory, and realizing that P is a Hermitian operator, allows the non BO rate R to be rewritten as:

$$R = (2\pi/\hbar) \int <(P/\mu)\chi_i| m_{i,f}^* |\chi_f> <\chi_f | m_{i,f} (P/\mu)\chi_i>$$
$$\delta(\varepsilon_f + E - \varepsilon_i)\, \rho(E)\, dE. \quad (34)$$

If the Fourier integral representation of the delta function is introduced and the sum over all possible final-state vibration-rotation states $\{\chi_f\}$ is carried out, the <u>total</u> rate R_T appropriate to this non BO case can be expressed as:

$$R_T = (2\pi/\hbar) \Sigma_f \int (1/2\pi\hbar) \int \exp[it(\varepsilon_f - \varepsilon_i + E)/\hbar]$$
$$<(P/\mu)\chi_i| m_{i,f}^* |\chi_f> <\chi_f | m_{i,f} (P/\mu)\chi_i> dt\, \rho(E)\, dE. \quad (35)$$

The next step is to replace $(\varepsilon_f + E) <\chi_f|$ by $<\chi_f| (T + V_f + E)$ and $(\varepsilon_i)| \chi_i>$ by $|(T + V_i)\, \chi_i>$ which reduces R_T to

$$R_T = (2\pi/\hbar)\, \Sigma_f \int (1/2\pi\hbar) \int \rho(E) <m_{i,f}(P/\mu) \exp(it(T + V_i)/\hbar)\, \chi_i|$$
$$|\chi_f> <\chi_f|\exp(it(E+ T + V_f)/\hbar)\, m_{i,f}\, (P/\mu)\chi_i> dt\, dE. \quad (36)$$

The sum $\Sigma_f |\chi_f><\chi_f| = 1$ can then be carried out to give

$$R_T = (2\pi/\hbar) \int (1/2\pi\hbar) \int \rho(E)$$
$$<m_{i,f}(P/\mu) \exp(it(T + V_i)/\hbar)\, \chi_i|\, \exp(it(E+ T + V_f)/\hbar)\, m_{i,f}\, (P/\mu)\chi_i> dt\, dE. \quad (37)$$

In this form, the rate expression looks much like that given for the photon absorption rate in Eq. (21), but with $m_{i,f}(P/\mu)$ replacing the molecule-photon interaction potential V. As in the absorption case, one can view (and even compute) R_T as the Fourier transform of the overlap of two time propagated functions:

(a) The first F_1 is the initial vibration-rotation state χ_i upon which the non BO perturbation $m_{i,f}\, (P/\mu)$ acts after which propagation on the neutral molecule's potential surface V_f is effected via $\exp(it(T + V_f)/\hbar)$.

(b) The second F_2 is the initial function χ_i which is propagated on the anion's surface V_i via $\exp(it(T + V_i)/\hbar)$ (producing, of course, $\exp(it\varepsilon_i/\hbar)\, \chi_i$) after which the perturbation $m_{i,f}(P/\mu)$ is allowed to act.

The time dependent overlap $<F_2 | F_1>$ is then Fourier transformed at energy $E = \varepsilon_i - \varepsilon_f$, and multiplied by the density of states $\rho(E)$ appropriate to the electron ejected with kinetic energy E.

It should be noted that to use the above time-domain formulas for computing rates, one would need an efficient means of propagating wave packets on the neutral and anion surfaces, and one, specifically, that would be valid for longer times than are needed in the optical spectroscopy case. Why? Because, in the non-BO situation, the $<F_2 | F_1>$ product is multiplied by $\exp(iEt/\hbar)$ and then integrated over time. In the spectroscopy case, $<F_2 | F_1>$ is multiplied by exp(-iωt) and integrated over time. However, in the former case, E corresponds to the (small) energy difference $E = \varepsilon_i - \varepsilon_f$, whereas $\hbar\omega$ is equal to the energy of the optical spectroscopic transition. Hence, short time propagators give sufficiently accurate $<F_2 | F_1>$ functions to use in the spectroscopy case, but longer time propagations will be needed in the non-BO case.

2.2.2 The Simplest Semi-Classical Approximation Fails

If one attempts to follow the photon absorption derivation and make the assumption that the kinetic energy operator T commutes with $V_{i,f}$ and with $m_{i,f}$ (n.b., T does commute with P/μ), the following expression is obtained for R_T:

$$R_T = (2\pi/\hbar) \int (1/2\pi\hbar) \int \rho(E) <m_{i,f}(P/\mu) \exp(it(V_i)/\hbar) \chi_i | \exp(it(E+ V_f)/\hbar) m_{i,f} (P/\mu)\chi_i> dt\, dE. \quad (38)$$

The Fourier integral over time can be carried out and one obtains

$$R_T = (2\pi/\hbar) \int \rho(E) <m_{i,f}(P/\mu) \chi_i | \delta(V_f+ E- V_i) m_{i,f} (P/\mu)\chi_i> dE. \quad (39)$$

For anions that are electronically bound, the anion's electronic energy V_i (Q) lies below the neutral molecule's electronic energy V_f (Q) as depicted in Figs. 2-5. Hence, because E is a positive quantity, there are no geometries for which the argument of the delta function in the above expression vanishes and, as a result, the non BO rate can not be cast in terms of shifted intersecting energy surfaces as can the photon absorption rate.

Therefore, the simplest classical treatment in which the propagator exp(it (T+V)/ħ) is approximated in the product form exp(it (T)/ħ) exp(it (V)/ħ) and the nuclear kinetic energy T is conserved during the "transition" produces a nonsensical approximation to the non BO rate. This should not be surprising because (a) In the photon absorption case, the photon induces a transition in the electronic degrees of freedom which subsequently cause changes in the vibration-rotation energy, while (b) in the non BO case, the electronic and vibration-rotation degrees of freedom must simultaneously interchange energy and momentum and/or angular momentum, which is impossible to do without the nuclei recoiling.

2.2.3 An Approximation that Works

The matrix element occurring in the non BO rate equation

$$M = \langle m_{i,f}(P/\mu) \exp(it(T + V_i)/\hbar)\, \chi_i \mid \exp(it(E+ T + V_f)/\hbar)\, m_{i,f}\ (P/\mu)\chi_i\rangle \quad (40)$$

can more fruitfully be handled by

a. replacing $\exp(it(T + V_i)/\hbar)\, \chi_i$ by $\exp(it(\varepsilon_i)/\hbar)\, \chi_i$, which is exact, and

b. approximating $\exp(it(E+ T + V_f)/\hbar)$ by $\exp(it(E+ T\,)/\hbar) \exp(it(V_f)/\hbar)$,

which is an approximation, thus achieving

$$M = \langle m_{i,f}(P/\mu) \exp(it(\varepsilon_i)/\hbar)\, \chi_i \mid \exp(it(E+ T\,)/\hbar) \exp(it(V_f\,)\, m_{i,f}\ (P/\mu)\chi_i\rangle. \quad (41)$$

Now introducing completeness relations in the forms

$$\int dp\ |p\rangle\langle p| = 1 \quad (42a)$$

$$\int dQ\ |Q\rangle\langle Q| = 1 \quad (42b)$$

and using $(T + V_i\,)\, \chi_i = \varepsilon_i\, \chi_i$, allows M to be written as

$$M = \int dQ' \int dQ \int dp \langle m_{i,f}(P/\mu) \exp(it(\varepsilon_i)/\hbar)\, \chi_i\ |Q'\rangle$$

$$\langle Q'|\ p\rangle \exp(it(E+ T_{class}\,)/\hbar) \langle p|Q\rangle\langle Q| \exp(it(V_f(Q)/\hbar\,)\, m_{i,f}\ (P/\mu)\chi_i\rangle. \quad (43)$$

Here T_{class} is the eigenvalue of the Q-coordinate's kinetic energy operator T in the momentum eigenbasis $\{|p\rangle\}$

$$T_{class} = (p^2/2\mu_Q) \quad (44)$$

and

$$<Q'|p> = (2\pi\hbar)^{-1/2} \exp(ipQ'/\hbar) \tag{45}$$

is the coordinate representation of the momentum eigenfunction along the active Q coordinate.

The integration over time can be carried out and gives the following expression fot the total rate:

$$R_T = (2\pi/\hbar) \int \rho(E) \int dQ' \int dQ \int dp <m_{i,f}(P/\mu)\, \chi_i\ |Q'> <Q'|\ p>$$
$$\delta(\ T_{class} + E + V_f\,(Q) - \varepsilon_i) \quad <p|Q> <Q|\ m_{i,f}\ (P/\mu)\chi_i> dE. \tag{46}$$

Since the energy E is restricted to match the state-to-state energy differences E = $\varepsilon_i - \varepsilon_f$, the integral over dE in Eq. (46) can be replaced by a sum over accessible final-state ε_f values multiplied by the spacing between neighboring such states ($dE_f = \varepsilon_f - \varepsilon_{f-1}$) :

$$R_T = (2\pi/\hbar)\, \Sigma_f\, \rho(\varepsilon_i - \varepsilon_f\,) dE_f \int dQ' \int dQ \int dp <m_{i,f}(P/\mu)\, \chi_i\ |Q'> <Q'|\ p>$$
$$\delta(\ T_{class} + V_f\,(Q) - \varepsilon_f) \quad <p|Q> <Q|\ m_{i,f}\ (P/\mu)\chi_i>. \tag{47}$$

In this form, which is analogous to Eq. (26) in the photon absorption case, the rate is expressed as a sum over the neutral molecule's vibration-rotation states to which the specific initial state having energy ε_i can decay of (a) a translational state density ρ multiplied by (b) the average value of an integral operator Δ whose coordinate representation is

$$\Delta(Q',Q) = \int dp\ |Q'>m_{i,f}\ (Q') <Q'|\ p>$$
$$\delta(\ p^2/2\mu_Q + V_f\,(Q) - \varepsilon_f) <p|Q>\ m_{i,f}\ (Q) <Q| \tag{48}$$

with the average value taken for the function

$$\psi = |(P/\mu)\chi_i> \tag{49}$$

equal to the anion's initial vibration-rotation state acted on by P/μ (in the harmonic approximation, $P/\mu\chi_i$ would yield a combination of functions of one higher and one lower quantum of vibration or rotation).

The integral operator Δ can be recast in a different form by carrying out the integration over the p-variable in Eq. (48) identifying

$$f(p) = (2\pi\hbar)^{-1/2} \exp(-ipQ/\hbar)\, (2\pi\hbar)^{-1/2} \exp(ipQ'/\hbar) \tag{50}$$

and $a = \varepsilon_f - V_f(Q)$. Doing so produces

$$\begin{aligned}
&\int_{-\infty}^{\infty} \delta(p^2/2\mu_Q - a) f(p)dp \\
&= \int_0^{\infty} \delta(p^2/2\mu_Q - a) f(p)dp + \int_{-\infty}^{0} \delta(p^2/2\mu_Q - a) f(p)dp \\
&= \int_0^{\infty} \delta(x-a) f(\sqrt{2\mu_Q x}) \frac{\mu_Q dx}{\sqrt{2\mu_Q x}} + \int_{\infty}^{0} \delta(x-a) f(-\sqrt{2\mu_Q x}) \frac{-\mu_Q dx}{\sqrt{2\mu_Q x}} \\
&= \int_0^{\infty} \delta(x-a) \{f(\sqrt{2\mu_Q x}) + f(-\sqrt{2\mu_Q x})\} \frac{\mu_Q dx}{\sqrt{2\mu_Q x}} \\
&= \{f(\sqrt{2\mu_Q a}) + f(-\sqrt{2\mu_Q a})\} \frac{\mu_Q}{\sqrt{2\mu_Q a}} \\
&= \frac{1}{2\pi\hbar} 2\mathrm{Cos}\{[Q' - Q]\sqrt{2\mu_Q(\varepsilon_f - V_f(Q))}/\hbar\} \frac{\mu_Q}{\sqrt{2\mu_Q(\varepsilon_f - V_f(Q))}}.
\end{aligned} \tag{51}$$

The quantity $\{2\mu_Q(\varepsilon_f - V_f(Q))\}^{1/2}$ is the classical momentum along the Q coordinate with energy ε_f moving on the <u>neutral</u> molecule's surface $V_f(Q)$, so μ_Q divided into this is the speed of movement at Q_0. Substituting this result back into the expression for R_T gives

$$R_T = (2\pi/\hbar)\sum_f \rho(\varepsilon_i - \varepsilon_f)dE_f \int dQ' \int dQ < m_{i,f}(P/\mu)\chi_i \,|\, Q' > \frac{1}{\pi\hbar}$$

$$\mathrm{Cos}\{[Q' - Q]\sqrt{2\mu_Q(\varepsilon_f - V_f(Q))}/\hbar\}$$

$$\frac{\mu_Q}{\sqrt{2\mu_Q(\varepsilon_f - V_f(Q))}} < Q \,|\, m_{i,f}(P/\mu)\chi_i > \tag{52}$$

Bearing in mind the discussion of the nature of the electronic non BO matrix elements $m_{i,f}(Q)$ given in Sec. I. C, the above rate expression can be further approximated by constraining Q' and Q to the region $Q'=Q=Q_0$ where the anion and neutral surfaces approach most closely:

$$d\{V_f(Q) - V_i(Q)\}/dQ = 0 \text{ at } Q = Q_0. \tag{53a}$$

If, as assumed, $m_{i,f}(Q)$ is significant only near Q_0, then we can approximate $m_{i,f}(Q)$ as

$$m_{i,f}(Q) = \delta(Q - Q_0)m^* \tag{53b}$$

where the quantity m* is the integral representing the total strength of the $m_{i,f}$ coupling concentrated at the geometry Q_0

$$m^* = \int m_{i,f}(Q)dQ. \tag{53c}$$

Introducing this approximation into Eq. (52) and using the fact that the argument of the Cos in Eq. (52) vanishes, allows R_T to be written in its simplest form as:

$$R_T = (2\pi/\hbar)\sum_f \rho(\varepsilon_i - \varepsilon_f)dE_f\{(P/\mu)\chi_i(Q_0)\}^2 >| m^* |^2 \frac{1}{\pi\hbar v_0} \tag{54}$$

where v_0 is the velocity along the Q-coordinate at the geometry Q_0:

$$v_0 = \frac{\sqrt{2\mu_Q(\varepsilon_f - V_f(Q_0))}}{\mu_Q} \tag{55}$$

and $|m_i^*|^2$ is the square of the integrated electronic non BO matrix element introduced above (n.g., $|m^*|^2$ has units of (gm cm^2/sec)^2.

3. INTERPRETATION OF RATE EXPRESSION

The semi-classical expression shown in Eq. (54) for the rate of ejection of electrons from a specified initial vibration-rotation state χ_i (Q) induced by non BO coupling to all accessible neutral-molecule-plus-free-electron final states (labeled f) gives this rate as:

1. A sum over all final vibration-rotation states ε_f lying below ε_i for which the geometry Q_0 is within the classically allowed region of the corresponding vibration-rotation wavefunction χ_f (Q) (so that v_0 is real) of
2. the modulus squared of the function $m^*(P/\mu)\chi_i$ evaluated at Q_0
3. multiplied by the state density $\rho(\varepsilon_i-\varepsilon_f)$ dE_f for the ejected electron and multiplied by $(2\pi/\hbar)(1/\pi\hbar)$, and finally
4. divided by the speed v_0 of passage through Q_0.

4. SUMMARY

The rate of ejection of electrons from anions induced by non BO couplings can be expressed rigorously as a Fourier transform of an overlap function between two functions

$$R_T = (2\pi/\hbar) \int (1/2\pi\hbar) \int \rho(E)$$

$$<m_{i,f}(P/\mu) \exp(it(T + V_i)/\hbar)\, \chi_i \,|\, \exp(it(E+ T + V_f)/\hbar)\, m_{i,f}\, (P/\mu)\chi_i> dt\, dE \tag{37}$$

one of which is the initial vibration-rotation function χ_i acted on by the non BO perturbation $m_{i,f}(P/\mu)$ and then propagated on the neutral molecule surface, the other being the initial χ_i propagated on the anion surface and then acted on by $m_{i,f}(P/\mu)$. In computer applications, it would be efficient to compute R_T in this manner whenever long-time surface propagation tools are applicable.

By introducing the simplest semi-classical approximation to the propagators, in which the nuclear motion kinetic energy is assumed to commute with the anion and neutral potential energy functions and with the non BO coupling operators, one obtains

$$R_T = (2\pi/\hbar) \int \rho(E) <m_{i,f}(P/\mu)\ \chi_i | \delta(V_f + E - V_i) | m_{i,f}(P/\mu)\chi_i> dE. \quad (39)$$

Unlike its success in treating the photon absorption rate expression, this simplest approximation produces a nonsensical expression in the present case because there are no geometries at which $(V_f + E - V_i) = 0$, as a result of which R_T is predicted to vanish. In the photon absorption situation, there are geometries at which the classical momentum is conserved (i.e., where the excited and shifted ground state surfaces intersect). In the non BO transition case, such geometries do not exist because the transition is not one in which the nuclear-motion momentum is conserved. Quite to the contrary, non BO transitions involve the simultaneous interchange of energy (from the nuclei to the electrons) and of momentum and/or angular momentum.

Improving on the semi-classical treatment of the vibration-rotation motion only slightly allows R_T to be recast in a form

$$R_T = (2\pi/\hbar)\sum_f \rho(\varepsilon_i - \varepsilon_f)dE_f \int dQ' \int dQ < m_{i,f}(P/\mu)\chi_i | Q' > \frac{1}{\pi\hbar}$$
$$\mathrm{Cos}\{[Q' - Q]\sqrt{2\mu_Q(\varepsilon_f - V_f(Q))}/\hbar\}$$
$$\frac{\mu_Q}{\sqrt{2\mu_Q(\varepsilon_f - V_f(Q))}} < Q | m_{i,f}(P/\mu)\chi_i > \quad (52)$$

that may prove computationally useful in cases where the geometry (Q)-dependence of the non BO electronic matrix element $m_{i,f}$ is known. In this expression, the rate is given in terms of the functions $m_{i,f}(P/\mu)\chi_i$, the density of state function ρ, the classical momentum on the neutral molecule's surface, and a Cos function whose argument is the classical action connecting the points Q' and Q via the momentum of the neutral molecule having energy ε_f. This Cos function will oscillate rapidly when this action is large, so its dominant contributions to the rate will arise for small momenta and/or small Q'-Q values.

Finally, by using what is known about the geometry dependence of the $m_{i,f}$ functions (i.e., that $m_{i,f}$ is strongly "peaked" near geometries Q_0 where the anion and neutral surfaces approach most closely), it is possible to further simplify the semi-classical equation for R_T

$$R_T = (2\pi/\hbar)\sum_f \rho(\varepsilon_i - \varepsilon_f)dE_f\{(P/\mu)\chi_i(Q_0)\}^2 >| m^* |^2 \frac{1}{\pi\hbar v_0} \tag{54}$$

to one that requires knowledge of the derivative of the initial-state vibrational wavefunction ($P/\mu\chi_i$) evaluated at Q_0, the speed v_0 at which classical motion on the neutral molecule surface passes through Q_0, the density of states ρ, and the magnitude of the integrated strength m* of $m_{i,f}$ at Q_0.

It should be emphasized that it is not the modulus $|\chi_i(Q_0)|^2$ that enters into the weighting function in Eq. (54), it is the derivative ($P/\mu\chi_i$) whose modulus squared enters. In contrast, in the photon absorption case, the rate involves, as given either in Eq. (28) or Eq. (30)

$$R_T = (2\pi/\hbar)\int d\Omega \int dS \int dd\, |\chi_i(s,d,\Omega)|^2\, |\mu_{i,f}(s,d,\Omega)|^2\, \delta(V_f(s,d) - V_i(s,d) - \hbar\omega). \tag{28}$$

$$R_T = (2\pi/\hbar)\int d\Omega \int dS \ |\chi_i(\mathbf{s},0,\Omega)|^2 \, |\mu_{i,f}(\mathbf{s},0,\Omega)|^2 \ | F_f(\mathbf{s}) - F_i(\mathbf{s})|^{-1} . \quad (30)$$

the modulus squared of χ_i itself. The qualitative difference in the two cases has to do with the inherent requirement that the nuclear-motion momentum and/or angular momentum change in non BO transitions while the same quantities are preserved in photon absorption events (in the semi-classical treatment).

ACKNOWLEDGMENTS

This work has been supported by NSF Grant CHE-9618904 and by proceeds of the Henry Eyring Endowed Chair. The author wishes to acknowledge his friend and colleague, Professor N. Y. Öhrn, for inspiring him to excellence in science, on the slopes, and in the back country.

REFERENCES

(1a) Lykke, K. R.; Neumark, D. M.; Andersen, T.; Trapa, V. J.; Lineberger, W. C. *J. Chem. Phys.* **1987**, *87* , 6842.

(1b) Neumark, D. M.; Lykke,K. R.; Andersen, T.; Lineberger, W. C. *J. Chem. Phys.* **1985**, *83*, 4364.

(1c) Andersen, T.; Lykke, K. R.; Neumark, D. M.; Lineberger, W. C. *J. Chem. Phys.* **1987**, *86*, 1858.

(2a) Marks, J.; Wentzel, D. M.; Comita, P. B.; Brauman, J. I. *J. Chem. Phys.* **1986**, *86*, 5284.

(2b) Meyer, F. K.; Jasinski, J. M.; Rosenfeld, R. N.; Brauman, J. I. *J. Am. Chem. Soc.* **1982**, *104*, 663.

(2c) Rosenfeld, R. N.; Jasinski, J. M.; Brauman, J. I. *J. Chem. Phys.* **1979**, *71*, 1030. Mead, R. D.; Lykke, K. R.; Lineberger, W. C.; Marks, J.; Brauman, J. I. *J. Chem. Phys.* **1984**, *81*, 4883.

(2d) Jackson, R. L.; Zimmerman, A. H.; Brauman, J. I. *J. Chem. Phys.* **1979**, *71*, 2088.

(2e) Zimmerman, A. H.; Reed, K. J.; Brauman, J. I. *J. Am. Chem. Soc.* **1977**, *99*, 7203.

(2f) Zimmerman, A. H.; Brauman, J. I. *J. Chem. Phys.* **1977**, *66*, 5823. Wight, C. A.; Beauchamp, J. L. *J. Am. Chem. Soc.* **1981**, *103*, 6501.

(4a) Acharya, P.; Kendall, R.; Simons, J. *SASP* **1984**, 84.

(4b) Acharya, P. K.; Kendall, R. A.; Simons, J. *J. Am. Chem. Soc.* **1984**, *106*, 3402.

(4c) Acharya, P. K.; Kendall, R.; Simons, J. *J. Chem. Phys.* **1985**, *83*, 3888.

(4d) Chalasinski, G.; Kendall, R. A.; Taylor, H.; Simons, J. *J. Phys. Chem.* **1988**, *92*, 3086.

(4e) O'Neal, D.; Simons, J. *J. Phys. Chem.* **1988**, *93*, 58.

(4f) Simons, J. *J. Chem. Phys.* **1989**, *91*, 6858.

(4g) Simons, J. *J. Am. Chem. Soc.* **1981**, *103*, 3971.

(5) Clary, D. L. *J. Phys. Chem.* **1988**, *92*, 3173.

(6) Berry, R. S. *J. Chem. Phys.* **1966**, *45*, 1228.

(7a) Simons, J. *J. Phys. Chem.* **1982**, *86*, 3615.

(7b) Taylor, H.; Simons, J. *J. Phys. Chem.* **1986**, *90*, 580.

(7c) Lax, M. *J. Chem. Phys.* **1952**, *30*, 1752.

(7d) Noda, C.; Zare, R. N. *J. Mol. Spectros.* **1982,** *95*, 254.

(8a) A good textbook treatment of how time correlation functions relate to photon absorption and other rates is given in D. A. McQuarrie, D. A. "Statistical Mechanics"; Harper and Row: New York, 1976.

(8b) Gordon, R. G. *Adv. Mag. Reson.* **1968**, *3*, 1.

(9a) Bergsma, J. P.; Berens, P. H.; Wilson, K. R.; Fredkin, D. R.; Heller, E. J. *J. Phys. Chem.* **1984**, *88*, 612.

(9b) Heller, E. J.; Brown, R. C. *J. Chem. Phys.* **1983**, *79*, 3336.

(9c) Heller, E. J. *J. Chem. Phys.* **1978**, *68*, 2066.

(9d) Heller, E. J. *J. Chem. Phys.* **1978**, *68*, 3891.

(10a) Landau, L. *Phys. Z. Sowjetunion* **1932**, *1*, 88.

(10b) Zener, C. *Proc. Roy. Soc. London* **1932**, *A137*, 696.

Density Matrix Treatment of Electronic Rearrangement

David A. Micha

Quantum Theory Project

University of Florida

Gainesville, Florida 32611–8435, USA

ADVANCES IN QUANTUM CHEMISTRY, VOLUME 35

0065-3276/99 $30.00

Contents

1. INTRODUCTION

The molecular dynamics of electronically excited systems, where electronic energy and charge transfer occur as a result of collisions or photoexcitations, requires a quantal treatment of the electronic rearrangement. The traditional approach to quantum molecular dynamics is to first generate the potential energy surfaces for fixed nuclei, and their couplings due to nuclear displacements, and to follow this by doing the quantal dynamics of the nuclear motions. (1–3) This procedure is very accurate when it can be done in detail, which is the case only for small molecular systems with small total energy. Otherwise the number of potential surfaces and their couplings become too numerous and complicated for detailed calculations.

An alternative to the traditional approach is to generate the electronic states as needed during the dynamics. This has been done for atomic collisions, where detailed calculations and comparisons with experimental results are possible.(4–8) General treatments of the coupling of electronic and nuclear motions in molecular systems can be done in a variety of formulations. In particular, Ohrn, Deumens and collaborators have implemented a general variational treatment in terms of coherent states for both electrons and nuclei, and have successfully applied it to collisional phenomena involving atoms and molecules.(9–11) Another approach developed from an early investigation of variational improvements to the time-dependent Hartree-Fock approximation applied to ion-atom collisions, and will be elaborated in the present contribution.(12–16) It consists of obtaining the wavefunctions of nuclei in the case of small deBroglie wavelengths, also called the eikonal approximation, and of allowing nuclear trajectories to evolve in effective potentials obtained from time-dependent many-electron wavefunctions.(17–23) Taking the limit of the eikonal approximation for vanishing wavelengths gives the classical description of nuclear motions, coupled to quantal electronic transitions. The eikonal approximation is more general because it allows for quantal phase interference of nuclear states, with their phases constructed from mechanical actions along trajectories. In this regard it is closely related to the semiclassical methods for wavefunctions and propagators introduced to deal with molecular collisions,(24, 25) and to wavepacket methods for quantum molecular dynamics.(26) The present treatment also relates to the recently revisited initial value representations (IVR), where transition probabilities and properties are obtained from semiclassical calculations integrating over initial values of coordinates and momenta.(27, 17, 28–30) Models have also been developed to describe electronic states coupled to nuclear motions in terms of Hamiltonian equations for electronic amplitudes.(31–33)

To properly describe electronic rearrangement and its dependence on both nuclear positions and velocities, it is necessary to develop a time-dependent theory of the electronic dynamics in molecular systems. A very useful approximation in this regard is the time-dependent Hartree-Fock approximation (34). Its combination with the eikonal treatment has been called the Eik/TDHF approximation, and has been implemented for ion-atom collisions.(21, 35–37) Approximations can be systematically developed from time-dependent variational principles.(38–41) These can be stated for wavefunctions and lead to differential equations for time-dependent parameters present in trial wavefunctions.

The treatment developed here is based on the density matrix of quantum mechanics and extends previous work using wavefunctions.(42–45) The density matrix approach treats all energetically accessible electronic states in the same fashion, and naturally leads to average effective potentials which have been shown to give accurate results for electronically diabatic collisions.(19) The approach is taken here for systems where the dynamics can be described by a Hamiltonian operator, as it is possible for isolated molecules or in models where environmental effects can be represented by terms in an effective Hamiltonian.

The following treatment starts with the complete quantal equations and introduces an eikonal representation which allows for a formally exact treatment. It shows how a time-dependent eikonal treatment can be combined with TDHF and a multiconfigurational extension in terms of density matrices. It then deals with the propagation of coupled fast and slow degrees of freedom by introducing a local interaction picture and describes the *relax-and-drive* computational procedure implemented in applications. (16, 15, 21, 46, 35–37) This procedure allows for the efficient calculation of coupled differential equations for functions with very different time scales, as is usual in molecular electronic rearrangement. The procedure is of general interest and is described here for propagation of a general density matrix.

2. DENSITY MATRIX TREATMENT

2.1. Equation of motion for the density operator

The density operator $\hat{\Gamma}(t)$ is a Hermitian and positive function of time, and satisfies the generalized Liouville-von Neumann (LvN) equation(47, 45)

$$i\hbar\frac{\partial\hat{\Gamma}}{\partial t} = [\hat{H}(t), \hat{\Gamma}(t)] + \mathcal{R}_D\hat{\Gamma} \tag{1}$$

which contains the Hamiltonian operator $\hat{H}$ for the system, in general a function of time, and a Liouville superoperator $\mathcal{R}_D$ which describes dissipative phenomena when the system interacts with a medium. This equation must be solved with the initial condition $\hat{\Gamma}_{in} = \hat{\Gamma}(t_{in})$, and with a normalization imposed here through the trace relation $tr(\hat{\Gamma}_{in}) = 1$.

This contribution considers systems which can be described with just the Hamiltonian, and do not need a dissipative term so that $\mathcal{R}_D = 0$. This would be the case for an isolated system, or in phenomena where the dissipation effects can be represented by an additional operator to form a new effective non-Hermitian Hamiltonian. These will be called here Hamiltonian systems. For isolated systems with a Hermitian Hamiltonian, the normalization is constant over time and the density operator may be constructed in a simpler way. In effect, the initial operator may be expanded in its orthonormal eigenstates $\Psi_n^{(in)}$ (density amplitudes) and eigenvalues w_n (positive populations), where *n* labels the states, in the form

$$\hat{\Gamma}_{in} = \sum_n w_n |\Psi_n^{(in)}\rangle\langle\Psi_n^{(in)}| \tag{2}$$

where the summation extends only over the populated states. The populations would follow from a set of initially known physical properties, as will be shown.

They add up to one. At later times, the density operator can be constructed from solutions of the time-dependent Schroedinger equation,

$$i\hbar\frac{\partial\Psi_n}{\partial t} = \hat{H}(t)\Psi_n(t) \tag{3}$$

with the initial conditions $\Psi_n(t_{in}) = \Psi_n^{(in)}$, which must be imposed for each state *n*. These time-dependent amplitudes can be constructed so that they are orthonormal. It follows then that at all times,

$$\hat{\Gamma}(t) = \sum_n w_n|\Psi_n(t)\rangle\langle\Psi_n(t)| \tag{4}$$

Although the density amplitudes satisfy the standard Schroedinger differential equation for quantal states, they can depend on initial statistical conditions and are more general than standard states.

The density amplitudes can usually be calculated more efficiently than the density operator because they depend on only one set of variables in a given representation although there are cases, such as shown below for the time-dependent Hartree-Fock density operator, where the advantages disappear and it is convenient to calculate the density operator. Expectation values of operators $\hat{A}(t)$ follow from the trace over the density operator, as

$$\langle A(t)\rangle = tr[\hat{\Gamma}(t)\hat{A}(t)]/tr[\hat{\Gamma}(t)] = \sum_n w_n\langle\Psi_n(t)|\hat{A}(t)|\Psi_n(t)\rangle \tag{5}$$

which is a weighted sum of expectation values from the density amplitudes.

2.2. Variational method for the density amplitudes

Computational strategies can be based on variational procedures using the Dirac-Frenkel time-dependent variational principle (TDVP). Introducing a short-hand notation so that

$$[\hat{H}(t) - i\hbar\partial_t]\Psi_n(t) = \hat{D}_t\Psi_n(t) = 0 \tag{6}$$

the TDVP is expressed, temporarily dropping the state subindex *n*, in terms of the action functional

$$\begin{aligned} A_t\left[\Psi^\dagger, \Psi\right] &= \int_{t_{in}}^{t} dt'\ \mathcal{L}_{t'}\left[\Psi^\dagger, \Psi\right] \\ \mathcal{L}_t\left[\Psi^\dagger, \Psi\right] &= \left(\ \langle\Psi(t)|\hat{D}_t\Psi(t)\rangle + \langle\hat{D}_t\Psi(t)|\Psi(t)\rangle\right) \\ &\quad /(2\ \langle\Psi(t)|\Psi(t)\rangle) \end{aligned} \tag{7}$$

where $\mathcal{L}$ is a Lagrangian functional, and the brackets indicate integration over electronic and nuclear variables, which are omitted here.

The variation condition $\delta\mathcal{A}_t = 0$ can be independently imposed for variations of Ψ and its adjoint. The condition of gauge invariance requires that trial functions have the form

$$\Psi(t) = \tilde{\Psi}(t)exp(-\frac{i}{\hbar}\mathcal{A}_t[\tilde{\Psi}^\dagger, \tilde{\Psi}]) \tag{8}$$

and leads to the variational equation (34)

$$\langle\delta'\tilde{\Psi}|\mathcal{D}_t\tilde{\Psi}\rangle = 0 \tag{9}$$

expressed in terms of a variation $\delta'\tilde{\Psi}$ orthogonal to the trial function so that $\langle\delta'\tilde{\Psi}|\tilde{\Psi}\rangle = 0$.

This procedure would generate the density amplitudes for each n, and the density operator would follow as a sum over all the states initially populated. This does not however assure that the terms in the density operator will be orthonormal, which can complicate the calculation of expectation values. Orthonormality can be imposed during calculations by working with a basis set of N states collected in the Nx1 row matrix $\mathbf{\Phi}(t)$ which includes states evolved from the initially populated states and other states chosen to describe the amplitudes over time, all forming an orthonormal set. Then in a matrix notation, $\Psi_n(t) = \mathbf{\Phi}(t)\mathbf{T}_n(t)$, where the coefficients $\mathbf{T}$ form 1xN column matrices, with ones or zeros as their elements at the initial time. They are chosen so that the square NxN matrix $\mathbf{T}(t) = [\mathbf{T}_n(t)]$ is unitary, to satisfy orthonormality over time. Replacing the trial functions in the TDVP one obtains coupled differential equations in time for the coefficient matrices,

$$[\mathbf{H}(t) - \mathbf{i\hbar}\langle\mathbf{\Phi}|\partial\mathbf{\Phi}/\partial t\rangle - i\hbar\mathbf{I}\partial_t]\mathbf{T}_n(t) = 0 \tag{10}$$

where $\mathbf{H}$ is the Hamiltonian matrix in the initial basis. Instead of working with this equation, it is more convenient to solve for the density matrix

$$\mathbf{\Gamma}(t) = \sum_n w_n\mathbf{T}_n(t)\mathbf{T}_n(t)^\dagger \tag{11}$$

which satisfies the matrix LvN equation

$$i\hbar\frac{d\mathbf{\Gamma}}{dt} = \mathbf{H}(t)\mathbf{\Gamma}(t) - \mathbf{\Gamma}(t)\mathbf{H}(t) \tag{12}$$

After solving this equation for the density matrix, its diagonalization provides the matrices of coefficients and the weights needed to reconstruct orthonormal density amplitudes. The density operator follows from $\hat{\Gamma}(t) = |\mathbf{\Phi}(t)\rangle\mathbf{\Gamma}(t)\langle\mathbf{\Phi}(t)|$

3. THE EIKONAL REPRESENTATION

3.1. The eikonal representation for nuclear motions

The dynamics of molecular systems frequently involves degrees of freedom (df's) with deBroglie wavelengths of different orders of magnitude. An example is the nuclear and electronic df's, where comparable forces and the larger masses of nuclei lead to larger nuclear momenta and therefore smaller wavelengths for them. Another example is that of rotational and vibrational motions in electronically adiabatic molecular collisions, where masses are comparable but forces have different magnitudes, giving shorter deBroglie wavelengths for rotations. In general we can label with q the long wavelength df's and with Q the short wavelength ones. It is well known that in the short wavelength limit the quantal equations of motion reduce to the classical ones. Therefore molecular motions couple the classical-like nuclear motions described by wavefunctions of the Q variables to the quantal electronic transitions described by wavefunctions of the q variables. It is possible to provide a rigorous description of both motions within an eikonal representation of the density operator.

The density operator in the coordinate representation is given by the functions $\Gamma(q, Q, q', Q', t)$, and can be expressed for Hamiltonian systems in terms of its amplitudes, which become the wavefunctions $\Psi_n(q, Q, t)$. It is convenient to introduce the formally exact eikonal representation,

$$\Psi_n(q, Q, t) = \chi_n(q, Q, t) exp[iS(Q, t)/\hbar] \tag{13}$$

where the phase factor S is a real function of the classical-like coordinates only, and is independent of the state index n. It can be chosen to extract from the wavefunctions their rapidly oscillating behavior over space and time so that the preexponential functions χ_n will change slowly with Q, and it will later be related to the mechanical action. The density function takes the form

$$\begin{aligned} \Gamma(q, Q, q', Q', t) &= \Lambda(q, Q, q', Q', t) exp\{i[S(Q, t) - S(Q', t)]/\hbar\} \\ \Lambda(q, Q, q', Q', t) &= \sum_n w_n \chi_n(q, Q, t) \chi_n(q', Q', t)^* \end{aligned} \tag{14}$$

For $Q = Q'$, this density function describes electronic motions for given nuclear positions, while for $Q = Q'$ it describes the quantal correlation of nuclear positions at time t, which should be small for classical-like variables. The equation of motion for the density function could be derived from the original LvN equation. Instead, it is more convenient to construct it from the wavefunctions. The phase factor and the preexponential factor are trial functions to be determined from the TDVP. The procedure followed here parallels that in ref. (23).

Introducing the Dirac "bra" and "ket" notation for operators and states of the electrons, while explicitly stating the nuclear coordinates, the operator $\hat{\Gamma}(Q, Q', t)$ is then expanded in the states $|\chi_n(Q)\rangle$. The Lagrangian functional becomes

$$\mathcal{L}'_t[\chi_n, \chi_n^*, S] = \frac{1}{2}[\int dQ\ \langle\chi_n(Q,t)|\hat{D}'_t\chi_n(Q,t)\rangle + c.c.]$$
$$\hat{D}'_t = [\frac{1}{2M}(\frac{\hbar}{i}\frac{\partial}{\partial Q} + \frac{\partial S}{\partial Q})^2 + \hat{H}_Q - (i\hbar\frac{\partial}{\partial t} - \frac{\partial S}{\partial t})] \tag{15}$$

where *c.c.* means the complex conjugate of the preceding term, *M* is a nuclear mass obtained by scaling nuclear coordinates, the partial derivative with respect to *Q* is a symbol for differentiation of many variables, and $\hat{H}_Q$ is a Hamiltonian operating over electronic coordinates for fixed nuclear coordinates. Varying the action functional with respect to the pre-exponential function χ_n, while keeping *S* fixed, the following equation is obtained,

$$\hat{D}'_t\chi_n(q, Q, t) = 0 \tag{16}$$

Assuming that all the solutions χ_n have been found, the equation for *S* follows by projecting on $\langle\chi_n|$, multiplying times the weights w_n and adding over *n* to form an average over the density operator, with the result

$$\frac{1}{2M}(\frac{\partial S}{\partial Q})^2 + V_{qu}\left(Q, \frac{\partial S}{\partial Q}, t\right) + \frac{\partial S}{\partial t} = 0 \tag{17}$$

Here a quantal potential has been introduced which implicitly depends on density amplitudes and their first and second derivatives with respect to *Q*, as

$$V_{qu}(Q,t) = \sum_n w_n[V_n(Q,t) + V'_n(Q,t) + V''_n(Q,t)] \tag{18}$$

with

$$V_n = \langle\chi_n|H_Q|\chi_n\rangle/\langle\chi_n|\chi_n\rangle$$

$$V'_n = \frac{i\hbar}{2M}\frac{\partial S}{\partial Q}\cdot\left[\langle\frac{\partial\chi_n}{\partial Q}|\chi_n\rangle - \langle\chi_n|\frac{\partial\chi_n}{\partial Q}\rangle\right]/\langle\chi_n|\chi_n\rangle \tag{19}$$

$$V''_n = -\frac{\hbar^2}{2M}\frac{1}{2}\left[\langle\frac{\partial^2\chi_n}{\partial Q^2}|\chi_n\rangle + \langle\chi_n|\frac{\partial^2\chi_n}{\partial Q^2}\rangle\right]/\langle\chi_n|\chi_n\rangle$$

where the first term in V_{qu} is equal to $\mathcal{V}(Q,t) = tr[\hat{\Gamma}(Q,Q,t)\hat{H}_Q]$ and can be identified as the Ehrenfest potential for the nuclear motions, while the second and third terms give quantal corrections.

The equation for S is recognized as a Hamilton-Jacobi equation for a mechanical action. The solution can be written in terms of a Lagrangian L_{qu} for the nuclear motions, introducing the path $\bar{Q}(t; Q_{in}, Q)$, starting initially at positions Q_{in} and ending at Q at time t, and the corresponding generalized velocities $\dot{\bar{Q}}$. The result is

$$
\begin{aligned}
S(Q,t) &= S_{in}(Q) + \int_{t_{in}}^{t} dt' \; L_{qu}(\bar{Q}, \dot{\bar{Q}}, t') \\
L_{qu}(\bar{Q}, \dot{\bar{Q}}, t) &= \dot{\bar{Q}}^2/(2M) - V_{qu}(\bar{Q}, t)
\end{aligned}
\tag{20}
$$

Therefore S can be constructed by solving the Lagrange equations for the nuclear variables. Alternatively it can be obtained from the Hamilton equations introducing the generalized momenta $P = \partial S/\partial Q$ and solving the equations

$$
\begin{aligned}
\partial P/\partial t = -\partial H_{qu}/\partial Q \;,\;\; \partial Q/\partial t = \partial H_{qu}/\partial P \\
H_{qu}(P,Q,t) = P^2/(2M) + V_{qu}(P,Q,t)
\end{aligned}
\tag{21}
$$

where H_{qu} is the Hamiltonian for classical motions containing quantal corrections in its potential, with the initial values $Q_{in} = Q(t_{in})$ and $P_{in} = P(t_{in})$, so that $Q = Q(t; Q_{in}, P_{in})$ and $P = P(t; Q_{in}, P_{in})$. The variational condition for χ_n can be stated for each path evolving from a set of initial values, as

$$
\begin{aligned}
&\langle \delta' \chi_n(Q,t) | [\frac{1}{2M}(\frac{\hbar}{i}\frac{\partial}{\partial Q} + \frac{\partial S}{\partial Q})^2 + \hat{H}_Q - H_{qu}] | \chi_n(Q,t) \rangle \\
&- \langle \delta' \chi_n(Q,t) | i\hbar \frac{\partial}{\partial t} | \chi_n(Q,t) \rangle = 0
\end{aligned}
\tag{22}
$$

which must be solve with the initial conditions $\chi_n(q, Q, t_{in}) = \chi_n^{(in)}(q, Q_{in})$. The Hamiltonian equations for P and Q, and the variational condition for χ_n provide together a formally exact set of coupled equations whose solution gives the time-evolution of the electronic states driven by nuclear motions. The present coupled equations generalize the ones previously presented in reference (21) to allow now for statistical weights in the quantal potential, which is the same for all the initially populated states n.

Initial conditions for the total molecular wavefunction with $n = I$ (including electronic, vibrational and rotational quantum numbers) can be imposed by adding elementary solutions obtained for each set of initial nuclear variables, keeping in mind that the χ_I and S depend parametrically on the initial variables

P_{in} consistent with conservation of total energy and momentum, for a given value of Q_{in}. For an initial molecular state I, constructed from coefficients a_I, the initial wavefunction is

$$\Psi_I^{(in)}(q, Q_{in}) = \int dP_{in}\, a_I(P_{in})\chi_I^{(in)}(q, Q_{in}; P_{in}) exp[iS^{(in)}(Q_{in}; P_{in})/\hbar] \tag{23}$$

From those coefficients it follows at later times that

$$\begin{aligned} \Gamma(q, Q, q', Q', t) = \sum_I w_I \int dP_{in}\, a_I(P_{in}) \int dP'_{in}\, a_I(P'_{in}) \\ \times\chi_I(q, Q, t; P_{in})^* \chi_I(q', Q', t; P'_{in}) \\ \times exp\{-i[S(Q, t; P_{in}) - S(Q', t; P'_{in})]/\hbar\} \end{aligned} \tag{24}$$

In the nuclear coordinate representation, the expectation values of an operator $\hat{A} = \hat{A}(Q, Q', t)$ follow from

$$\begin{aligned} \langle A(t)\rangle = \int dQ \int dQ' \hat{A}(Q, Q', t)\hat{\Gamma}(Q', Q, t) \\ = \int dP_{in} \int dQ \int dP'_{in} \int dQ' F(P_{in}, Q, P'_{in}, Q', t) \\ = \int dP_{in} \int dQ_{in} \int dP'_{in} \int dQ'_{in} \\ \times J(t; Q_{in}) J'(t; Q'_{in}) F'(P_{in}, Q_{in}, P'_{in}, Q'_{in}, t) \end{aligned} \tag{25}$$

where the second and third members of this equation show how it is possible to obtain the average by integration over initial conditions using the Jacobian transformation from space to initial coordinates, $J(t; Q_{in}) = det[\partial Q/\partial Q_{in}]$. This provides a convenient expression for calculations in a semiclassical limit. Integration over the space of nuclear coordinates has been replaced by integration over initial coordinates within a region of space from which bundles of trajectories emerge as streamlines. For each set of initial values, the integrand can be obtained over time and added to form the integrals at each time. This is similar in spirit to the initial value representation (IVR) recently developed by several authors, and is closely related to a procedure employed in the calculation of scattering transition integrals.(17, 18)

A similar procedure may be followed to calculate a transition probability $P(J \to K)$ into a final state Ψ_K^0 at time t_{fn}; this is obtained from the operator $\hat{A} = |\Psi_K^0\rangle\langle\Psi_K^0|$ and the trace over the density operator $\hat{\Gamma}_J(t)$ which evolves from a $\hat{\Gamma}_J^{(in)}$ with the weights $w_I = \delta_{IJ}$, giving $P(J \to K) = \langle\Psi_K^0|\hat{\Gamma}_J(t_{fn})|\Psi_K^0\rangle$.

The variational procedure for the states χ_I can be implemented in many different ways; a very general formalism introduces complex parameters $z(t)$, which can be identified with expansion coefficients or state rotation angles, as done in a development of simplectic differential equations for the variables $z(t)$ and their complex conjugate $z^*(t)$ from the TDVP, and in particular further developed for molecular systems.(10)

The treatment presented so far is quite general and formally exact. It combines the eikonal representation for nuclear motions and the time-dependent density matrix in an approach which could be named as the Eik/TDDM approach. The following section reviews how the formalism can be implemented in the eikonal approximation of short wavelengths for the nuclear motions, and for specific choices of electronic states leading to the TDHF equations for the one-electron density matrix, and to extensions of TDHF.

3.2. The eikonal/time-dependent Hartree Fock approximation and extensions.

The time-dependent eikonal approximation for the density operator can be introduced by analogy to the treatment for stationary states. Here the steps leading to the working equations are first outlined for a general system. For nuclear motions with short de Broglie wavelengths $\lambda = h/P$, it is assumed that the χ_I vary little over a wavelength so that

$$|\langle\chi_I|\frac{\partial\chi_I}{\partial Q}\rangle/\langle\chi_I|\chi_I\rangle| << \lambda^{-1} \tag{26}$$

and

$$|\langle\chi_I|\frac{\partial^2\chi_I}{\partial Q^2}\rangle/\langle\chi_I|\frac{\partial\chi_I}{\partial Q}\rangle| << \lambda^{-1} \tag{27}$$

from which it follows that the quantum potential and Hamiltonian for the nuclei are approximated by

$$\begin{aligned} V_{qu} &\approx \mathcal{V}(Q,t) = \sum_I w_I V_I(Q) \\ H_{qu} &\approx \mathcal{H}(P,Q,t) = P^2/(2M) + \mathcal{V}(Q,t) \end{aligned} \tag{28}$$

and that the variational condition for each χ_I is approximated by

$$\begin{aligned} &\langle\delta'\chi_I(Q,t)|[\frac{\hbar}{i}\frac{P}{M}\frac{\partial}{\partial Q} + \hat{H}_Q - W(P,Q,t)]|\chi_I(Q,t)\rangle \\ &-\langle\delta'\chi_I(Q,t)|(i\hbar\frac{\partial}{\partial t})_{q,Q}|\chi_I(Q,t)\rangle = 0 \\ &W = \mathcal{H}(P,Q,t) + i\frac{\hbar}{2M}\frac{\partial P}{\partial Q} \end{aligned} \tag{29}$$

where it is made explicit what coordinates are fixed during time differentiation. The imaginary term in the function W describes the divergence of streamlines. Using that

$$P/M = \dot{Q}$$
$$\dot{Q}\frac{\partial}{\partial Q} + (\frac{\partial}{\partial t})_{q,Q} = (\frac{\partial}{\partial t})_q \tag{30}$$

and the trial solution form

$$\chi_I[q, Q(t), t] = \eta_I(q, t) exp[i \int_{t_{in}}^{t} dt' \ W(t')/\hbar] \tag{31}$$

the working variational functional for the new time-dependent electronic wavefunction is simply

$$\langle \delta' \eta_I(t) | [\hat{H}_{Q(t)} - (i\hbar \frac{\partial}{\partial t})_q] | \eta_I(t) \rangle = 0 \tag{32}$$

where the electronic Hamiltonian has been shown to depend on time through $Q(t)$. This variational condition, implemented for a parametrized trial function, gives differential equations for state parameters that must be solved coupled to the Hamiltonian equations for the nuclear motions, to construct the density operator in a self-consistent way. This wavefunction is generated after specifying its initial value, $\eta_I^{(in)}(q) = \eta_I(q, t_{in})$. The density operator and averages can be obtained directly from the wavefunctions $\eta_I(q, t)$. To proceed, it is necessary to specify the form of the wavefunctions η_I, and to apply the variational condition to their parameters.

Molecular rearrangement resulting from molecular collisions or excitation by light can be described with time-dependent many-electron density operators. The initial density operator can be constructed from the collection of initially (or asymptotically) accessible electronic states, with populations w_I . In many cases these states can be chosen as single Slater determinants formed from a set of orthonormal molecular spin orbitals (MSOs) $\{i_m\}$ as $I = (i_1 ... i_m ... i_N)$. Describing the N-electron system by wavefunctions of the position and spin variables $(\vec{r}_1, \zeta_1, ... \vec{r}_n, \zeta_n, ... \vec{r}_N, \zeta_N) = (1, ..., n, ... N)$ in a concise notation, a determinantal wavefunction is constructed from N MSOs, $\{\psi_i(\vec{r}, \zeta)\}$, as

$$\eta_I^{(in)}(q) = \langle 1, 2, ..., N | D_I^{(in)} \rangle = (N!)^{-1/2} det[\psi_{i_m}(n)] \tag{33}$$

from which follows the N-electron density operator

$$\hat{\Gamma}^{(in)} = \sum_I w_I | D_I^{(in)} \rangle \langle D_I^{(in)} | \tag{34}$$

This in turn leads to the initial one-electron density matrix by contraction over electron variables,(48)

$$\begin{aligned}\rho^{(in)}(1,1') &= N\int d(2)...d(N)\Gamma^{(in)}(1,2,...,N;1',2,...,N)\\ &= \sum_I w_I \rho_I^{(in)}(1,1')\end{aligned} \tag{35}$$

The one-electron density matrix at later times,

$$\rho(1,1',t) = \sum_I w_I \rho_I(1,1',t) \tag{36}$$

is obtained from the LvN equation by contraction over all electron variables except one, and reduces to the well known time-dependent Hartree-Fock equation for each term

$$\hat{F}\hat{\rho}_I - \hat{\rho}_I\hat{F} = i\hbar\partial\hat{\rho}_I/\partial t \tag{37}$$

where $\hat{F} = \hat{H}^{(1)} + \hat{G}[\hat{\rho}_I]$ is the Fock operator written as a sum of a one-electron term $\hat{H}^{(1)}$, plus the HF self–consistent potential energy operator $\hat{G}[\hat{\rho}_I]$.

The average effective potential needed in the Hamiltonian equations is now

$$\mathcal{V}^{(HF)} = V_{cc} + \sum_I w_I tr[\hat{\rho}_I(\hat{H}^{(1)} + \hat{F})]/2 \tag{38}$$

where the first term is the core-core interaction potential and the second term is a sum over all initially populated states. The TDHF equations for each $\hat{\rho}_I$ and the classical Hamiltonian equations must be solved simultaneously and provide the Eik/TDHF approximation.

The TDHF equations must be solved without recourse to perturbation theory for problems of electronic rearrangement driven by nuclear motions, where couplings are strong. This can be done for each initial state I introducing basis sets and solving matrix TDHF equations. Given a general basis set of one-electron orbitals $\{\phi_p(t)\}$, with overlap integrals $\langle\phi_p|\phi_q\rangle = S_{pq}$, and expanding MSO's with spin state $\gamma(\zeta)$ in this basis to obtain

$$\psi_i^\gamma(\vec{r},t) = \sum_p \phi_p(\vec{r},t)c_{ip}^\gamma(t) \tag{39}$$

where the coefficients are complex valued, the density operator in this basis is

$$\hat{\rho}(t) = \sum_{pq} |\phi_p\rangle P_{pq}(t)\ \langle\phi_q| \tag{40}$$

with P_{pq} the (*pq*)–element of the one-electron density matrix **P**. The TDHF equation for the density matrix is then

$$i\hbar\dot{\mathbf{P}} = \mathbf{S}^{-1}(\mathbf{F} - i\hbar\mathbf{\Omega})\mathbf{P} - \mathbf{P}(\mathbf{F} - i\hbar\mathbf{\Omega})^{\dagger}\mathbf{S}^{-1} \tag{41}$$

where $\mathbf{S} = \langle\phi|\phi\rangle$ is the overlap matrix, and

$$\mathbf{\Omega} = \langle\phi|\frac{\partial}{\partial t}\phi\rangle \tag{42}$$

is an orbital coupling due to the motion of the nuclei, both given in a matrix notation where $|\phi\rangle$ is a $1\mathrm{x}N_B$ row matrix of basis functions. This matrix can lead to spurious couplings of asymptotic electronic states, and must be eliminated with a transformation to a basis of travelling atomic functions (TAFs) ξ_μ; MO's are then expanded as linear combinations of TAF's , or LCTAFs,

$$\begin{aligned}\psi_i^{\gamma}(\vec{r},t) &= \sum_{\mu} \xi_\mu(\vec{r},t)c_{\mu i}^{\gamma}(t)\\ \xi_\mu(\vec{r},t) &= \chi_\mu(\vec{r})T_m(\vec{r},t)\end{aligned} \tag{43}$$

where here χ_μ is an AF centered at nuclear position $\vec{R}_m(t)$, and

$$T_m(\vec{r},t) = \exp\{im_e[\vec{v}_m(t)\cdot\vec{r} - \int_{t_{in}}^{t} dt' v_m^2(t')/2)]/\hbar\} \tag{44}$$

is an electron translation factor, a function of the velocity vector of nucleus *m*. It may also be considered an eigenvalue of the translation (or boost) multiplicative operator T_m , which gives $|\xi_\mu\rangle = \hat{T}_m|\chi_\mu\rangle$. Details of computational aspects with these basis sets can be found in references.(21, 36)

The multiconfiguration extension of TDHF, or MCTDHF, can be developed also starting from Slater determinants

$$D_K(1,2,...,N,t) = (N!)^{-1/2}det[\psi_{k_m}(n,t)] \tag{45}$$

where $K = (k_1,...,k_m,...,k_N)$ is an ordered set among $N_B > N$ MSOs, that can be combined into configuration spin functions (CSFs) $\Phi_K(1,2,..,N,t)$ to construct eigenstates of the spin operators. The multiconfiguration function

$$\eta_I(1,2,...,N,t) = \sum_{K=1}^{N_C} \Phi_K(1,2,...,N,t)C_{KI}(t) \tag{46}$$

depends parametrically on the expansion coefficients C_{KI}, and on the LCTAFs parameters within the MSOs. It is a sum over a selected subset of N_C configurations, usually smaller than the total number of configurations which can be constructed from the complete active space of MSOs. Placing the trial function into the TDVP, a variation of the expansion coefficients gives the set of coupled equations for the C_{KI} coefficients. This can be rewritten introducing the elements $(\mathbf{M}_I)_{KL} = C_{KI}C^*_{LI}$ of the configurational density matrix. In a matrix notation, using the equation for the coefficients and its adjoint, one finds that

$$\mathbf{H}\mathbf{M}_I - \mathbf{M}_I\mathbf{H} = i\hbar d\mathbf{M}_I/dt \tag{47}$$

where $\mathbf{H} = [H_{LK}]$.

The variation of the MSOs can be done for each initial state I with Lagrange multipliers ϵ_{kl} introduced to impose the orthonormality of the MSOs. Details of the derivation may be found in reference (23). The resulting equations generalize the TDHF Fock operator and differential equation for the MSOs. Introducing the matrix notation $\psi_k(1,t) = \phi(1,t)\mathbf{c}_k(t)$, where the coefficients form a N_Bx1 column, and the set of the N_BxN_B one-electron density matrices $\gamma^{(kl)} = \mathbf{c}_k\mathbf{c}_l^\dagger$, as well as a compact notation involving supermatrices where each element is itself a matrix, it is possible to define the $N_B^2 \times N_B^2$ supermatrix $\underline{\mathbf{F}} = [\bar{\mathbf{F}}^{(lk)}]$ and a a density supermatrix of dimensions $N_B^2 \times N_B^2$,

$$\underline{\gamma}(t) = \underline{\mathbf{c}}\ \underline{\mathbf{c}}^\dagger = [\mathbf{c}_l\mathbf{c}_k{}^\dagger] \tag{48}$$

With this, a compact equation is found for the coefficients of the MSO expansions in the atomic basis, for each initial state I, as

$$\underline{\mathbf{F}}\,\underline{\gamma}_I - \underline{\gamma_I}\,\underline{\mathbf{F}} = i\hbar d\underline{\gamma_I}/dt \tag{49}$$

Once the Fock operators have been constructed from a set of MSOs, this matrix equation is linear in its unknowns. Its coefficients are dependent on time in a way determined by the forces driving the electrons. These forces are the nuclear Coulomb potentials in molecular collisions or dynamics, but they could also be weak external fields.

Averages of properties require integrals over CSFs which can readily be written for one- and two-electron operators, insofar the Slater determinants and the MSOs are orthonormal by construction, in terms of one- and two-electron density matrices.

4. THE PROPAGATION OF COUPLED FAST AND SLOW DEGREES OF FREEDOM

4.1. Propagation in a local interaction picture

The large disparity between nuclear and electron masses has another important effect. For thermal or hyperthermal energies up to about 10,000 eV, nuclear velocities are small compared with electronic ones. As a result, the oscillations in time of electronic states are much faster that those of the nuclear variables. Since both degrees of freedom are coupled, it is not efficient to solve their coupled differential equations by straightforward time-step methods. Instead it is necessary to introduce propagation procedures suitable for coupled equations with very different time scales: short for electronic states and long for nuclear motions. The following procedure introduces a unitary transformation at sequences of intervals to create a local interaction picture for propagation over time.

The calculation of the density operators over time requires integration of the sets of coupled differential equations for the nuclear trajectories and for the density matrix in a chosen expansion basis set. The density matrix could arise from an expansion in many-electron states, or from the one-electron density operator in a basis set of orbitals for a given initial many-electron state; a general case is considered here. The coupled equations are

$$\partial P/\partial t = -\partial \mathcal{H}/\partial Q\ , \quad \partial Q/\partial t = \partial \mathcal{H}/\partial P \tag{50}$$

and

$$\mathbf{H\Gamma} - \mathbf{\Gamma H} = i\hbar d\mathbf{\Gamma}/dt \tag{51}$$

and must be solved for the initial conditions at $t = t_{in}$: $Q_{in} = Q(t_{in})$ and $P_{in} = P(t_{in})$, and $\mathbf{\Gamma}_{in} = \mathbf{\Gamma}(t_{in})$. To start with, the matrices and trajectory variables are assumed known at a time t_0; the density matrices are first obtained as they relax over the interval $t_0 \leq t \leq t_0 + \Delta t$ while keeping the nuclear variables fixed. They are the solutions of the equations

$$\mathbf{H}_0\mathbf{\Gamma}^0(t) - \mathbf{\Gamma}^0(t)\mathbf{H}_0 = i\hbar d\mathbf{\Gamma}^0/dt \tag{52}$$

where it is shown that the density matrices change with time as they relax from their (non-stationary) values at t_0. The initial conditions are $\mathbf{\Gamma}^0(t_0) = \mathbf{\Gamma}_0$. Since the Hamiltonian matrix is now constant in time, these coupled equation are simple first order differential equations with constant coefficients, and can be integrated by diagonalizing the matrix of coefficients. The results are sums of rapidly oscillating functions in time, reflecting the rapid electronic motions. Relaxation occurs in two ways: by transitions between electronic configurations, and by rearrangements of the molecular orbitals.

In reality the nuclei are moving and one must account for the driving effect of their displacement and velocity changes within the interval $t_0 \leq t \leq t_1$. Provided this is small, and insofar as the nuclear motions are slower than the electronic ones, one can assume that the driving effect will only be corrections to the relaxing densities; this can be verified by shortening the time interval and repeating the calculations. The corrected densities are obtained writing

$$\begin{aligned} \mathbf{\Gamma}(t) &= \mathbf{\Gamma}^0(t) + \mathbf{U}^0(t)\mathbf{\Gamma}'(t)\mathbf{U}^0(t)^\dagger \\ \mathbf{U}^0(t) &= exp[-\frac{i}{\hbar}\mathbf{H}_0(t-t_0)] \end{aligned} \tag{53}$$

for the density matrix, where $\mathbf{U}^0$ defines a unitary transformation to a local interaction picture at each time t_0.

Replacing this in the LvN equation, it is found that

$$\begin{aligned} i\hbar\frac{d\mathbf{\Gamma}'}{dt} &= [\mathbf{V}, \mathbf{\Gamma}_0] + [\mathbf{V}, \mathbf{\Gamma}'] \\ \mathbf{V}(t) &= \mathbf{U}^0(t)[\mathbf{H}(t) - \mathbf{H}_0]\mathbf{U}^0(t)^\dagger \end{aligned} \tag{54}$$

Here the matrix $\mathbf{V}$ contains the effect of the nuclear displacements; therefore the inhomogeneous first term to the right is a driving term; the second term to the right is of second order in the driving effect, and could be dropped in calculations. Formally, the solution for the configuration density matrix correction is

$$\begin{aligned} \mathbf{\Gamma}'(t) &= \mathbf{\Delta}'(t) + (i\hbar)^{-1}\int_{t_o}^{t} dt' \, [\mathbf{V}(t'), \mathbf{\Gamma}'(t')] \\ \mathbf{\Delta}'(t) &= (i\hbar)^{-1}\int_{t_o}^{t} dt' \, [\mathbf{V}(t'), \mathbf{\Gamma}_0] \end{aligned} \tag{55}$$

where the driving term $\mathbf{\Delta}'$ can be obtained from a quadrature, and the second term can usually be neglected.

With the new density matrices known up to time t_1, it is possible to advance the nuclear positions and momenta by integrating their Hamilton equations. This completes a cycle which can be repeated to advance to a later time t_2. This sequence based on relaxing the density matrix for fixed nuclei and then correcting it to account for nuclear motions has been called the *relax-and-drive procedure*, and has been numerically implemented in several applications.(16, 15, 21, 46, 35–37)

4.2. The relax-and-drive computational procedure

Straightforward stepwise integration of the coupled Hamiltonian and LvN differential equations would be computationally inaccurate, because the fast electronic oscillations would demand very small time-steps, while the slow nuclear motions would need many steps to give converged final results. The accumulation of round-off errors would lead to large inaccuracies. An alternative is to separately do some of the integrations by quadratures. An obvious approach would be to use a perturbation expansion around the initial density matrix $\mathbf{\Gamma}_0$, but this requires small quadratures because the density matrix relaxes rapidly for fixed nuclei. An alternative solution which works well is to make a first order perturbation corrections to the relaxing (time-dependent) density matrix.

The correction to the relaxing density matrix can be obtained without coupling it to the differential equations for the Hamiltonian equations, and therefore does not require solving coupled equations for slow and fast functions. This procedure has been successfully applied to several collisional phenomena involving both one and several active electrons, where a single TDHF state was suitable, and was observed to show excellent numerical behavior. A simple and yet useful procedure employs the first order correction $\mathbf{\Gamma}'(t) = \mathbf{\Delta}'(t)$ and an adaptive step size for the quadrature and propagation. The density matrix is then approximated in each interval by

$$\mathbf{\Gamma}(t) = \mathbf{\Gamma}^0(t) + \mathbf{U}^0(t)\mathbf{\Delta}'(t)\mathbf{U}^0(t)^\dagger \tag{56}$$

with the first term describing relaxation and the second one giving the driving effect.

To advance from t_0 to $t_1 = t_0 + \Delta t$, the nuclear trajectory is first advanced to the time $t_{1/2} = t_0 + \Delta t/2$ and the relaxing density $\mathbf{\Gamma}^0(t)$ is calculated at this time; then the correction

$$\mathbf{\Delta}'(t_1) = (i\hbar)^{-1} \int_{t_0}^{t_1} dt' \, [\mathbf{V}(t'), \mathbf{\Gamma}_0] \tag{57}$$

is obtained with the (easily improved) approximation

$$\mathbf{V}(t) = \mathbf{U}^0(t)[\mathbf{H}(t_{1/2}) - \mathbf{H}(t_0)]\mathbf{U}^0(t)^\dagger \tag{58}$$

which allows an analytical integration of each matrix element. This is finally followed by recalculation of the nuclear trajectory and full density matrix at time t_1. To assure an accurate propagation, the step size Δt is varied to keep the density matrix correction within high and low tolerances within the interval, in accordance with

$$\varepsilon_{low} \leq \| \mathbf{\Delta}'(t) \| \, / \, \| \mathbf{\Gamma}^0(t) \| \leq \varepsilon_{high} \tag{59}$$

and the normalization is checked. This leads to an efficient adaptation of the step size, so that for example in a collision it will start large, will then decrease, and later increase again after the interaction forces have disappeared. The propagation accuracy can also be verified by reversing the propagation direction in time.

An alternative relax-and-drive procedure can be based on a strictly unitary treatment where the advance from t_0 to t_1 is done with a norm-conserving propagation such as provided by the split-operator propagation technique.(49, 50) This however is more laborious, and although it conserves the norm of the density matrix, it is not necessarily more accurate because of possible inaccuracies in the individual (complex) density matrix elements. It can however be used to advantage when the dimension of the density matrix is small and exponentiation of matrices can be easily done.(51, 52)

5. CONCLUSION

A formulation of electronic rearrangement in quantum molecular dynamics has been based on the Liouville-von Neumann equation for the density matrix. Introducing an eikonal representation, it naturally leads to a general treatment where Hamiltonian equations for nuclear motions are coupled to the electronic density matrix equations, in a formally exact theory. Expectation values of molecular operators can be obtained from integrations over initial conditions.

Approximations have been reviewed in the case of short deBroglie wavelengths for the nuclei to derive coupled quantal-semiclassical computational procedures, by choosing different types of many-electron wavefunctions. Time-dependent Hartree-Fock and time-dependent multiconfiguration Hartree-Fock formulations are possible, and lead to the Eik/TDHF and Eik/TDMCHF approximations, respectively. More generally, these can be considered special cases of an Eik/TDDM approach, in terms of a general density matrix for many-electron systems.

The Eik/TDDM approximation can be computationally implemented with a procedure based on a local interaction picture for the density matrix, and on its propagation in a relax-and-drive perturbation treatment with a relaxing density matrix as the zeroth-order contribution and a correction due to the driving effect of nuclear motions. This allows for an efficient computational procedure for differential equations coupling functions with short and long time scales, and is of general applicability.

6. ACKNOWLEDGMENTS

The author thanks the National Science Foundation and the Office of Naval Research of the USA for partial support. He has been motivated in this work by enlightening conversations with Yngve Ohrn.

REFERENCES

(1) Herzberg, G. *Spectra of Diatomic Molecules, 2nd. edition*; Van Nostrand: Princeton, NJ, 1950.

(2) Mott, N. F.; Massey, H. S. W. *The Theory of Atomic Collisions*, Third ed.; Clarendon: Oxford, 1965, chap. 19.

(3) Nikitin, E. E.; Umanskii, S. Y. *Theory of Slow Atomic Collisions*; Springer-Verlag: Berlin, 1984.

(4) *Atomic and Molecular Collision Theory*; Gianturco, F., Ed.; Plenum: New York, 1980.

(5) Kimura, M.; Lane, N. F. *Adv. At. Mol. Opt. Phys.* **1990**, *26*, 79.

(6) Fritsch, W.; Lin, C. D. *Phys. Rep.* **1991**, *202*, 1.

(7) Bransden, B. H.; McDowell, M. R. C. *Charge Exchange and the Theory of Ion-Atom Collisions*; Clarendon: Oxford, England, 1992.

(8) Errea, L. F.; Harel, C.; Jouin, H.; Mendez, L.; Pons, B.; Riera, A. *J. Phys. B* **1994**, *27*, 3603.

(9) Deumens, E.; Diz., A.; Taylor, H.; Ohrn, Y. *J. Chem. Phys.* **1992**, *96*, 6820.

(10) Deumens, E.; Diz., A.; Longo, R.; Ohrn, Y. *Rev. Mod. Phys.* **1992**, *96*, 917.

(11) Longo, R.; Deumens, E.; Ohrn, Y. *J. Chem. Phys.* **1993**, *99*, 4554.

(12) Gazdy, B.; Micha, D. A. *Phys. Rev. A* **1986**, *33*, 4446.

(13) Micha, D. A.; Gazdy, B. *Phys. Rev. A* **1987**, *36*, 539.

(14) Gazdy, B.; Micha, D. A. *Phys. Rev. A* **1987**, *36*, 546.

(15) Runge, K.; Micha, D. A.; Feng, E. Q. *Intern. J. Quantum Chem. Symposium* **1990**, *24*, 781.

(16) Feng, E. Q.; Micha, D. A.; Runge, K. *Intern. J. Quantum Chem.* **1991**, *40*, 545.

(17) Micha, D. A. *J. Chem. Phys.* **1983**, *78*, 7138.

(18) Stodden, C. D.; Micha, D. A. *Intern. J. Quantum Chem. Symposium* **1987**, *21*, 239.

(19) Cohen, J. M.; Micha, D. A. *J. Chem. Phys.* **1992**, *97*, 1038.

(20) Cohen, J. M.; Micha, D. A. *J. Chem. Phys.* **1993**, *98*, 2023.

(21) Micha, D. A.; Runge, K. *Phys. Rev. A* **1994**, *50*, 322.

(22) Micha, D. A. *Int. J. Quantum Chem.* **1994**, *51*, 499.

(23) Micha, D. A. *Int. J. Quantum Chem.* **1996**, *60*, 109.

(24) Marcus, R. A. *J. Chem. Phys.* **1971**, *54*, 3965.

(25) Miller, W. H. *Adv. Chem. Phys.* **1974**, *25*, 69.

(26) Heller, E. J. *J. Chem. Phys.* **1975**, *62*, 1544.

(27) Miller, W. H. *J. Chem. Phys.* **1970**, *53*, 3578.

(28) Herman, M. F.; Kluk, E. *Chem. Phys.* **1984**, *91*, 27.
(29) Heller, E. J. *J. Chem. Phys.* **1991**, *95*, 9431.
(30) Campolieti, G.; Brumer, P. *J. Chem. Phys.* **1992**, *96*, 5969.
(31) Meyer, H.-D.; Miller, W. H. *J. Chem. Phys.* **1979**, *70*, 3214.
(32) Olson, J. A.; Micha, D. A. *J. Chem. Phys.* **1984**, *80*, 2602.
(33) Deumens, E.; Ohrn, Y. *J. Phys. Chem.* **1988**, *92*, 3181.
(34) McWeeny, R. *Methods of Molecular Quantum Mechanics*, 2nd ed.; Academic: San Diego, CA, 1989.
(35) Micha, D. A.; Runge, K. *Chem. Phys. Lett.* **1995**, *238*, 132.
(36) Runge, K.; Micha, D. A. *Phys. Rev. A* **1996**, *53*, 1388.
(37) DaCosta, H. F. M.; Micha, D. A.; Runge, K. *J. Chem. Phys.* **1997**, *107*, 9018–27.
(38) Dirac, P. A. M. *Proc. Cambridge Philos. Soc.* **1930**, *26*, 376.
(39) Frenkel, J. *Wave Mechanics. Advanced General Theory*; Clarendon: Oxford, England, 1934.
(40) Kerman, A. K.; Koonin, S. *Ann. Phys. (N.Y.)* **1976**, *100*, 332.
(41) Kramer, P.; Saraceno, M. *Geometry of the Time-Dependent Variational Principle in Quantum Mechanics*; Springer-Verlag: New York, 1981.
(42) Tolman, R. C. *The Principles of Statistical Mechanics*; Clarendon: Oxford, England, 1938.
(43) Fano, U. *Rev. Mod. Phys.* **1957**, *29*, 74.
(44) Lowdin, P. O. *Intern. J. Quantum Chem. Symposium* **1982**, *16*, 485.
(45) Blum, K. *Density Matrix Theory and Applications, 2nd. edition*; Plenum: New York, NY, 1996.
(46) Micha, D. A.; Feng, E. Q. *Comp. Phys. Comm.* **1994**, *90*, 242.
(47) von Neumann, J. *Mathematical Foundations of Quantum Mechanics*; Princeton Univ.: Princeton, New Jersey, 1955.
(48) Lowdin, P. O. *Phys. Rev.* **1955**, *97*, 1490.
(49) Feit, M. D.; Fleck, J. A.; Steiger, A. *J. Comput. Phys.* **1982**, *47*, 412.
(50) Leforestier, C.; et al. *J. Comput. Phys.* **1991**, *94*, 59.
(51) Fernandez, F. M.; Micha, D. A. *J. Chem. Phys.* **1992**, *97*, 8173.
(52) Kosloff, R. *Annu. Rev. Phys. Chem.* **1994**, *45*, 145.

Theoretical and Experimental Studies of the Benzene Radical Cation: Effects of Selective Deuteration

S. Lunell[a], J. W. Gauld[a], R. M. Kadam[b], Y. Itagaki[b] and A. Lund[b]

[a]*Department of Quantum Chemistry, Uppsala University, Box 518, S-75120 Uppsala, Sweden*

[b]*Department of Physics and Measurement Technology, Linköping University, S-58183 Linköping, Sweden*

Abstract

Geometries, hyperfine structure, and relative stabilities of the different positional isomers of monodeuterated benzene cations have been studied theoretically by density functional theory, using the B3-LYP functional, and experimentally by ESR and ENDOR spectroscopy. A comparison between theoretical and experimental results at 30 K gives acceptable agreement, but further experiments on multiply deuterated species should improve the analysis by making the effects of deuteration larger.

ADVANCES IN QUANTUM CHEMISTRY, VOLUME 35

0065-3276/99 $30.00

1. Introduction

The use of selective deuteration is a powerful tool in electron spin resonance (ESR) experiments, in order to establish unequivocal assignments of experimental spectra of radicals. The reason for this is, as is well known, the difference in magnetic properties between the deuteron and the proton, which can be exploited to distinguish chemically inequivalent hydrogens in the molecule.

In addition to the magnetic differences between the deuteron and proton, however, their mass difference may also cause observable effects. A well known example is found in the theory of chemical reactions, where the so called kinetic isotope effects (KIE's) are an important source of information about reaction mechanisms. Also in the field of ESR, such effects may arise, although these have been much less studied than the KIE's.

A particularly interesting case is when a set of hydrogens which are chemically equivalent in the unionized molecule become inequivalent in the positive ion. Obvious examples are Jahn-Teller active molecules, but the same phenomenon may be found also in Jahn-Teller inactive systems. Since deuteration for practical reasons must be done before ionization, it may happen that a single deuterated molecule may produce several inequivalent isomers of the radical cation, e.g., upon irradiation. This will obviously influence the recorded ESR spectrum.

We have previously in a number of papers [1-5] investigated these effects for both the Jahn-Teller inactive molecule *n*-butane [1] and the Jahn-Teller active molecules ethane, cyclopropane, and cyclohexane [2-5]. The choice of systems was largely dictated by the availability of experimental results [5-8]. New experiments being performed on selectively deuterated benzene have motivated a closer theoretical study of this system, and a first presentation of these investigations is given in the present paper.

The benzene radical cation is an archetype of a Jahn-Teller active system. Removal of one electron from the highest occupied MO, the doubly degenerate e_{1g} orbital, leads to a Jahn-Teller unstable ion, which distorts from D_{6h} to D_{2h} symmetry. The ground state of the ion can be either $^2B_{3g}$, corresponding to a compressed geometry (two C—C bonds shorter than the four others), or $^2B_{2g}$, corresponding to an elongated geometry (two C—C bonds longer than the four others). A definite assignment of the ground state of the radical cation to the compressed ($^2B_{3g}$) state could be made less than a decade ago by comparing experimental hyperfine coupling parameters obtained from matrix-isolation ESR experiments with theoretical values obtained from correlated *ab initio* calculations [9,10].

In the present study, we focus on the effects of substituting one of the protons by a deuteron. While giving only one isomer of the unionized molecule, this produces two inequivalent isomers of the Jahn-Teller distorted ion; one isomer where the deuteron occupies one of the two sites on the C_2 symmetry axes (H_1 or H_4) and one where it occupies one of the four equivalent remaining sites (H_2, H_3, H_5 or H_6). The effects on the ESR spectrum will below be illuminated both theoretically and experimentally.

2. Theoretical Method

Density functional theory (DFT) calculations were carried out using the Gaussian 94 [11] suite of programs. Optimized geometries and harmonic vibrational frequencies were obtained using the hybrid Becke exchange functional (B3) [12, 13], as implemented [14] in Gaussian 94 [11], and the correlation functional due to Lee, Yang and Parr (LYP) [15], in combination with the 6-31G(d) and 6-311G(d,p) basis sets. The unrestricted procedure (UB3-LYP) was used throughout (the symbol U is hereafter neglected for simplicity).

In our previous study of the benzene radical cation [9], it was shown that computational methods which do not include correlation predict an erroneous ground state of the ion. Since then, the rapid development in the field of density functional theory (DFT) has resulted in computational methods which are of comparable speed to *ab initio* Hartree-Fock calculations, while often being of comparable accuracy to high-level correlated procedures. In particular, the DFT method B3-LYP, has been found to give reasonable geometries and properties for a range of chemical systems [16], which motivates its use in the present context.

The hyperfine coupling constant (hfcc) calculations were performed by carrying out single point calculations at the B3-LYP/6-311+G(2df,p) level, based on the B3-LYP/6-31G(d) optimized geometries, i.e., B3-LYP/6-311+G(2df,p)//B3-LYP/6-31G(d).

The inequality of the C—H bonds in the radical cation implies that all C—H bonds do not have the same force constants. In a simplistic approximation, the zero-point vibrational energy (ZPVE) of a C—H stretching vibration will be proportional to $(k/m_H)^{1/2}$, where k is the force constant of the C—H bond and m_H is the mass of the hydrogen nucleus. The effect on the ZPVE of replacing one proton by a deuteron will hence depend on the deuteration site, such that the ZPVE will be lowered more if the deuteron occupies a site with a larger force constant, i.e. a shorter bond. This, in general, means a site with low unpaired spin density.

At low enough temperatures, only that substitutional isomer which has the lowest ZPVE will be populated. At higher temperatures, however, other isomers may also be found, so that a superposition of several spectra is observed in an ESR experiment. Previously, we have found [1-4] that the abundances of these other isomers are well predicted by a Boltzmann distribution based on the differences in ZPVE. Since the 6-31G(d) and 6-311G(d,p) basis sets gave very similar geometries in the present case (see below), the vibrational frequencies and ZPVE for the mono-deuterated isomers were calculated only at the B3-LYP/6-31G(d) level.

3. Experimental Method

Benzene, benzene-d_1, $CFCl_3$ and CF_3CCl_3 were obtained commercially and were not further purified. Solutions of c:a 0.3 - 1 volume % of benzene in $CFCl_3$ were prepared in suprasil quartz tubes of 4 mm outer diameter on a vacuum line. The samples were degassed and sealed under vacuum (< 10^{-4} Torr). Polycrystalline samples were prepared by rapid freezing in liquid nitrogen. The samples were irradiated at 77 K for 5 minutes at an approximate dose rate of 250 G/min. using the radiation from an X-ray tube with a W anode operated at 70 kV and 20 mA.

The ESR spectra were recorded with a Bruker ER200 spectrometer at X band. A few samples showing clear orientational dependence were further investigated by recording the ESR spectra for different angles of rotation about the tube axis. The true polycrystalline nature of the samples obtained by rapid cooling was similarly tested. The ENDOR data of the polycrystalline samples were obtained with a Bruker ER250 ENDOR accessory and an ENI RF 500 amplifier. ENDOR spectra were obtained using RF modulation depths of c:a 200 kHz. Sample temperatures above 77 K were regulated with a Bruker VT4111 temperature controller.

4. Theoretical Results

The results of the geometry optimizations are shown in Table 1. With some minor deviations, the geometrical parameters agree well with those of ref. [9]. One can also note that, as already mentioned, the geometries obtained using the 6-31G(d) and 6-311G(d,p) basis sets, respectively, are very similar.

Table 2 shows the hyperfine coupling constants obtained at the B3-LYP/6-311+G(2df,p)//B3-LYP/6-31G(d) level. The calculations reiterate the results of reference [9], that the largest spin density is on H_1, which is consistent with the

Table 1: Optimized Geometrical Parameters[a] for the Benzene Radical Cation Ground State ($^2B_{3g}$), Obtained Using B3-LYP with Various Basis Sets.

	Basis Set	
	6-31G(d)	6-311G(d,p)
R1	1.432	1.429
R2	1.372	1.369
R3	1.087	1.085
R4	1.085	1.083
—	121.6	121.6
—	119.2	119.2
—	119.4	119.4

[a] Bond lengths in Ångströms, angles in degrees.

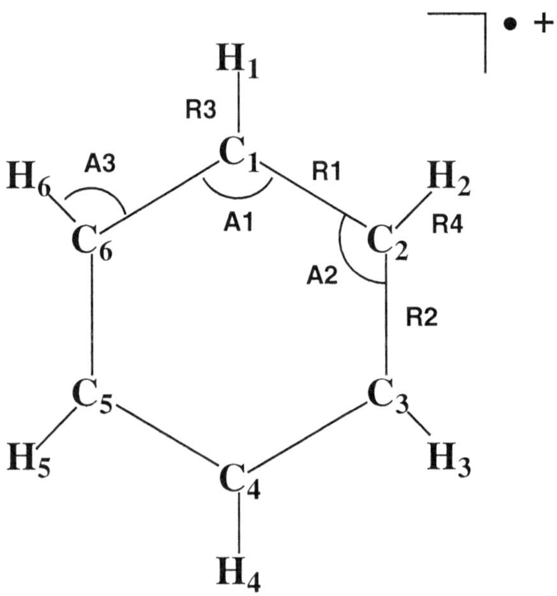

Table 2: Calculated and Experimental Hydrogen Hyperfine Coupling Constants (in Gauss) for the Benzene Radical Cation Ground State ($^2B_{3g}$).

	B3-LYP/6-311+G(2df,p) //B3-LYP/6-31G(d)[a]	SDCI[b]	Exptl
H_1 (2H)	−10.2	−9.6	−8.9[a] (−8.2[c])
H_2 (4H)	−1.4	−1.4	−1.9[a] (−2.4[c])

[a] This work. [b] From ref [9]. [c] From ref [17].

Table 3: Vibrational Frequency Components[a] in the High- and Low-Frequency regions (cm^{-1}) and Total ZPVE[a] (kJ mol^{-1}) for the Non- and Mono-Deuterated Isomers of the Benzene Radical Cation.

	Frequency range		
Deuterated Position	Low (0 - 2000 cm^{-1})	High (2000 - 3500 cm^{-1})	ZPVE (kJ mol^{-1})
None	23673.17	19401.97	257.6
C_1	23107.23	18560.39	239.6
C_2	23067.34	18560.96	239.4

[a] Calculated at the B3-LYP/6-31G(d) level, see text.

information from Table 1 that the C_1—H_1 bond is weaker (longer) than the C_2—H_2 bond. One can, however, note that all of the coupling constants are negative, i.e. are caused by spin polarization, implying that the unpaired spin in principle is localized to the carbon-carbon bonding π-system. The effects of ionization on the C—H bonds are therefore rather small, which is also reflected in the small difference (0.002 Å) between the C_1—H_1 and C_2—H_2 bonds in the cation.

The ZPVE's for the two different substitutional isomers are given in Table 3. As expected from the above arguments, a slightly larger lowering of the ZPVE is obtained when the deuterium is bound to the low spin density site C_2 than when it is bound to the high spin density site C_1. One can however, note that the difference is very small, only 0.2 kJ mol^{-1} (see further Section 6, Discussion).

5. Experimental Results

A. ESR RESULTS

The ESR spectrum of $C_6H_6^{\bullet+}$ trapped in $CFCl_3$ at 15 K is shown in Figure 1a and agrees with that reported previously [18]. The principal values of the hyperfine coupling were obtained from previous ESR and ENDOR measurements [17, 18]. The best agreement with experiment was obtained with the axes oriented as in Table 4. In the latter study, the simulated ENDOR spectra were insensitive to the orientation of the tensor axes, however, and the assignment was made on the basis of molecular orbital calculations [9]. The tensor data are reproduced here for convenience (see Table 4).

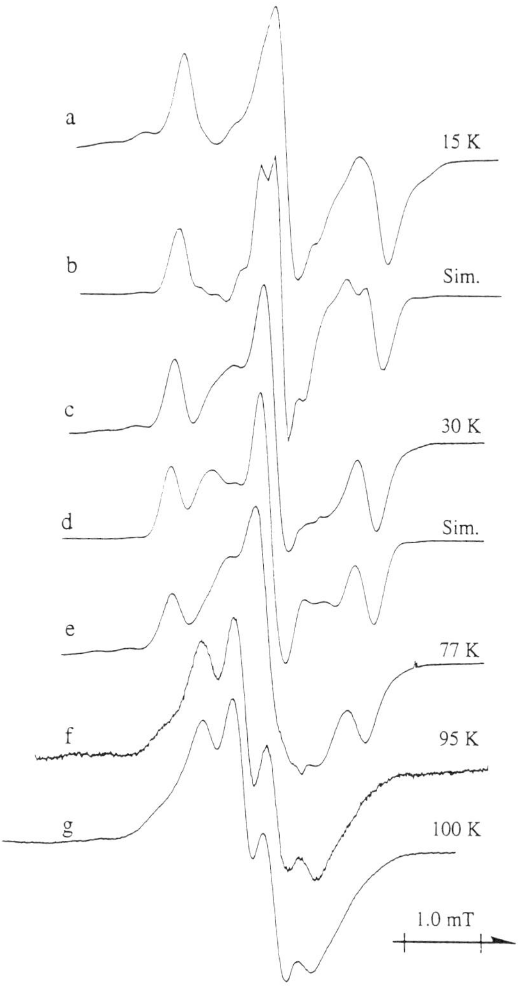

Figure 1. ESR spectra at some temperatures of X-irradiated polycrystalline frozen solutions (c:a 1 mol%) of C_6H_6 and C_6H_5D, respectively. a) $C_6H_6^{\bullet+}$ at 15 K. b) simulated spectrum by using the parameters shown in Table 4. c) $C_6H_5D^{\bullet+}$ at 30 K in $CFCl_3$. d) simulation using hyperfine coupling data shown in Table 4 for a mixture of component 1 : component 2 of $C_6H_5D^{\bullet+}$. The ratio 1:2 = 0.38:0.62 obtained from the component analysis of Figure 2 was employed.

Table 4: The g and hyperfine tensors (in Gauss) used in the ESR and ENDOR simulations of $C_6H_5D^{•+}$ in $CFCl_3$ at 30 K.

tensors	principal components	principal value	x	y	z
$C_6H_5D^{•+}$	g_x	2.0029	1	0	0
	g_y	2.0029	0	1	0
	g_z	2.0023	0	0	1
A (1, 4)	A_x	–13.66	1	0	0
	A_y	–3.86	0	1	0
	A_z	–9.23	0	0	1
A (3,5,6)	A_x	–2.91	±0.3771	0.9262	0
	A_y	0.00	–0.9262	±0.3771	0
	A_z	–2.91	0	0	1
A D(2)	A_x	–0.45	±0.3771	0.9262	0
	A_y	0.00	–0.9262	±0.3771	0
	A_z	–0.45	0	0	1

The polycrystalline ESR spectrum at 30 K of $C_6H_6^{•+}$ trapped in a $CFCl_3$ matrix is shown in Figure 1c. The spectrum is composed of a triplet component similar to that in Figure 1a. Additional features are present about the center, indicating a second component. The two components were assigned to two forms of $C_6H_5D^{•+}$ with the deuterium at a high spin density position, i.e., C_1 or C_4 (component 1) and with the deuterium at a low spin density position, C_2, C_3, C_5 or C_6 (component 2), respectively, shown below. Simulations of the ESR spectra of the components were carried out based on the ENDOR data in Table 4. The two components shown in absorption mode in Figure 2a have triplet and doublet structures. The experimental and fitted spectra, the latter obtained with a least squares procedure [19], are shown in absorption mode in Figure 2b. The intensity ratio between component 1 and

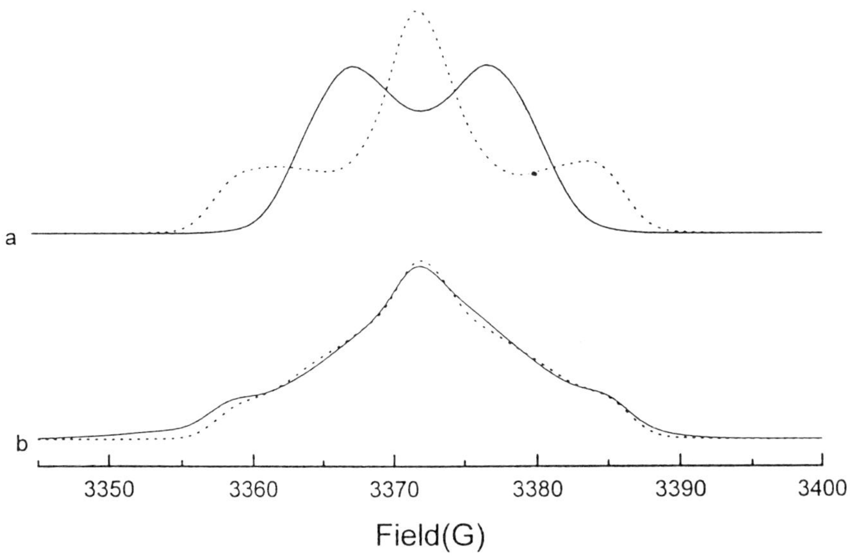

Figure 2. Component analysis of an X-irradiated polycrystalline frozen solution (c:a 1 mol%) of C_6H_5D in $CFCl_3$. a) simulated components 1 (dashed) and 2 (solid) using hyperfine coupling data from ENDOR (Table 4). b) experimental (solid) and fitted (dashed) ESR spectra of $C_6H_5D^{\bullet+}$ in absorption mode. The ratio component 1: component 2 = 0.38:0.62 was obtained by a least squares analysis.

2 was estimated as 0.38:0.62 with a calculated (statistical) error of ±0.0035. The actual error is probably larger (see Discussion).

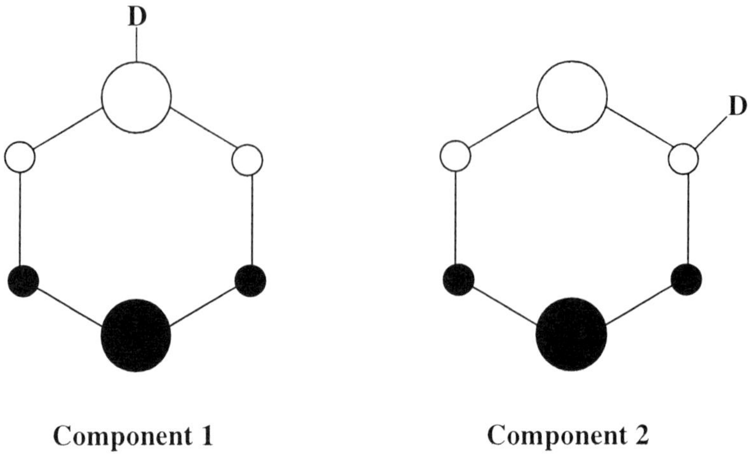

On increasing the temperature all the protons become equivalent by dynamic averaging.

B. ENDOR RESULTS

ENDOR measurements were performed at 30 K in the case of $C_6H_5D^{\bullet +}$ in a $CFCl_3$ matrix. An ENDOR spectrum obtained by sweeping the radio frequency between 1 - 35 MHz, with the magnetic field locked to the central position of the ESR spectrum, is shown in Figure 3. The strong signal in the region of 13.80 MHz is due to the matrix radical. Note that the line marked with * is the second harmonic of the strongest ENDOR lines of the matrix radical due to the non linearity of the radio frequency power amplifier. The ENDOR signal outside this region corresponds to the principal values of the hyperfine coupling arising from ring protons and deuterium nuclei. It is seen from the figure that at a radio frequency $\nu > 15$ MHz, the ENDOR spectrum shows four hyperfine transitions denoted as A^+, B^+, C^+, and D^+ (high frequency transitions ν^+). The low frequency pairing lines are observed only for C^+ and D^+. The other two pairing lines A^- and B^- are too weak and they are buried in the strong

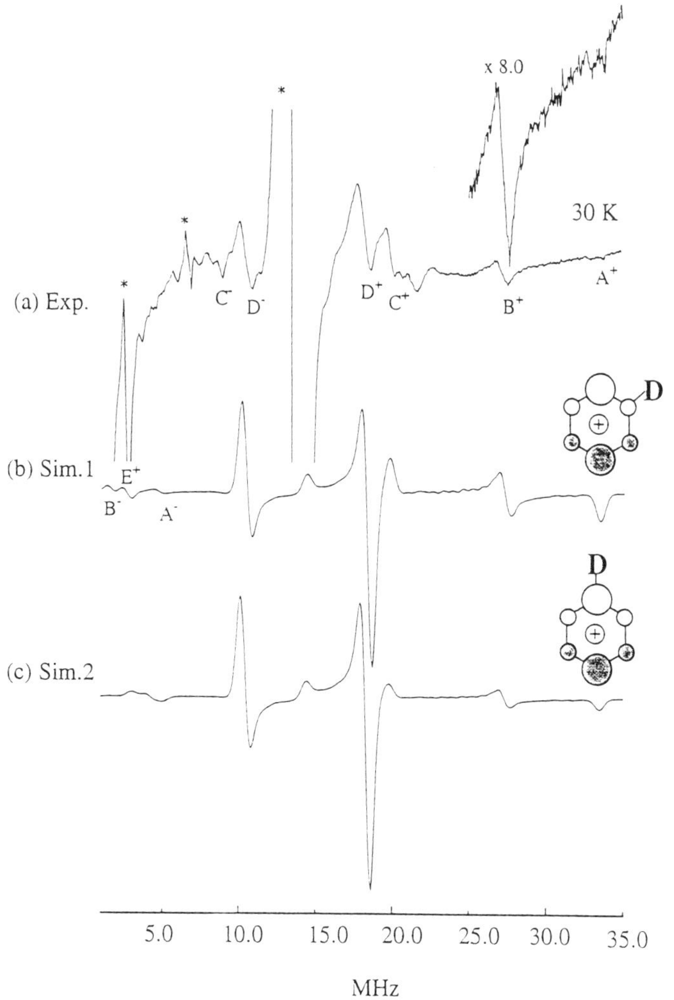

Figure 3. ENDOR spectrum of $C_6H_5D^{\bullet+}$ radical cation. (a) obtained experimentally by saturating central ESR transitions (b) simulated spectrum with H_1 and H_4 at high spin density positions and D_2, H_3, H_5 and H_6 at low spin density positions. The hyperfine coupling constants used in the ENDOR simulation are listed in Table 4. (c) simulated spectrum with H_1 and D_4 at high spin density positions and H_2, H_3, H_5 and H_6 at low spin density positions.

matrix signal. These low frequency ENDOR transitions are further confirmed by ENDOR spectral simulations. These ENDOR transitions are centred around the proton Larmor frequency, near $\nu_H \pm A_i/2$ MHz. The ENDOR signals have the splitting of 38.80 MHz, 26.00 MHz, 11.10 MHz and 8.15 MHz and they correspond to the hyperfine coupling constants of 13.66 G, 9.23 G, 3.86 G and 2.91 G, respectively. The pair of transitions centred around the Larmor frequency of deuterium (ν_D = 2.20 MHz) with a separation of 0.45 MHz is also buried in the strong matrix signal. The direction cosines, the hyperfine coupling constants and the g values used in the ESR and ENDOR simulations are given in table 4.

6. Discussion

As can be seen from Table 2, the agreement between measured and calculated isotropic hyperfine coupling parameters is good, confirming previous interpretations [9] of ESR data.

In order to assess the analysis given in Figures 1 and 2 in terms of the components 1 and 2, the theoretically predicted abundances of the two components were calculated at 30 K, using the ZPVE values from Table 3. Including the statistical factors, the ratio obtained from the ZPVE's is 0.31:0.69, implying that substitution at the H_2 position is slightly favoured compared to the purely statistical probability, as expected from the difference in bond strengths bewtween C_1—H_1 and C_2—H_2. In a comparision with the experimentally estimated ratio of 0.38:0.62, the agreement between theory and experiment can be considered as acceptable, considering the difficulties with overlapping signals from the matrix radicals in the experiment and the small magnitude of the ZPVE differences in the calculations.

There are some interesting aspects of the results which deserve additional study. A close scrutiny of Table 3 reveals that the difference in ZPVE between the two isomers of the mono-deuterated benzene cation does only to a very minor extent (0.5 cm^{-1}) stem from the C—H (C—D) stretching vibrations, as one would have anticipated from the simplistic arguments given in Section 2. Instead, virtually the whole difference (40 cm^{-1}) comes from the low-frequency vibrations, i.e. predominantly bending vibrations.

One can here compare with other similar studies on selectively deuterated hydrocarbons [1-4]. On one hand, the abundances of the different substitutional isomers of cations of the aliphatic (non-cyclic) hydrocarbons ethane [4] and *n*-butane [1] fully follow the trends anticipated from the stretching frequencies, namely that deuteration at a short (strong) bond is more favourable that at a weaker bond, by about 1.5 kJ mol^{-1} in the case of mono-deuteration. In the case of cyclopropane, on the other hand, the changes in the ZPVE for the C—H stretching vibrations do follow the anticipated trends, but the effect on the low-frequency region is larger and opposite in direction, so that deuteration actually is favoured at the *weaker* bonds, by about 0.5 kJ mol^{-1} in the case of mono-deuteration. The benzene cation seems in this context to occupy an intermediate position – the changes in the different coupled vibrational modes cancel almost completely, leading to a ZPVE difference of only 0.2 kJ mol^{-1} between the two isomers in the mono-deuterated case.

Additional experimental studies on multiply deuterated benzene cations would give more information, by enlarging the effects on the ZPVE and also by introducing new structural features in the experimental spectra which can facilitate their interpretation. This would enable a more detailed and more accurate analysis, both theoretically and experimentally. Such experiments will hopefully be carried out in the near future.

Acknowledgments

This work was supported by the Swedish Natural Science Research Council (NFR) and the Göran Gustafsson Foundation. A grant of computer time on the T3E of the National Supercomputer Centre (NSC) in Linköping is gratefully acknowledged. It is a pleasure to dedicate this paper to Yngve Öhrn as a token of appreciation for all contacts over the years, in Uppsala as well as in Gainesville.

References

[1] S. Lunell, L. A. Eriksson, L. Worstbrock, J. Am. Chem. Soc. 113 (1991) 7508.

[2] L. A. Eriksson and S. Lunell, J. Am. Chem. Soc. 114 (1992) 4532.

[3] S. Lunell and L. A. Eriksson, Int. J. Quantum. Chem., Quant. Chem. Symp. 26 (1992) 575.

[4] L. A. Eriksson and S. Lunell, J. Phys. Chem. 97 (1993) 12215.

[5] P. Wang, M. Shiotani and S. Lunell, Chem. Phys. Letters. 292 (1998) 110.

[6] M. Lindgren, A. Lund and G. Dolivo, Chem. Phys. 99 (1985) 103; M. Lindgren and A. Lund, J. Chem. Soc., Faraday Trans. I 83 (1987) 1815.

[7] K. Matsuura, K. Nunome, M. Okazaki, K. Toriyama and M. Iwasaki, J. Phys. Chem. 93 (1989) 6642.

[8] M. Iwasaki, K. Toriyama and K. Nunome, Chem. Phys. Letters. 111 (1984) 309.

[9] M.-B. Huang and S. Lunell, J. Chem. Phys. 92 (1990) 6081.

[10] Due is a certain non-uniqueness in the symmetry assignments in D_{2h} symmetry, some authors, e.g. in ref [9], instead designate the compressed ground state as $^2B_{2g}$ and the elongated one as $^2B_{1g}$. In the present paper, we follow the convention set by the Gaussian program [11] and designate the ground state as $^2B_{3g}$.

[11] M.J. Frisch, G.W. Trucks, H.B. Schlegel, P.M.W. Gill, B.G. Johnson, M.A. Robb, J.R. Cheeseman, T.A. Keith, G.A. Petersson, J.A. Montgomery, K. Raghavachari, M.A. Al-Laham, V.G. Zakrzewski, J.V. Ortiz, J.B. Foresman, J. Cioslowski, B.B. Stefanov, A. Nanayakkara, M. Challacombe, C.Y. Peng, P.Y. Ayala, W. Chen, M.W. Wong, J.L. Andres, E.S. Replogle, R. Gomperts, R.L. Martin,

D.J. Fox, J.S. Binkley, D.J. DeFrees, J. Baker, J.P. Stewart, M. Head-Gordon, C. Gonzalez and J.A. Pople, Gaussian 94, Gaussian Inc., Pittsburgh PA (1995).

[12] A.D. Becke, J. Chem. Phys. 98 (1993) 1372.

[13] A.D. Becke, J. Chem. Phys. 98 (1993) 5648.

[14] P.J. Stephens, F.J. Devlin, C.F. Chabalowski and M.J. Frisch, J. Phys. Chem. 98 (1994) 11623.

[15] C. Lee, W. Yang and R.G. Parr, Phys. Rev. B 37 (1988) 785.

[16] See, for example, C. W. Bauschlicher Jr., Chem. Phys. Lett. 246 (1995) 40.

[17] M. Iwasaki, K. Toriyama and K. Nunome, J. Chem. Soc., Chem. Commun. (1983) 320.

[18] R. M. Kadam, R. Erickson, K. Komaguchi, M. Shiotani and A. Lund, Chem. Phys. Lett., 290 (1998) 371.

[19] Modernized version of the ASESK program written by T. Vänngård, 1961, cf. A. Lund, J. Phys. Chem., 76 (1972) 1411.

A Theoretical Study of the $[Fe_2(\mu - S_2)(P(o - C_6H_4S)_3)_2]^{2-}$ Electronic Spectrum

Marshall G. Cory, Krassimir K. Stavrev* and Michael C. Zerner*

Quantum Theory Project

University of Florida

Gainesville, Florida 32611

ABSTRACT

The absorption spectrum of the $[Fe_2(\mu\text{-}S_2)(P(o\text{-}C_6H_4S)_3)_2]^{2-}$ complex was examined within the Intermediate Neglect of Differential Overlap model parametrized for spectroscopy using a methodology developed earlier for 2-Fe ferredoxins. We demonstrate here that the low energy UV-visible absorption spectrum of this complex can be interpreted as originating from sulfur-to-iron charge-transfer transitions. The results obtained favor the model of antiferromagnetically coupled $d^5 - d^5$ Fe atoms over other possibilities, for example the formation of a diamagnetic $[Fe(d^6)\text{-}S_2^0\text{-}Fe(d^6)]$ complex.

ADVANCES IN QUANTUM CHEMISTRY, VOLUME 35

0065-3276/99 $30.00

Introduction

In this work we examine the low energy UV-visible absorption spectrum of the $\left[Fe_2(\mu - S_2)(P(o - C_6H_4S)_3)_2\right]^{2-}$ complex, Figure 1, whose synthesis, structure, and properties have recently been reported.[1] The complex contains a $[Fe - S - S - Fe]^{4+}$ core and is a structural isomer of the 2-Fe $[Fe - (\mu - S)_2 - Fe]^{2+}$ ferredoxin. The electronic structure of the disulfide complex is, however, unknown, and can be associated with either an antiferromagnetically (AF) coupled $\left[Fe(d^5) - S_2^{2-} - Fe(d^5)\right]^{4+}$ system, or with a diamagnetic $\left[Fe(d^6) - S_2^0 - Fe(d^6)\right]^{4+}$ core.[2,3] The Mössbauer spectrum of this compound shows that the two Fe atoms are equivalent, but does not distinguish between the two alternative electronic distributions, while the observed resonance shifts in proton NMR favor the high-spin model.[1] Further, the assignment of the absorption band origins of this complex have yet to be made.[1] These bands have a variety of possible origins including inter- or intra-metal *d-d* transitions, metal-to-ligand and ligand-to-metal $M \leftrightarrow L$ charge-transfer (CT) bands, or possibly ligand $\pi \rightarrow \pi^*$ transitions.[2,3] We intend to clarify this picture.

In an earlier work,[4] we have proposed a theoretical procedure for the spectroscopy of antiferromagnetically (AF) coupled transition-metal dimers and have successfully applied this approach to the electronic absorption spectrum of model 2-Fe ferredoxin. In this work we apply this same procedure to the $\left[Fe_2(\mu - S_2)(P(o - C_6H_4S)_3)_2\right]^{2-}$ complex in order to better understand the electronic structure of this compound. As in our previous work[4] we base our analysis on the Intermediate Neglect of the Differential Overlap model parameterized for spectroscopy (INDO/S), [5-8] utilizing a procedure outlined in detail in Reference 4.

Structure and Method

The structure of the $\left[Fe_2(\mu - S_2)(P(o - C_6H_4S)_3)_2\right]^{2-}$ complex is presented in Figure 1. The structure we used was obtained from x-ray crystallography,[1] symmetrized such that the environments of the two Fe atoms are the same. The local symmetry of each Fe atom is C_{3v}, the overall symmetry of the complex is C_i. To aid in the interpretation of the computed results we oriented the complex such that the z-axis passes through the center of mass and the projections of the two $P - Fe - S$ line segments onto the z-axis are maximized.

The complex contains 72 atoms with 244 valence electrons distributed in 226 valence atomic orbitals. In order to reduce the computational effort, and to assess the contribution of the ligand π-orbitals to the overall spectrum, we examined a "reduced" model, see Figure 2, in which the benzene rings of the ligands are replaced by -HC=CH- groups. This model compound consists of

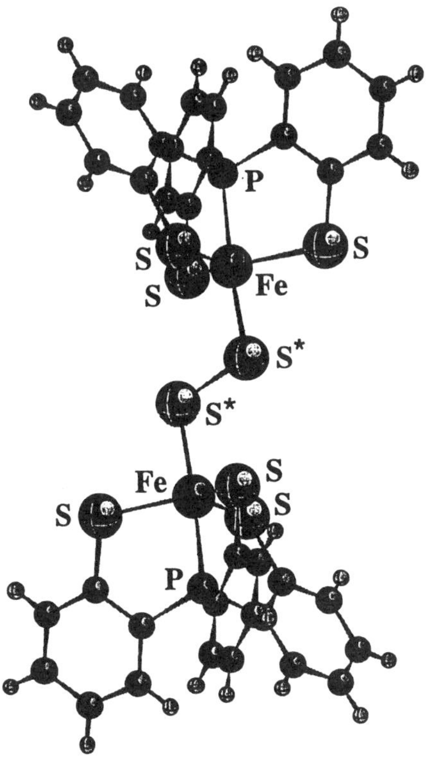

Figure 1. The original 72-atom structure of the $[Fe_2(\mu - S_2)(P(o - C_6H_4S)_3)_2]^{2-}$ complex, from Reference 1. S^* stands for a bridging, inorganic sulfur. The coordinates around each Fe atom have been symmetrized.

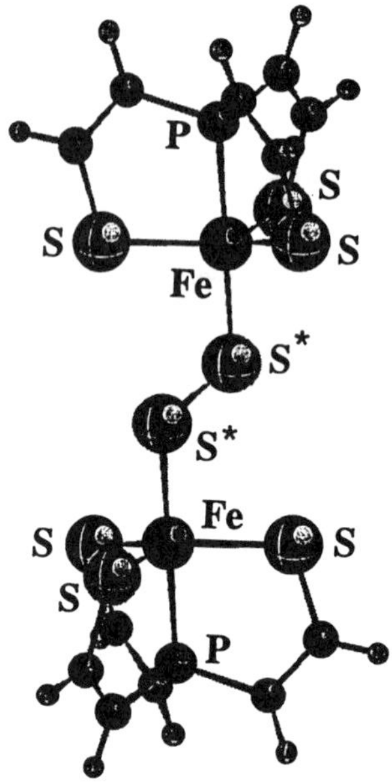

Figure 2. A reduced 36-atom model of the $\left[Fe_2(\mu - S_2)(P(o - C_6H_4S)_3)_2\right]^{2-}$ complex; the benzene rings are substituted by ethylene groups, see Figure 1.

36 atoms, 136 valence electrons, and 118 valence atomic orbital (AO) basis functions. Thus we partially preserve the π-character of the ligands and reduce the computational effort by about a factor of 8.

The electronic spectrum of the complex was calculated using the INDO/S model within the Restricted Open-shell Hartree-Fock (ROHF) approximation[9] as implemented in the ZINDO program.[10] Excitations from the highest multiplicity reference state have been calculated using the single excitation Configuration Interaction (CIS) procedure, and the calculated absorption spectrum fitted to a Lorentzian band convolution over the computed states. (See the appendix of Reference 11.) For the case of AF coupling it is computationally difficult to obtain the states of lower electronic spin multiplicity (with singlets being the lowest) due to the large number of configuration-state-functions (CSF) necessary to properly describe the reference function and the excited states. However, we can derive the singlet CT spectrum from the spectrum obtained for the highest multiplicity, i.e. $2S$+1=11 or 11-tet (undectet), via first order perturbation theory.[4] In Reference 4 an analysis of the terms contributing to the ligand-to-metal charge transfer transition for cases in which each d-orbital is singly occupied suggested that

$$\begin{aligned} {}^{1}\Delta E(delocal) &= {}^{11}\Delta E(delocal) - 5\bar{K}_{dd} + 5\bar{K}_{dL''} \\ &= {}^{11}\Delta E(delocal) - E(shift) \end{aligned} \tag{1}$$

where ${}^{1}\Delta E(delocal)$ refers to calculations in which the molecular orbitals of d-atomic character are not localized (as is the normal case for high-spin ferromagnetic calculations), $\bar{K}_{dd}$ is an average exchange integral between d-atomic-orbitals, and $\bar{K}_{dL''}$ is the average exchange between ligand and metal d-orbitals. $\bar{K}_{dd}$ can be estimated from atomic spectroscopy, i.e. $\bar{K}_{dd} = \frac{5}{196}\left(F^2(dd) + F^4(dd)\right) = 2525\ \text{cm}^{-1}$ using $F^2(dd) = 61\,000\ \text{cm}^{-1}$ and $F^4(dd) = 38\,000\ \text{cm}^{-1}$.[12] $\bar{K}_{dL''}$ can be estimated from these calculations (with L'' a sulfur ligand) as 340 cm^{-1}. This yields

$$\begin{aligned} {}^{1}\Delta E(delocal) &= {}^{11}\Delta E(delocal) - 12\,625\ \text{cm}^{-1} + 1\,700\ \text{cm}^{-1} \\ &= {}^{11}\Delta E(delocal) - 10\,925\ \text{cm}^{-1} \end{aligned} \tag{2}$$

In other words, we estimate the spectrum of the AF-singlet ground state by applying a shift, to the calculated undectet spectrum.[4] For this case we use an empirical shift of 13 500 cm^{-1} so as to yield maximum coincidence between the calculated and observed spectra, a value easily obtained, for example, if we had used the Slater-Condon integrals from the Fe^+ ion in equation (1) rather than the values for neutral Fe.

The oscillator strengths for the reported spectra were calculated as

$$f^{I}_{osc} = \tfrac{2}{3}\Delta E_I |\langle 0|\vec{\mu}|I\rangle|^2 \tag{3}$$

thus

$$^{1}f^{I}_{osc} = \frac{^{11}\Delta E_I - E(shift)}{^{11}\Delta E_I}\ ^{11}f^{I}_{osc} \tag{4}$$

where f^{I}_{osc} is the computed oscillator strength of state *I*. Results for the $\left[Fe\left(d^6\right) - S_2^0 - Fe\left(d^6\right)\right]^{4+}$ core have also been obtained using single and double excitations CI (CISD) from the $\left[Fe\left(d^5\right) - S - S - Fe\left(d^5\right)\right]^{4+}$ reference state and compared with the latter's high-spin state. Lower multiplicity reference states have not been considered for use in the calculation of the spectrum as such calculations are computationally expensive and the transition energies are not expected to deviate substantially from those derived from the high-spin case.[4]

Results and Discussion

We summarize our calculations on the 72- and 36-atom structures in Tables 1 and 2 respectively. ROHF energies were obtained for different electron distributions corresponding to S_z values for the two Fe atoms of $\left(\frac{5}{2}, \frac{5}{2}\right)$, $\left(\frac{5}{2}, \frac{3}{2}\right)$, and $\left(\frac{3}{2}, \frac{3}{2}\right)$. These configurations are distinguished by assigning ten, eight or six electrons, respectively, to the open-shell.

For the UHF calculations the undectet implies $S_z = 5$ (septet implies $S_z = 3$) and the electrons are assigned as suggested by the ROHF calculation. For the $S_z = 0$ case the electrons are assigned to be AF coupled. The Projected Unrestricted Hartree-Fock (PUHF) energies correspond to the fully projected UHF values with $S = S_z$.[13] A brief overview of the PUHF method is given below, a more detailed description can be found in Reference 4.

The UHF wavefunction can be written as

$$^{2S_z+1}\Psi^{UHF}_{s_z} = \sum_{s=s_z}^{s_z+N_\beta} \omega_s \sum_{k=0}^{N_\beta} C_k(S, S_z, n)\ ^{2S+1}\psi^{k}_{s_z} \tag{5}$$

where N_β is the number of β-spin electrons and the ω_s and $C_k(S, S_z, n)$ are those of Sasaki and Ohno,[14] and the projector[15] as

$$\widehat{O}_s = \prod_{s_z \le l \ne s} \frac{\widehat{S}^2 - l(l+1)}{s(s+1) - l(l+1)} \tag{6}$$

Thus yielding

$$^{2S+1}\Psi_{S_z} = \widehat{O}_s\ ^{2S_z+1}\Psi^{UHF}_{s_z} \tag{7}$$

The procedure we follow is very similar to that of Harriman[16–18]. Again, as in the case of the 2-Fe ferredoxin model,[4] these two methods, i.e. ROHF and

Table 1. The 72-atom model examined by different theoretical methods. The energy differences (ΔE in *kcal/mol*) are calculated with respect to the lowest SCF energy. q(Fe) stands for Mulliken population charges on the Fe atoms; q(S) and SS(b.i.) are the Mulliken population charges and the bond index for the bridging S atoms, respectively; ΔE_Q is the calculated Mössbauer quadrupole splitting constant [*mm/sec*]. The PUHF spin states are those projected from the UHF wavefunction with $S = S_z$.

Method	Multiplicity	ΔE	q(Fe)	q(S)	SS(b.i.)	ΔE_Q
ROHF	11	60.50	1.91	-0.61	0.971	1.43
	9	59.86	1.88	-0.59	0.934	2.01
	7	59.49	1.84	-0.58	0.978	2.57
UHF	11	43.55	1.91	-0.61	0.971	1.45
	7	29.05	1.83	-0.56	0.978	3.02
	1	29.32	1.83	-0.56	0.978	3.02
PUHF	11	29.74	1.91	-0.61	0.971	1.40
	7	8.04	1.83	-0.57	0.978	3.03
	1	0.00	1.83	-0.57	0.978	3.03
RHF	1	186.06	1.82	-0.43	1.245	1.27
Exp.	1	—	—	—	1 or 2[a]	2.04[b]

a) See text.
b) Reference 1.

Table 2. The 36-atom model examined by different theoretical methods. The energy differences (ΔE in *kcal/mol*) are calculated with respect to the lowest SCF energy. q(Fe) stands for Mulliken population charges on the Fe atoms; q(S) and SS(b.i.) are the Mulliken population charges and the bond index for the bridging S atoms, respectively; ΔE_Q is the calculated Mössbauer quadrupole splitting constant [*mm/sec*]. The PUHF spin states are those projected from the UHF wavefunction with $S = S_z$.

Method	Multiplicity	ΔE	q(Fe)	q(S)	SS(b.i.)	ΔE_Q
ROHF	11	52.40	1.93	-0.61	0.971	1.41
	9	52.08	1.89	-0.60	0.934	2.04
	7	51.96	1.85	-0.59	0.976	2.70
UHF	11	34.14	1.92	-0.61	0.971	1.42
	7	19.58	1.84	-0.57	0.978	3.03
	1	19.64	1.84	-0.57	0.978	3.02
PUHF	11	26.67	1.92	-0.61	0.971	1.39
	7	6.09	1.84	-0.57	0.978	3.03
	1	0.00	1.84	-0.57	0.978	3.02
RHF	1	179.47	1.84	-0.44	1.237	1.27

PUHF, yield the lowest energies as well as reasonable Mössbauer quadrupole splitting constants, ΔE_Q.

We see from Tables 1 and 2 that the closed-shell Restricted Hartree-Fock (RHF) procedure does not provide a good description of the singlet state. The RHF energies are the highest of the methods we utilize for both the 72- and 36-atom structures. Even when followed by a reasonably sized CI, the RHF results are not comparable in their descriptive quality to those of the open-shell calculations. This argues against a closed-shell $d^6 - d^6$ diamagnetic electronic description for this complex. The lowest energy we obtained was for the fully projected UHF singlet state which corresponds to an AF $\left(\frac{3}{2}, -\frac{3}{2}\right)$ electronic spin-distribution between two d^5 Fe atoms. This observation is consistent with a very recent experiment finding that the monomer 1-Fe compound is of 3/2 spin.[19] The ferromagnetic septet spin-states lie close in energy, with a predicted AF Heisenberg exchange splitting constant of approximately -450 cm^{-1}[1] when calculated from the septet and singlet PUHF energy differences.

Figure 3 summarizes two important molecular orbital (MO) structures for the states of interest, $\left(\frac{5}{2}, \frac{5}{2}\right)$ and $\left(\frac{3}{2}, \frac{3}{2}\right)$. We note again that the local symmetry around the Fe atoms is C_{3v}, and our orbital diagram reflects this. Shown are the $Fe(III) - Fe(III)$ ferromagnetic situations (FM), although in both cases we calculate the AF structures, with half the electrons of β spin, of lower energy. Both FM and AF structures are considerably lower in energy than the all-paired $d^6 - d^6$ situations formally $(Fe(II) - Fe(II))$. Note that the singly occupied molecular orbital energies derive from an open-shell ROHF operator and do <u>not</u> correspond directly to ionization processes as do those obtained from the closed-shell RHF operator (Koopmann's approximation). The doubly occupied HOMO's are composed of sulfur bridging orbitals and placing electrons from these orbitals into the singly occupied lower lying d-orbitals correspond to excited charge transfer states.

The Mulliken charges of the iron and sulfur atoms, and especially the bond index between the bridging sulfurs (see Tables 1 and 2), favor a bonding description of a single $S^* - S^*$ bond that stems from the $\left[Fe - S_2^{2-} - Fe\right]^{4+}$ core electronic distribution, although with a considerable amount of S→Fe charge transfer. In the case of a diamagnetic $\left[Fe\left(d^6\right) - S_2^0 - Fe\left(d^6\right)\right]$ complex[2, 3] the two sulfurs would be formally neutral and possess a bond index for the sulfur bridge of 2, as in the diatomic molecule, itself. This is not observed in any of the calculations we have performed although the closed-shell RHF $d^6 - d^6$ singlet calculation favors this description most, as might be expected.

Tables 1 and 2 indicate that the results obtained for the 72- and 36-atom models are essentially the same. This is indicative of the aromatic rings in the ligands having no significant effect on the calculated energy differences, charges, or magnetic splittings. We do not expect there to be a marked difference in the

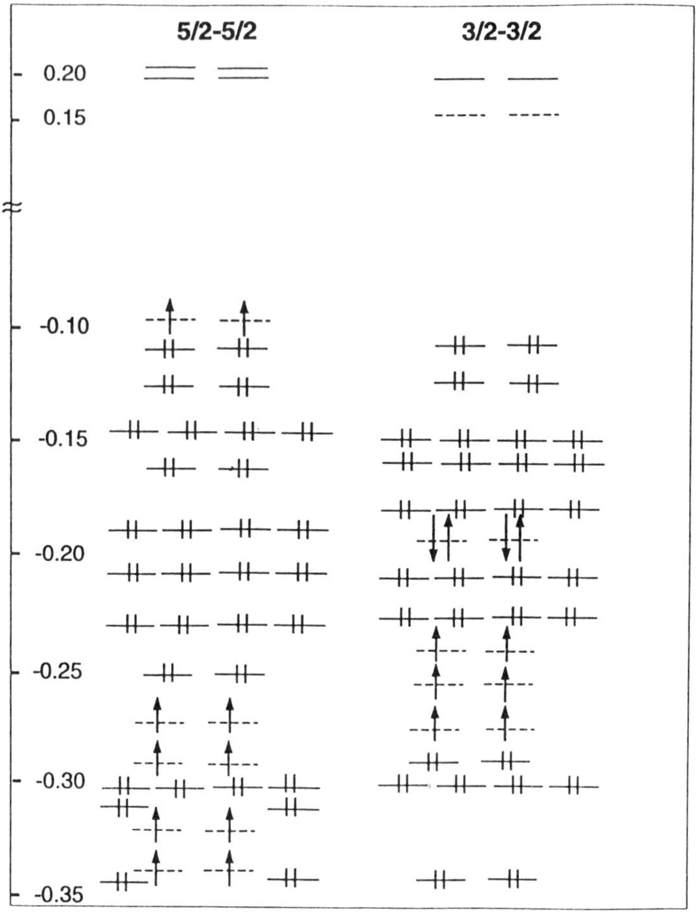

Figure 3. Molecular-orbital diagrams as obtained by the ROHF method. Dashed lines indicate MOs dominated by the metal *d*-orbitals, the solid lines stand for doubly occupied or virtual ligand orbitals. Orbitals which are close in energy are presented as degenerate; the average deviation from degeneracy is approximately 0.01 a.u. In the case of a septet state (S=3), the singly occupied open-shell orbitals come from a separate Fock operator and their orbital energies do not relate to ionization potentials as do the doubly occupied MOs (i.e. Koopmann's approximation). For these reasons, the open-shell orbitals appear well below the doubly occupied metal orbitals. Doubly occupying these gives rise to excited states, see text.

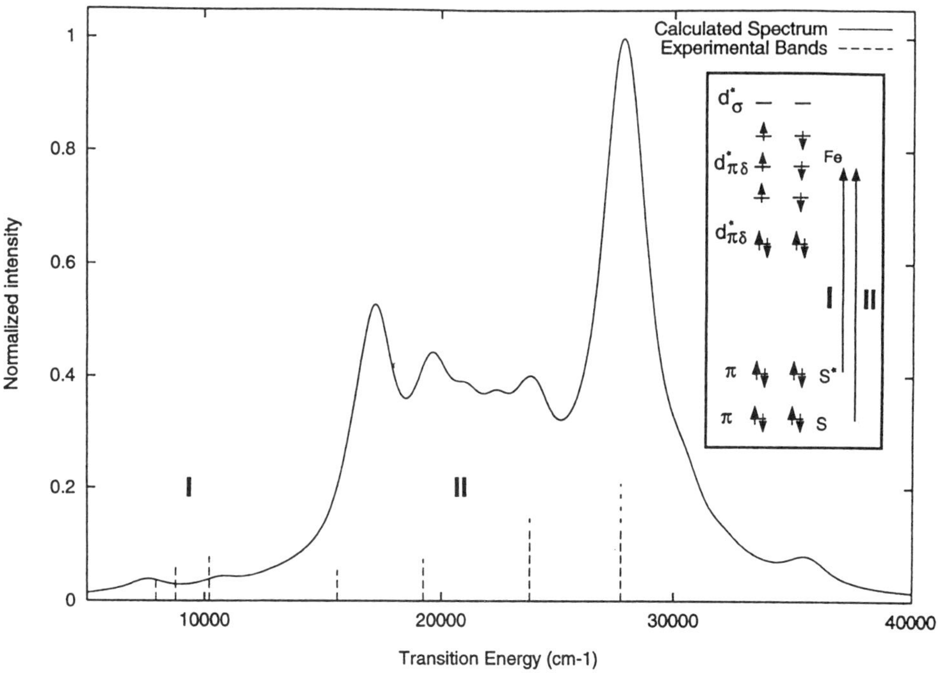

Figure 4. The calculated spectrum of the complex after a Lorentzian band convolution. Region **I** is dominated by bridging-sulfur-to-iron CT transitions, while region **II** is mostly due to organic-sulfur-to-iron electron transitions. Regions **I** and **II** are explained in a MO diagram. The vertical lines correspond to the experimental bands observed in the absorption spectrum of the $\left[Fe_2(\mu - S_2)(P(o - C_6H_4S)_3)_2\right]^{2-}$ complex, from Reference 1.

CT and *d-d* spectroscopy of the two model complexes. Indeed, we calculate the low energy absorption spectra for the two model structures to be nearly the same and we have found no effect on the electronic structure of the core complex that is due to the more distant carbon atoms. Furthermore, the frontier orbitals are those originating from the [Fe-S-S-Fe] core, see Figure 4, as well as the organic sulfurs, with only marginal contributions from the P atoms. In the molecular orbital picture the assignment of the excitations observed in the calculated spectrum, Figure 4, are CT bands that occur in two different frequency regions. Region **I** is of lower energy, the near infrared (IR), and is dominated by electronic transitions from MOs localized to the sulfur bridge into MOs principally of iron *d*-character. The bands of region **II** are in the visible region and are mostly from organic sulfur MOs into the Fe d_{π}^{*} orbitals with an admixture of Fe d_{δ}^{*}. The calculated spectrum correlates well with that of experiment and this lends further support to the argument for the model of AF coupled Fe atoms over that of the simple closed-shell diamagnetic picture. In addition, we note that region **I** (which we assigned to CT from the bridging sulfurs to the Fe atoms) disappears in the monomer, which has a thiolate group substituted for the disulfide bridge.[19]

It has been previously noted [1–3] that the near IR region might be an important signature for biological molecules with similar $Fe - S - S - Fe$ structures, and we agree. The P-cluster of nitrogenase is a promising candidate with one such Fe-S-S-Fe bridge.[20, 21] The M-clusters of the same enzyme also contain a number of inorganic sulfur bridges. We are currently studying such systems[22] and we believe that the present analysis can aid in the elucidation of the electronic structure of these complex transition metal systems.

Acknowledgment

We would like to thank Dr. Stephen Koch, SUNY-Stony Brook for providing the X-ray structure of the complex and for the valuable discussions. This work was supported in part through a grant from the Office of Naval Research.

Dedication

"This paper is dedicated to our friend Nils Yñgve Öhrn, interested in everything, and always there to help, advise, discuss and argue. It is difficult to imagine a better colleague with whom to share the day."

References

[1] Franolic, J. D.; Millar, M.; Koch, S. A. *Inorg. Chem.* **1995**, *34*, 1981.

[2] Sellmann, D.; Mahr, G.; Knoch, F. *Angew. Chem. Int. Ed. Engl.* **1991**, *30*, 1477.

[3] Sellmann, D.; Mahr, G.; Knoch, F. *Inorg. Chim. Acta* **1994**, *224*, 35.

[4] Cory, M. G.; Stavrev, K. K.; Zerner, M. C. *Int. J. Quant. Chem.* **1997**, *63*, 781.

[5] Pople, J. A. Beveridge, D. L. *Approximate Molecular Orbital Theory*, 1st ed.; McGraw-Hill: New York, 1970.

[6] Ridley, J. Zerner, M. C. *Theor. Chim. Acta* **1973**, *32*, 111.

[7] Bacon, A. D. Zerner, M. C. *Theor. Chim. Acta* **1979**, *53*, 21.

[8] Zerner, M. C.; Loew, G. H.; Kirchner, R. F.; Müller-Westerhoff, U. T. *J. Amer. Chem. Soc.* **1980**, *102*, 589.

[9] Edwards, W. D. Zerner, M. C. *Theor. Chim. Acta* **1987**, *72*, 347.

[10] Zerner, M. C. *ZINDO Program, University of Florida, Gainesville Florida.*

[11] Cory, M. G.; Hirose, H.; Zerner, M. C. *Inorg. Chem.* **1995**, *34*, 2969.

[12] Slater, J. C. *Quantum Theory of Atomic Structure, Vol. I*, 1st ed.; McGraw-Hill: New York, 1960.

[13] Cory, M. G. Zerner, M. C. *in preparation.*

[14] Sasaki, F. Ohno, K. *J. Math. Phys.* **1963**, *4*, 1140.

[15] Löwdin, P. O. *Phys. Rev.* **1955**, *97*, 1509.

[16] Harriman, J. E. *J. Phys. Chem.* **1964**, *40*, 2827.

[17] Hardisson, A. Harriman, J. E. *J. Phys. Chem.* **1967**, *46*, 3639.

[18] Phillips, D. H. Schug, J. C. *J. Phys. Chem.* **1974**, *61*, 1031.

[19] Koch, S. A. *private communication.*

[20] Mouseca, J.-M.; Noodleman, L.; Case, D. A. *Inorg. Chem.* **1994**, *33*, 4819.

[21] Mouseca, J.-M.; Noodleman, L.; Case, D. A. *Coord. Chem. Rev.* **1995**, *144*, 199.

[22] Stavrev, K. K. Zerner, M. C. *Angew.Chem. (35/1): Chem. Res. J.* **1996**, *2*, 83.

Index

D

I

J

K

L

M

P

Q

Y

Z

ISBN 0-12-034835-7

90018

9 780120 348350